THE IMMUNOLOGY OF PARASITIC INFECTIONS

A Handbook for Physicians, Veterinarians, and Biologists

by
Omar O. Barriga, D.V.M., Ph.D.
Department of Veterinary Pathobiology
The Ohio State University

University Park Press
Baltimore

UNIVERSITY PARK PRESS
International Publishers in Science, Medicine, and Education
300 North Charles Street
Baltimore, Maryland 21201

Typeset by Maryland Composition Company, Inc.

Manufactured in the United States of America by the
Maple Press Company

Library of Congress Cataloging in Publication Data
Barriga, Omar O.
The immunology of parasitic infections.
Includes index.
1. Parasitic diseases—Immunological aspects.
2. Veterinary parasitology. I. Title. (DNLM:
1. Parasitic diseases—Immunology. 2. Parasitic
diseases—Veterinary. WC 695 B275i)
RC119.B3   616.9′6079   80-26809
ISBN 0-8391-1621-7

Contents

Preface

The intelligent man must often ask himself why, in the dawn of the 21st century, there are still 800 million of his fellow men infected with hookworms, 300 million with amebiasis, and 200 million with schistosomiasis; why 400 million people are at risk of contracting malaria and 50 to 150 million actually acquire it every year; why 10 million human beings harbor the agent of Chagas' disease and 35 million are under its menace; why livestock is so scarce and horses nonexistent in vast regions of Africa covered by luxurious vegetation; why the United States, where the control of parasitic diseases has reached a high degree of efficiency, lost 1.2 billion dollars per year in the decade 1951–1960 as a result of animal parasitism.

The possible answers are numerous, but the basic reason may reside in the fact that animal parasites are much older than mankind itself: the parasitic protonematodes originated with the insects, the protocestodes with the crustacea, and the digenetic prototrematodes with the mollusks. The parasitic protozoa are probably much older than the parasitic helminths.

Throughout more than 600 million years, and accompanying the vertebrates since their phylogenetic origins, the parasitic species had the time and opportunity to genetically select and to adapt to the peculiarities of their hosts in a very effective manner. We do not know the number of species (undoubtedly high) that succumbed in this process, but those that survived must have taken advantage of every biological "loophole" that facilitated their continuity.

Nature, free of our anthropocentric bias, for millenia took proper care that the parasitic as well as the host species had a fair opportunity to persist in time. The relatively scarce pathogenicity of contemporary animal parasites and the comparative inefficiency of the defense mechanisms of the hosts are eloquent proofs of the perfection of the host-parasite relationship as we know it nowadays. It is not surprising, then, that the efforts of the professionals of human or animal health have had meager results in their attempts to destroy such an old association that has already come successfully through so many vicissitudes.

The advent of the lymphoid system in the vertebrates and its further refinement in the homoiotherms must have represented a formi-

dable obstacle that relatively few parasitic species were able to con-
quer. Nevertheless, it is astounding that those species are able to
coexist with a physiological mechanism such as the immune system,
the primordial purpose of which, if we yield to a pinch of teleology,
is precisely to rid the organism from exogenous or endogenous invaders. This coincidence seemed so unnatural that only 50 years ago, as
many years after the celebrated experiments of Pasteur in Pouilly-le-
Fort, Hegner directed the attention of the scientific community to the
fact that the host-parasite association in blood protozoa was similar
to that existing in bacterial infections. Until fairly recently, even individuals with training in biomedical sciences believed that the immune
responses to animal parasites should be peculiar to these organisms
and different from the reactions to other pathogenic agents.

The first attempt to present the immunity against parasitic animals
in a comprehensive manner was the monumental work by Taliaferro,
published in 1929. Twelve years later, Culbertson repeated the feat by
reviewing the advances achieved in that period in a text that even today
can be read with benefit. The prodigious advancement of immunology
in the last decades (the number of papers in immunology rose from 220
in 1940 to 7660 in 1970) and the increasing amount of investigations
related to the immunity against animal parasites allowed almost 30
years to elapse before a team of 35 authors dared gather the information
on the discipline in two comprehensive volumes (Jackson, Herman,
and Singer, 1969–1970). More recent works that attempted to give detailed and erudite accounts of the state of the art have been edited by
Soulsby (1972) and by Cohen and Sadun (1976), with the collaboration
of numerous specialists.

Although the fine details of the immunology to parasites have gone
far beyond the possibilities of exhaustive examination by nonspecial-
ized readers, the study of the immune response against animal parasites
is becoming more and more important for the professionals connected
with human and animal health, and for biologists. Problems such as
the natural refractoriness to malaria in some human populations, the
mechanisms responsible for the lethal anaphylaxis in the terminal
phases of piroplasmosis, the differences in the evolution of infections
by leishmanias, the significance of the EVI antibodies in Chagas' disease, the precise identification of hydatid disease and visceral larva
migrans, the correct interpretation of the presence of specific agglutinins in bovine trichomoniasis, new possibilities for the prevention of
sequelae in schistosomiasis, and the effective prophylaxis of respiratory nematodiases have already been or are in the process of being
solved by immunological studies.

In addition, parasite immunology is contributing considerably to revealing the phylogenetic relationships among diverse parasitic organisms and is constantly discovering new facets of the fascinating host-parasite association. Finally, in recent years, parasites have been demonstrated to constitute quite adequate probes for exploring diverse peculiarities of the immune system.

Thus the time has come to write a text that explains the basic principles and the practical implications of the immunity against animal parasites, for physicians, veterinarians, biologists, and the students of these disciplines who do not have the time, the vocation, or the need to pursue advanced studies of the subject. This book is directed to them and also, as an introduction to the specialty, to the parasitologists and immunologists who have not had training in the complementary discipline. I think that even some colleagues may welcome the identification of some of the numerous areas that are in most urgent need of research. The book places emphasis on the parasitic infections occurring in the Americas. I hope that the necessary lack of depth in a text of this nature is duly compensated by a coordinate treatment that emphasizes principles and applications rather than fine details.

For the benefit of those readers who need to refresh their knowledge of immunology, two chapters have been specially included: one that summarizes the most important concepts in the discipline, and another that briefly explains the molecular mechanisms of the immunological tests commonly used in clinical parasitology. For those who do not remember the parasitological aspects, the most relevant characteristics of the natural history of the parasites that are not transmitted directly have been mentioned at the beginning of each chapter. Those readers who may desire to deepen their knowledge in any specific field should find abundant opportunity to do so in the bibliographic references at the end of each chapter. Necessarily, a large proportion of these are review papers. Nevertheless, they cite the original publications on which the opinions expressed in the text are based.

Finally, perhaps in no field of human activity is it as true as in the academic life that "no man is an island." My own training has had the benefit of the contribution of too many people to name here. It is only fair, however, to mention Professor Amador Neghme, of the University of Chile, who initiated me in the fascinating task of teaching and doing research in the biomedical sciences and encouraged my interest in the humanities; Drs. Norman D. Levine and Diego R. Segre, of the University of Illinois, who were patent examples of superior scholarship and unequaled kindness; and Dr. E. J. L. Soulsby, citizen of the world, and lately at the University of Cambridge, with whom I had the honor to work for five unforgettable years in Philadelphia. My gratitude goes to all of them. Anything good in this book is a reflection of their teachings and examples; anything incorrect is my exclusive responsibility.

Omar O. Barriga

SOURCES OF INFORMATION

Cohen, S., and Sadun, E. H. (eds.). 1976. Immunology of Parasitic Infections. Blackwell Scientific Publications, Oxford.

Culbertson, J. T. 1941. Immunity against Animal Parasites. Columbia University Press, New York.

Jackson, G. J., Herman, R., and Singer, I. (eds.). 1969–70. Immunity to Parasitic Animals. Appleton-Century-Crofts, New York.

Soulsby, E. J. L. (ed.). 1972. Immunity to Animal Parasites. Academic Press, Inc., New York.

Taliaferro, W. H. 1929. The Immunology of Parasitic Infections. The Century Co., New York.

To
  Inés,
  Omar Jr.,
  and Alvaro,
for their
  love and
  support.

The
Immunology
of Parasitic
Infections

Chapter 1

Review of
Basic Concepts of Immunology

The advent of the vertebrates in the parade of evolution brought into existence a group of animals far more complex than earth had seen before; prolificacy alone was no longer able to secure their continuity in time. The coordination and preservation of these novel organisms demanded systems much more sophisticated than those previously existing; some were acquired by perfecting old physiological mechanisms and a few were virtually newly created. The immune system was among the latter.

The primordial function of the immune system is the preservation of the biochemical identity of the individual, by detecting and reacting against exogenous or endogenous invaders (pathogens and tumors, respectively). In the classic writings, immunity has often been described as homologous to defense. Even accepting that the general evolutionary "purpose" of immunity must have been to provide the vertebrates with an efficient mechanism to fight disease, the immune reactions must be recognized as automatic responses that occur whenever the required conditions happen, regardless of their end result. Thus, in real life, we find that some immune responses are deleterious for the invader, others are injurious for the host, and, finally, others are indifferent for either partner.

This chapter discusses briefly the main features of the immune system in order to form a basis for understanding its participation in the production, course, and control of parasitic infections, as well as its possible use for practical applications. There is currently a high degree of interest in immunology, and the field is undergoing explosive growth; therefore a large number of textbooks on the subject has been written in the last few years. Most of them are quite adequate, so that selecting references for further study becomes almost a matter of personal preference. The interested reader may benefit from consulting the book by Benacerraf and Unanue (1979), which provides an excellent introduction to general immunology. The textbook edited by Fudenberg et al. (1978) is a more comprehensive work that presents up-to-date information on the basic and clinical immunology of humans. The manuals by Tizzard (1977) and the Olsen and Krakowka (1979) are particularly appropriate for those interested in the immunology of domestic animals.

STRUCTURE AND GENERAL FUNCTION OF THE IMMUNE SYSTEM

The structural basis of the immune system is the lymphohematopoietic tissue, and the main reactive cell is the lymphocyte. A general view of the immune system indicates that it was acquired only by the vertebrates and that, among them, it evolved to reach the highest complexity in the birds and, particularly, in the mammals.

The structures that form the immune system are generally described at three levels: level 1, the stem cell compartment (fundamentally the bone marrow), where new lymphocytes are produced; level 2, the central or primary lymphoid organs (thymus and bursa), where the bone marrow lymphocytes acquire their capacity to respond to immunological stimulation; and level 3, the peripheral or secondary lymphoid organs (e.g., spleen, lymph nodes, Peyer's patches), where the lymphocytes are stimulated and respond to their corresponding antigens.

The lymphocytes that acquire immune competence in the thymus are known as thymus dependent or thymus derived, or, simply, *T lymphocytes* or *T cells*. Once differentiated in the thymus, they migrate to the peripheral lymphoid tissue and become particularly abundant in the blood, lymph, and paracortical zones of the lymph nodes. The lymphocytes that mature in the bursa of Fabricius are called bursa dependent or bursa derived, or *B lymphocytes* or *B cells,* and are found principally in the bone marrow and in the germinal centers of the lymph nodes. The human spleen contains approximately equal proportions of T and B cells. The anatomical equivalent of the bursa has not been identified in mammals, but there is evidence that the bone marrow exerts bursa-equivalent functions in these vertebrates.

The recognition of a substance as foreign or "nonself" by the immune system is limited to macromolecules that have a certain degree of complexity (*antigens*). Practically all foreign proteins with a molecular weight of 10,000 or more, numerous "nonself" polysaccharides of 60,000 or more, and the combinations of extraneous proteins with polysaccharides are effective antigens. A considerable number of substances of low molecular weight (*haptens*) are also able to stimulate an immune response when they are conjugated with an antigenic protein. Oligosaccharides, lipids, and nucleic acids, as well as synthetic compounds, occasionally behave as haptens.

In practically all cases, the recognition of the antigenic substance, either free or constituting part of a more complex structure (cell, virus), begins with its ingestion and "processing" by macrophages. There is no consensus yet on what happens to the antigen during this processing, but apparently those portions of the antigenic molecule that will be recognized as foreign (*antigenic determinants*) are distributed on the surface of the macrophage and become easily accessible to the lymphocytes.

The next step in the chain of the immune response is still obscure and has been the subject of numerous studies and interpretations. Simply explained, the native antigen is able to prime the effector lymphocytes genetically predetermined to react with it, but this stimulation is not enough to initiate their multiplication. The macrophage-processed antigen, however, triggers a subpopulation of T lymphocytes called *regulatory cells*; at least functionally, these are divided into a "promoter" or "helper" set and an "inhibitory" or "suppressor" set. The helper cells appear within the first hours of the immune response and produce soluble substances that stimulate the proliferation of the effector lymphocytes already primed by the antigen. The suppressor cells become functional 3 or 4 days after the initiation of the immune response and set a limit to the multiplication of the effector lymphocytes, which, otherwise, would continue multiplying like a neoplasia.

Following the sequential stimulation by the native antigen and by the product of the helper cells, the effector B lymphocytes begin to divide in rapid succession for several generations until they differentiate to plasma cells, which are active producers of antibody. During this proliferation, some daughter cells remain at the stage of small lymphocytes as "memory cells." They may persist for years in the body and, when stimulated by a subsequent dose of the same original antigen, they initiate a more rapid, intense, and effective immunological response (*secondary* or *anamnestic* response) than occurred on the first contact with the antigen (*primary* response). A few antigens constituted by numerous repeated monomeric units (such as the polysaccharides of pneumococcus or the lipopolysaccharides of enterobacteria) are able to stimulate B lymphocytes without the assistance of helper cells; the antibodies produced in these cases are virtually pure immunoglobulin M. This observation may be particularly relevant to parasitology, since helminths contain abundant polysaccharides.

It is not clear at the moment whether effector T lymphocytes are as dependent on the activity of helper cells to proliferate as are the B lymphocytes. At any rate, subsequent to the introduction of the antigen in the host, the effector T lymphocytes divide for several generations until becoming mature cells. When these cells enter in contact again with the same antigen that stimulated their formation (either because it remained in the body or it was acquired de novo), they release a number of humoral substances collectively known as *lymphokines*.

Certain materials called *mitogens* are able to stimulate the proliferation of T cells, B cells, or both in the absence of antigen or of helper cells. Several of them are proteins extracted from plants, but substances with mitogenic activity have been recently reported to have been found in some parasites. At least in theory, these substances might be able to enhance or abolish the response to a particular antigen, by expanding or preempting the population of lymphocytes predetermined to react with it.

The immune response can be artificially enhanced by the administration of compounds generically called *adjuvants,* together with the antigen. One of the most widely used is Freund's adjuvant, which contains mineral oil and an emulsifier to assure its intimate mixture with the antigen. Dead *Mycobacterium* may or may not be added (complete or incomplete adjuvant, respectively). Because of the severe local inflammation produced by Freund's adjuvant, other adjuvants (e.g., aluminum hydroxide) are preferred for clinical applications. Adjuvants not only increase the immune responses to an antigen, but can also favor the production of one particular manifestation of immunity over the others: Freund's complete adjuvant facilitates the expression of cell-mediated immunity; *Bordetella pertussis* and *Ascaris* extracts stimulate the formation of immunoglobulin E antibodies; bacterial endotoxins enhance antibody production. Also, substances that normally do not produce a detectable immune response can be made antigenic by inoculating them along with potent adjuvants.

The development of the lymphoid system occurs early in ontogeny: small lymphocytes appear in the peripheral blood of human fetuses at 7–8 weeks of gestation and their lymph nodes are populated already from the fourth month on. However, the ability to respond to different antigens (which is genetically predetermined for each lymphocyte) is attained at various times around birth: lambs born with congenital toxoplasmosis present specific antibodies produced prenatally, whereas lambs infected with *Haemonchus contortus* do not develop protective immunity if they are under 4–6 months of age. Before acquiring immune competence, lymphocytes are particularly sensitive to very large or very small doses of antigens: experimentally, it has been verified that administration of antigens in these ranges often induces lack of immunological responses to subsequent conventional doses of the same antigen (*immunological tolerance*). This inhibition of the responses

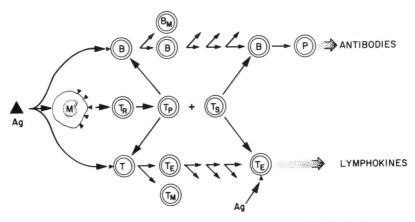

Figure 1.   Schematic representation of the cellular events entailed in the immune response. The newly acquired antigen (▲) primes competent B and T lymphocytes (**B** and **T**) and is ingested and processed by appropriate macrophages (**M**). The macrophage-processed antigen subsequently stimulates two subpopulations of regulator T cells (T_R): promoter T lymphocytes (T_P), which induce the multiplication of the immunocompetent cells beginning a few hours after the introduction of the antigen; and suppressor T lymphocytes (T_S), which inhibit their proliferation starting on the third or fourth day. The stimulated B lymphocytes divide for several generations, generating memory B cells (B_M), and finally mature to plasma cells (**P**) that are active producers of antibodies. The stimulated T lymphocytes also divide several times, generating memory T cells (T_M) and effector T cells (T_E); these latter will produce lymphokines on reencounter with the homologous antigen.

usually passes after a few weeks, but it may persist indefinitely if the presence of the antigen continues. Conventional amounts of antigen, on the other hand, may accelerate the acquisition of immune competence for the homologous antigens. Although there is little solid evidence yet, some observations suggest that these phenomena of tolerance or rapid maturation might occur in the course of parasitic infections in very young hosts.

From the account above (summarized in Figure 1), it is clear that immunity can be expressed in two general ways: by formation of antibodies, or by production of lymphokines. Traditionally, these two branches of immunity have been called humoral and cell-mediated immunity, respectively.

HUMORAL IMMUNITY

Antibodies are glycoproteins that constitute the group known as immunoglobulins (Ig). Two major characteristics define the immunoglobulins: (1) chemically, they possess a structure with two heavy and two light polypeptide chains, arranged in a typical fashion (see below); and

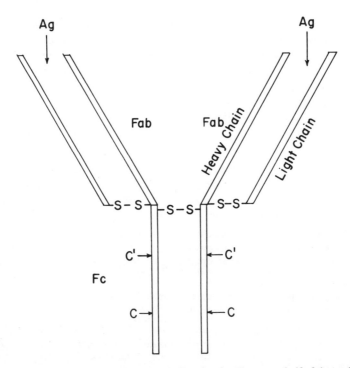

Figure 2.   Schematic representation of an IgG molecule. The upper half of the molecule is the Fab fragments and the lower half is the Fc fragment. The sites for combination with the antigen (**Ag**), with the first component of the complement (**C′**) and with cells (**C**) have been indicated by arrows. The IgG molecule very closely resembles the basic units found in the other immunoglobulin classes.

(2) biologically, their production is stimulated by an antigen with which they react specifically. Since the lymphocytes are genetically predetermined to react with only one antigen and to produce the corresponding antibody, by definition immunoglobulins must be antibodies. Conventionally, however, the term *antibody* is reserved for those immunoglobulins with a known specificity, i.e., the immunoglobulins whose corresponding antigen we know.

Certain chemical and physicochemical peculiarities have permitted the grouping of the diverse immunoglobulins into five classes: IgG, IgM, IgA, IgE, and IgD. The first three have been further categorized into subclasses. IgG is the most abundant in the serum and was the earliest to be identified as a molecular species; for this reason, most of the structural studies of immunoglobulins refer to it.

The basic unit of the immunoglobulins is formed by two polypeptide chains of about 450 amino acids each (heavy chains) joined to two polypeptide chains of about 200 amino acid each (light chains) (Figure

2). The portion of the heavy chains that is not accompanied by light chains can be separated from the rest of the molecule and crystallized; because of this characteristic it has been called the *Fc* (crystallizable) fragment. This part of the molecule is the part responsible for a number of important biological properties of some classes of immunoglobulins, such as the ability to cross the placenta and the digestive epithelium of diverse mammalian species, the capacity to activate the complement system, and the faculty to attach to macrophages or to cells that contain vasoactive amines. The portion of the immunoglobulin molecule that contains a heavy chain associated with a light chain is known as the *Fab* (antigen-binding) fragment, since the sequence of amino acids in both chains is such that it complements the particular chemical structure of the corresponding antigen and reacts with it.

The light chains may correspond to either of two different structural types (lambda and kappa) that may be found in any of the five classes of immunoglobulins. The heavy chains may belong to any of five structural classes: gamma, mu, alpha, epsilon, or delta. The class of the heavy chains present in an immunoglobulin molecule determines the class of that immunoglobulin; thus two gamma heavy chains with two lambda or with two kappa light chains constitute an IgG molecule.

The different classes of immunoglobulins possess peculiar physicochemical and biological characteristics, which have been summarized in Table 1. Among the former, the molecular weight and the coefficient of sedimentation (measured in Svedberg units) are the most popular.

Immunoglobulin M

IgM is formed by five basic immunoglobulin units joined by the free end of their Fc fragments. This region also possesses a characteristic polypeptide (*J chain*) that, despite previous beliefs, does not seem to be essential for the polymerization of the five basic units. The IgM molecule has 10 antigen binding sites, but it has been demonstrated numerous times that it only binds five molecules of antigen with efficiency; when the whole molecule is separated into five basic units, each unit is monovalent for the antigen. In these circumstances, all the antibody activities that require two or more antigen binding sites (e.g., agglutination) disappear. This peculiarity is of common use in laboratories: the decrease of the agglutinating activity of a serum after treatment with the reducing agent 2-mercaptoethanol (which disassembles the IgM molecule) is an approximate assessment of the activity of IgM antibodies in that serum.

IgM is found predominately in the intravascular compartment. Large amounts of IgM antibodies found in secretions in cases of in-

Table 1. Some characteristics of the human immunoglobulins

Characteristic	Immunoglobulin				
	IgM	IgG	IgA	IgE	IgD
Molecular weight	900,000	150,000	160,000[a]; 400,000[b]	190,000	180,000
Sedimentation coefficient	19 S	7 S	7 S[a]; 11 S[b]	8 S	7 S
Serum concentration (mg/ml)	0.6–2.0	8–16	1.5–4.0	Traces	Traces
Percentage in intravascular space	76	45	42	51	75
Half-life (days)	5	23	6	2	3
Heat stability (56°C; 30 min)	yes	yes	yes	no	yes?
Mercaptoethanol stability	no	yes	no	no	yes?
Complement activation	yes	yes	no[c]	no[c]	no
Crosses placenta or digestive epithelium	no	yes	no	no	no
Sensitizes homologous skin	no	?	no	yes	no
Sensitizes heterologous skin	no	yes	no	no	no
Attaches to macrophages	no?	yes	no	no	no
Attaches to mast cells/basophils	no	yes	no	yes	no

[a] Seric.
[b] Secretory.
[c] By the alternative pathway when aggregated.

fection of the mucous membranes suggest that this immunoglobulin may be produced locally in response to antigens that enter through the mucosae, as occurs with IgA.

IgM is the first immunoglobulin to appear during ontogenic development (usually a little before birth) and constitutes the first antibodies produced on inoculation of an antigen. Except in the case of polysaccharide antigens, IgM antibodies are usually transient and are replaced by IgG antibodies after a few weeks. Commonly, the IgM antibodies either do not exhibit secondary responses or have anamnestic responses that are little more intense than the original response. These antibodies are particularly efficient in reacting with particulate antigens (reactions of agglutination) and in activating the complement system.

Immunoglobulin G

IgG is the most abundant and ubiquitous of the immunoglobulins: it is found in almost identical proportions in the intravascular and extravascular compartments; it predominates in the milk and the intestine of ruminants; and it is found in considerable concentration in all secretions.

IgG appears in the organism at or a little after birth, and the corresponding antibodies are detectable in the serum about a week after inoculation of a potent antigen. Commonly the antibodies disappear in a few weeks, but they may persist indefinitely if the antigenic stimulation is maintained. On reinoculation of an antigen, IgG antibodies are responsible for virtually all of the secondary response. IgG is the immunoglobulin transferred from mother to offspring: through the placenta in primates; through the placenta and milk in rodents and carnivores; and through the colostrum in large herbivores and pigs.

The combination of IgG antibodies with the macrophages by their Fc fragments and with the antigen by their Fab fragments approximates the cells to the antigen and promotes phagocytosis (*opsonizing activity*). A subclass of IgG (IgG_1), identified so far in several rodents, dogs, monkeys, and possibly most domestic animals and humans, but not in rabbits, has the ability to attach by its Fc fragment to the mast cells and basophils of the same species that produced it (*homocytotropy*). This immunoglobulin behaves similarly to the IgE, but, unlike it, IgG_1 is found in high concentration in the serum, is stable to reducing agents and to heat (56°C for 30 min), and remains attached to the cells only for a few hours. The IgG of humans and many animals (but not of ungulates) has a similar affinity for the mast cells and basophils of a species different from that in which it was produced (*heterocytotropy*). The biological significance of this phenomenon is difficult to perceive, but it has been utilized in a highly sensitive serological test (the passive cutaneous anaphylactic test).

IgG antibodies are particularly efficacious in reacting with soluble macromolecules and small particles (viruses); they are especially effective in reactions of precipitation and neutralization of toxins and viruses. They agglutinate and activate the complement system, but with less efficiency than the IgM antibodies at the same molar concentration.

Immunoglobulin A

IgA occurs as monomers, dimers, or higher polymers of the basic unit. The monomeric form predominates in the internal compartments of the human body (but not in other animals), whereas the dimeric association is the common form found in all organic secretions. Seric IgA appears to be produced anywhere in the body, but secretory IgA is synthesized fundamentally by the lymphoid tissue located under the epithelium of the mucous membranes and exocrine glands. The dimers are apparently assembled within the respective plasma cells, and a J chain is added at the same location. Another protein, the *secretory component*, which affords particular resistance to enzymatic digestion, is appended during the transport of the dimer through the epithelium.

IgA antibodies are not opsonizing and do not activate the complement by the classic pathway; however, their agglomeration is effective in initiating the alternative pathway. For a long time, it was suspected that IgA antibodies had a particularly valuable participation in the protection against infections acquired through the mucous membranes, but the corresponding mechanism was unknown. It has been recently demonstrated that secretory IgA antibodies can neutralize viruses and prevent the adherence of bacteria to the mucosae by binding the respective receptors on the pathogen. On several occasions it has been shown that the production of secretory IgA antibodies requires local stimulation by the antigen; its parenteral inoculation has produced circulating IgG antibodies instead.

Immunoglobulin E

IgE is found in minute amounts in virtually all organic fluids. A large proportion of the IgE present in external secretions is synthesized locally, as is the case with IgA. IgE antibodies, or *reagins*, are the main (in humans, maybe the only) antibodies responsible for immediate-type hypersensitivity or Type I allergy. The penetration of certain antigens (*allergens*) in the body causes the preferential formation of IgE antibodies, which attach to mast cells and basophils by their Fc fragments. Combinations of these antibodies with subsequent doses of the same antigen will bridge two nearby molecules of antibody, thus initiating a chain of reactions that results in the release of vasoactive amines from the cells.

Immunoglobulin D

IgD is a relatively new discovery. It has been found (together with IgM) on the surface of B lymphocytes and is believed to be an element for the recognition of the respective antigens by the cell.

CELL-MEDIATED IMMUNITY

It has been known for a long time that manifestations of immunity such as delayed-type hypersensitivity, the primary rejection of grafts, the elimination of some tumors, and certain autoimmune diseases occurred in the absence of specific antibodies but were transferable by lymphoid cells. It was thought that the effects registered in these cases were due to the direct activity of the lymphoid cells on the target tissues, and the corresponding mechanism was called *cellular* or *cell-mediated immunity*. Studies in the last decade have shown that the effects of cell-mediated immunity are actually due to the action of humoral substances, collectively known as *lymphokines,* that are released by primed effector T cells on reencountering the homologous antigen. The original nomenclature has been conserved, however, as a matter of convenience.

At present, there are at least a score of biological effects that are attributed to lymphokines: the responsible substance, however, has been partially isolated and characterized only on a few occasions. Ingenious laboratory assays have indicated that many of these effects are attributable to different substances, but it is still possible that a single lymphokine can exert more than one activity or that a single action requires the presence of more than one lymphokine.

As in the case of the antibodies, lymphokines are produced only when an appropriate lymphocyte reacts with the homologous antigen. Once the lymphokines are released from the lymphocyte, however, the activity of most of them is nonspecific and they will operate in their particular characteristic manner regardless of the antigen that was involved. In most of the known cases, the lymphokines are proteins from 20,000 to 80,000 daltons that appear to act mostly at short range and have a half-life measurable in minutes or hours when they are in the body.

The lymphokines (or their biological effects) most relevant to parasitic infections are the following:

1. *Migration inhibitory factors*, which prevent the spread of wandering cells in vitro. One of these factors acts on the leucocytes of the buffy coat and two others affect the macrophages. Of these latter, one is independent of the antigen, but the other requires its presence to operate.
2. *Macrophage-stimulating factors*. One factor enhances the production of energy, the phagocytosis, and the digestion by macrophages; another promotes the same activities, but only against cells that possess the antigen that induced its production.
3. *Chemotactic factors*, which attract inflammatory cells to the site of their release. Independent factors that call for macrophages, lymphocytes, polymorphonuclears, and eosinophils have been identified.
4. *Mitogenic factors*. One factor stimulates the proliferation of non-primed lymphocytes (and of other cultured cells) and another promotes the division of primed B lymphocytes. The latter actually performs the function of the helper cells.
5. *Cytostatic factors*. One factor detains or delays the proliferation of the cells in a culture and another (*lymphotoxin*) lyses them.

Rigorous proofs of the biological significance of lymphokine production are still lacking, but few authors doubt that they may have a preponderant role in the actions of defense against some invaders. An exercise in imagination will illustrate the possible activities of the lymphokines in this regard. The initial nonspecific inflammation caused by most pathogenic agents is bound to attract to the area some of the circulating T lymphocytes, which will be primed by the structures of the pathogen or by its products. If the aggressor is still present after a few days, the renewed contact of its antigens with the already primed and mature lymphocytes will cause the production of lymphokines. The chemotactic factors will attract macrophages, neutrophils, and eosinophils with the ability to attack the germ, and new lymphocytes to amplify the immune reaction. The mitogenic factors, in the meanwhile, will stimulate the proliferation of local lymphocytes (and macrophages?) to provide an immediate increase of the reactive cells in the zone. At the same time, the migration inhibitory factors prevent the wandering macrophages and leucocytes from leaving the area, collaborating in this manner to recruit defensive elements. While a reassuring numeric superiority is being achieved, macrophage-stimulating factors make sure that these cells operate at peak efficiency and cytostatic factors enter to cooperate in the destruction of the aggressor. As a further assurance, the mitogenic factor for primed B lymphocytes ensures that antibodies are produced, even if helper cells are absent.

Certain lymphoid cells that are able to kill cultured cells on contact have been described recently. They look like lymphocytes but do not show the distinctive characteristics of either B or T cells; because of their biological activity, they have been named *K* (killer) *cells*. K cells do not require prior priming by the antigen to exert their lethal effects. Some of them have receptors for the Fc portion of the IgG, and the target cell must be coated with its corresponding antibody to facilitate the approximation of the killer cell; others attack the target cell without need for an antibody.

At this moment, K cells cannot be included within the humoral or the cellular responses, and their participation in parasitic infections is unknown. Their mode of action, however, suggests that they might operate in the reactions of defense against protozoa.

THE COMPLEMENT SYSTEM

Of the several nonspecific humoral factors of defense against pathogens that have been described, none is as important and general as the complement system. The complement consists of nine proteins (or protein complexes) of the serum that are self-activating by a sequence of reactions that must be initiated by an external element.

The activation of the complement by the "classic pathway" begins when one molecule of IgM antibody or two adjacent molecules of IgG antibodies bind their corresponding antigen. The combination with the antigen uncovers a site between two adjacent Fc fragments for the combination of the first component of the complement, C1. Subsequent to its attachment on the immunoglobulin molecules, C1 acquires enzymatic activity and cleaves the two next components: C4 and C2. A fragment is released from each of these factors and the remaining fragments combine between them to form the enzyme $\overline{C42}$, or C3 convertase. This enzyme can attach to biological membranes and, after its formation, the activating reactions can proceed in the absence of antigen, antibody, or $\overline{C1}$. C3 convertase acts on the next component in the series, C3, and cleaves it into two portions: one is released and the other binds local structures. Those $\underline{C3}$ fragments that attach in the vicinity of $\overline{C42}$ form a new enzyme, $\overline{C423}$, that acts on C5, splitting it into two portions. Again, one of the portions is released and the other attaches to local structures and presents a site of combination for C6 and C7. The trimolecular complex so formed, C567, in turn, forms a site for the attachment of C8, and the combination with this complement constitutes a new site for the binding of C9.

In summary, during the activation of the complement by the classic pathway several sesile compounds that attach to local structures are produced. Of these, only C56789 is fairly long-lived and able to cause damage to the biological structures to which it is attached. Also, free fragments derived from C4, C2, C3, and C5 are formed; they exhibit biological activities that are mentioned below.

A few years ago, a new method of activation called the "alternative pathway" was identified; this may be identical to the old and never completely understood "properdin system." The alternative pathway consists of the activation of C3 and the subsequent components of the chain, in the absence of C1, C4, and C2. In this case, the three initial components of the classic pathway are replaced by other proteins that operate in a sequence not completely worked out yet. Unlike the classic pathway, the alternative pathway works well only with concentrated serum and requires only the presence of magnesium (instead of calcium and magnesium) ions. The initiation of the cascade is triggered by a polysaccharide of the wall of yeast cells (zymosan), by inulin, by gram-negative endotoxins, by a factor from cobra venom, by trypsinlike enzymes, or by products of some parasites. Conglomerates of IgG, IgA, and IgE can also initiate the alternative activation, but an antigen-antibody reaction is not necessary. Since the reactions of both pathways are identical beyond C3, C56789 and free C3 and C5 fragments are also produced in this case.

Several components of both pathways are destroyed by heat (56°C for 30 min), aging at room temperature, or incubation with NH_3. These procedures, particularly heating of a serum, are often utilized to eliminate or to investigate the participation of the complement in a given reaction.

The complement system is a powerful one and is able to exert a number of biological functions in the organism. The complex C56789 has the ability to disturb the lipidic phase of the biological membranes to which it attaches; by this mechanism it can destroy diverse classes of cells (e.g., erythrocytes, lymphocytes, platelets, protozoa) and pathogenic agents (gram-negative bacteria, viruses). Several cells of the body (B lymphocytes, macrophages, polymorphonuclears, nonprimate platelets, primate erythrocytes) bear receptors for the combined C3 fragment; binding of C3 on an antigen facilitates the contact of the antigen with these receptored cells, to initiate the humoral response in the case of B lymphocytes or to proceed to phagocytosis in the case of macrophages and polymorphonuclears. In the case of erythrocytes, this phenomenon has been used to devise a sensitive procedure of serological diagnosis (immune adherence test). The free C3 and C5 fragments have the ability to release vasoactive amines from mast cells and basophils (*anaphylotoxic activity*) and to attract polymorphonuclears to the area. Anaphylactic and chemotactic activities are also attributed to the free C4 and C2 fragments.

All these actions contribute to damage local structures and to cause inflammation: both may be effective deterrents of the proliferation and survival of the aggressor, but at the same time they may cause damage of the host tissues.

Intense formation of combined C3 and, secondarily, C4 may lead to the production of IgM or IgG antibodies against these fragments. The anticomplement antibodies are usually called *immunoconglutinins* and their detection in a subject presumes that extensive complement activation has taken place.

MECHANISMS OF TISSUE INJURY

Since the immunological reactions are automatic responses, on occasion they may be injurious for the host rather than protective. This has been frequently reported in parasitic diseases in the last few years, and the number of examples is likely to increase as more research is done. In 1963, Coombs and Gell proposed a classification of the injury-producing immune reactions (hypersensitivities or allergies) based on their main mechanism of production. Because of its adaptability to clinical applications, this division is being adopted increasingly by the specialists.

Type I Hypersensitivity, or Immediate-type Allergy, or Reaginic Anaphylaxis

Reactions mediated by the release of vasoactive substances from mast cells and basophils, as a consequence of the combination, on the cell membrane, of IgE antibodies with their corresponding antigen are called Type I hypersensitivity, immediate-type allergy, or reaginic anaphylaxis. An ever-increasing number of biologically active materials have been described in connection with the degranulation of mast cells and basophils (e.g., histamine, slow-reacting substance of anaphylaxis, serotonin, eosinophil chemotactic factor of anaphylaxis, platelet aggregating factor, bradykinin, heparin, prostaglandins). The most evident manifestations of this allergy, however, are dilation and hyperpermeability of the small vessels, contraction of hollow organs with muscle walls (large vessels, respiratory passages, intestine) and hypersecretion of mucous surfaces (bronchii, intestine). The particular signs in a patient will depend on the location of the target organs and on the predominant mediator produced, which are fairly typical for each species. The reaction may also be local or systemic, according to the diffusion of the allergen in the body.

Type II Hypersensitivity, or Cytotoxic Reaction

Reactions involving the combination of IgG or IgM antibodies with antigens of the cell membrane are known as Type II hypersensitivity or cytotoxic reactions. Free antigens or haptens that are adsorbed by the cell membranes are equally effective. Commonly, destruction of the cell by activation of the complement or by promotion of phagocytosis follows. In a few cases, the binding of antibodies stimulates the functions of the cells (antithyroid autoantibodies) or protects them from other deleterious manifestations of immunity (antitumor antibodies). Transfusional reactions, antibody-dependent K cell activity, and some drug-induced anemias belong to this group.

Type III Hypersensitivity, or Immune Complex Deposition

Type III hypersensitivity, or immune complex deposition, occurs when antigen-antibody complexes containing complement-activating, precipitating antibodies, are produced in the presence of a moderate excess of antigen. The complexes that precipitate in the tissues and are not removed by macrophages initiate the activation of the complement locally. Besides the injurious effects of the complement itself, the lysosomal enzymes of the neutrophils attracted to the area contribute to intensify the damage. If antigen-antibody complexes remain in the circulation, the disturbance of the local irrigation favors their further deposition and the lesion becomes chronic.

The localized Arthus reaction, systemic serum sickness, and a variety of immune complex diseases are examples of Type III allergy.

Type IV Hypersensitivity, or Delayed-type Hypersensitivity, or Cell-mediated Immunity

Type IV or delayed-type hypersensitivity are typical cell-mediated reactions in which the multiple deleterious activities taking place in the inflamed area also affect the host's cells. Strictly speaking, antibody-independent K cell activity does not belong to this group, but some authors prefer to include it here in order to keep the classification simple and manageable.

GENERAL MECHANISMS OF DEFENSE AGAINST PARASITIC INFECTIONS

The simple observation of the world around us shows that not all individuals are equally susceptible to all parasites, or affected to the same extent by them, or uniformly effective in controlling their proliferation. These three properties are brought about by essentially different mechanisms that are described as natural resistance, resilience, and acquired resistance, respectively.

Natural, Constitutive, Innate, Genetic, or Nonspecific Resistance

These terms, used by various authors to convey a similar idea, all refer to the insusceptibility of certain host species to the invasion and proliferation of some parasite species. Natural resistance may be absolute or relative according to whether the host is completely insusceptible to the parasitic invasion or only exhibits a reduced suitability to the survival of the parasite. It exists prior to the first encounter with the parasite, is not enhanced by previous infections, and is a consequence of structural and physiological features inherent to the host's species. The often-used expression *"natural immunity"* in reference to this phenomenon seems inappropriate since, by definition, the immune system does not play any role in natural resistance. *"Genetic resistance"* pretends to emphasize the inheritability of the causative mechanisms, but fails to separate this process from the acquired resistance that certainly also has a genetic background. *"Nonspecific resistance"* may be an oversimplification when applied to natural resistance to parasitic infections, since constitutive resistance to *Plasmodium knowlesi* and *Plasmodium vivax* appears to be quite specific (see Chapter 3) and the same may prove true when more information about other infections becomes available.

Abundant evidence has been produced that natural resistance to many microorganisms depends on the existence of physical barriers (e.g., skin, epithelia of mucous membranes, mucous secretions, washing action of tears and urine), chemical factors (lysozyme, interferon, properdin), physicochemical circumstances (pH, temperature, redox potential), or biological conditions (nonspecific phagocytosis, nonspecific inflammation, associated flora). Many of these mechanisms are likely to affect animal parasites also, particularly protozoa, but these organisms are peculiar in the sense that they have a defective regulatory biochemistry and must rely on the host to provide required physiological signals. Throughout this text are abundant examples (e.g., "Nematodes of the Digestive Tract") that critical portions of the development of parasites are closely regulated by stimuli from the host. This fact brings a new possibility of natural resistance: the host may interrupt the initial infection by failing to supply the necessary signals. Current experimental evidence indicates that this mechanism may be a major factor of natural resistance in parasitic infections.

Resilience

Resilience is a term popularized rather recently that expresses the long-known fact that some infected individuals compensate for the damage produced by a parasite better than other members of the same species. Some studies, for example, have shown that sheep with hemoglobin A develop less anemia than sheep with hemoglobin B under a similar infection with *Haemonchus contortus*; discordant results have also been reported, however. Although the mechanisms responsible for a high resilience are unknown at present, this physiological property appears to be inherited and to be independent of the ability to produce an effective immune response.

Acquired, Inducible, Immunological, or Specific Resistance

Acquired resistance is characteristically mediated by an immune response to parasite antigens (antibodies or cell-mediated immunity), often assisted by elements peripheral to the immune system (phagocytes, complement, K cell activity). In contraposition to natural resistance, by definition acquired resistance does not exist (to a detectable degree at least) prior to the first encounter with the parasite and usually is enhanced by previous infections. The expression *"acquired immunity"* is frequently used interchangeably with *"acquired resistance."* The former term actually refers to any manifestation of immunity elicited by an antigen, whether it is protective or not; the latter is properly restricted to only those immune reactions that are protective for the host.

The efficacy of immunological resistance covers a wide spectrum, from its virtual absence to the total elimination of the parasitic population (the latter is commonly referred to as *sterilizing immunity*). In most parasitic infections, and probably as a consequence of a lengthy evolutionary process, the host's immunity is able to reduce the growth of the parasitic population but not to exterminate the parasites. At the same time, the residual parasites provide the stimulation to keep the immunity at an effective level; if they are eliminated, the resistance wanes shortly afterward. This peculiar form of protective immunity is traditionally called *premunition*.

Customarily, acquired immunity is classified as *active* when the host produces the immune response (by infection or by immunization) or *passive* when the host receives the immune elements (antibodies or effector T cells) from some other source (by maternal transfer or by needle inoculation). The recent advances in immunology have made possible the production of *adoptive immunity* in which an individual receives and accepts primed lymphocytes from a histocompatible donor.

SOURCES OF INFORMATION

Benacerraf, B., and Unanue, E. R. 1979. Textbook of Immunology. Williams & Wilkins Company, Baltimore.

Coombs, R. R. A., and Gell, P. G. H. (eds.). 1963. Clinical Aspects of Immunology. Blackwell Scientific Publishing, Oxford.

Fudenberg, H. H., Stites, D. P., Caldwell, J. L., and Wells, J. V. (eds.). 1978. Basic and Clinical Immunology. 2nd Ed. Lange, Los Altos, Cal.

Olsen, R. G., and Krakowka, S. 1979. Immunology and Immunopathology of Domestic Animals. Charles C Thomas, Publisher, Springfield, Ill.

Tizzard, I. R. 1977. An Introduction to Veterinary Immunology. W. B. Saunders Company, Philadelphia.

Chapter 2

Inducers of
Immunity and Its Verification

THE PARASITIC ANTIGENS

Since parasites are species that are little related to their hosts phylogenetically, it is natural to expect that many of their macromolecules will be recognized as foreign, and therefore antigenic, by the host. In fact, parasites are sources of numerous and potent antigens that elicit the corresponding immune responses on the part of the lymphatic system of the host.

In the case of parasitoses of the blood and the tissues, the parasitic antigens have an excellent chance to achieve the close contact with the reticuloendothelial system of the host that is conventionally considered an essential condition for an immune response to occur. In many parasitoses restricted to lumenal organs, however, the mucous membranes appear to allow the passage of antigenic substances in quantity and quality enough to stimulate the immune system of the host. The finding of specific circulating antibodies in healthy carriers of *Entamoeba histolytica*, in asymptomatic cases of infections by *Trichomonas vaginalis,* in several intestinal cestodiases, and in trichuriases of lower animals testifies to this regard. In none of these cases does the union of host and parasite reach sufficient intimacy to justify the introduction of important quantities of antigens in the internal environment of the host. As is the case in the alimentary allergies, however, substances released by the parasites in the lumen of the organ appear to be absorbed by the host without undergoing important alterations.

The chemical study of many parasitic antigens has demonstrated the variability that was expectable: some of them are proteins, many are proteins conjugated to carbohydrates, a considerable number are polysaccharides, and complexes of lipids with either proteins or polysaccharides have been identified on a few occasions. In certain cases, it has been found that the corresponding antigenic determinant is actually a hapten bound to parasitic (or even host) proteins: the allergen in the oral secretion of fleas is a hapten that conjugates with the collagen of the host's skin; infections of *Ascaris* produce antibodies to the hapten phosphorylcholine of the cell membrane of many microorganisms; the hapten commonly called Forssman antigen has been verified in *Trichinella spiralis, Schistosoma mansoni, Hymenolepis nana,* and other helminths.

Some of these substances are part of the morphological structures of the parasites and have been named *somatic* antigens, *structural* antigens, or *endoantigens*. When the extract of a parasite is injected experimentally into a laboratory animal to produce an antiserum against the respective parasite, the somatic antigens act as potent antigens that give origin to a large variety of antibodies, in high concentration. It is possible, nevertheless, that these antigens have a more reduced role in natural infections, since the parasite must be phagocytized (subsequent to its destruction in the case of large organisms) for the structural materials to be processed in such a physical form that they could effectively stimulate the immune system. It has been debated for some time whether the cuticle of the nematodes is or is not antigenic in natural infections. Recent biochemical and immunological findings suggest that it is.

Other antigens that are the product of the physiological activity of the parasites have been called *metabolic* antigens, *excretion-secretion (E-S) products*, or *exoantigens*. A large number of these substances are enzymes, such as those produced during the invasion or the migration of the parasites in the tissues of the host, or in connection with their moults, or during the suction of blood, or in relation to other activities of the parasites. At least 16 antigens with enzymatic activity have been demonstrated in extracts of *S. mansoni*.

A third group of antigens are the *soluble antigens* or *S antigens* (Wilson, 1978), which are antigenic substances found free in the tissues or fluids of infected hosts. At present, it is not completely clear whether these antigens are parasite metabolic products, somatic substances released by dying parasites, or even host materials altered by the parasitism. At any rate, they may have an important part in the defense against the parasite, in the pathology of the infection, or in the evasion of the immune response by the parasitic organism.

Besides the substances characteristic of the parasite, numerous antigens demonstrated by inoculation of parasitic extracts into laboratory animals actually correspond to materials derived from the original host of the parasite. Kagan injected an extract of hydatid cyst of sheep into rabbits and verified that nine of the 19 antigen-antibody systems produced were equally evidenced with ovine serum instead of hydatid material. Capron found that, of 21 antigens identified in extracts of *S. masoni* obtained from hamsters, four were also present in the tissues of noninfected hamsters and five were shared by the snail intermediate host. Recently, Goodger reported that an antigen of *Babesia argentina* remained firmly bound to host fibrinogen after the process of purification. Some of these host antigens are undoubtedly substances acquired by the parasite during its activities in the host and carried in its intestine or adsorbed on its membranes; others, however, may be materials shared by the tissues of the host and of the parasite that arouse during evolution. Some investigators believe that some parasites might have the ability to synthesize substances similar to those of the host in response to the current act of parasitism. The importance of sharing or adsorbing host antigens for the survival of the parasitic species is mentioned throughout the book and is summarized in Chapter 7.

Despite the elevated number of antibody systems and expressions of cell-mediated immunity that a host can produce as a consequence of a parasitic infection, only those immunological processes able to alter some vital function of the parasite, to destroy it in vivo, or to modify its habitat in some drastic manner will confer resistance to the infection. The cross-reactivity between *Ascaris suum* and *Toxocara canis* is a good illustration for this concept: both parasites share numerous antigens, and infections of laboratory animals with either of them produce sera that react in vitro with both species. Infections with *A. suum*, however, do not protect against later challenges with *T. canis* (the reverse is probably also true). Evidently *A. suum* and *T. canis* possess common substances that act as potent antigens, but none of them is so vital to the respective parasite that its reaction with the corresponding antibodies would impair the persistence of the organism in its host.

The antigens that elicit protective immunological reactions have been called *functional* or *protective* antigens. Many experiments have brought consensus among the specialists that most of these antigens are of metabolic origin, at least in the metazoan parasites. In fact, it seems logical that antibodies directed against enzymes of the parasite should be able to neutralize them and to alter the physiology of the invader. On the other hand, antibodies to inert structures of metazoa, which chemically may not be susceptible to the activity of the complement, seem to have few possibilities of hurting the parasite! In the protozoa, it is possible that antibodies against metabolically inactive structures (if such a thing exists among these organisms) still may stimulate phagocytosis.

Experimentally, it has been demonstrated many times that immunization with metabolic products of parasites, often obtained from a determined developmental stage, frequently results in a remarkable resistance to the homologous infection. Inoculation of somatic antigens, on the contrary, rarely has achieved comparable results.

The metabolic antigens are often very characteristic of the organisms that produced them, and the corresponding antibodies have an exquisite specificity that allows them to distinguish among different strains of the same parasitic species or even among diverse developmental stages of the same parasite. The somatic antigens, on the contrary, are amply distributed in nature, and they are frequently shared by different species and by diverse genera. Not uncommonly, they are present even in more than one phylum. Experimental studies seem to indicate that the cross-reactivity is mainly associated with the polysaccharide antigens. This phenomenon is not unexpected since, because of their inherent chemical characteristics, the polysaccharides do not enjoy the possibilities of stereochemical (and therefore antigenic) variation that are typical of the proteins. It is thus easy to assume that a particular polysaccharide spatial configuration repeats over and over in different zoological (and botanical?) groups, so that the antibodies directed against it would not be able to distinguish among these groups. Examples of this idea abound: specificity for the blood group agglutinogen A2 has been detected in *Ascaris, Trichinella,* and *Clostridium*; *Ascaris* and *Trichinella* possess specificities characteristic of *Diplococcus* (pneumococcus) polysaccharides; the Forssman antigen has been found in several helminths. Since the cross-reactivity of some antigens may be due to identical polysaccharide moieties conjugated to different proteins, removal of the carbohydrate portion with specific enzymes may abolish the cross-reactions (Torres and Barriga, 1975). This method does not seem to have been tested yet.

For reasons of procedural convenience, the antigenic preparations used with immunodiagnostic purposes in parasitology are generally extracts of whole parasites, with tremendous predominance of the somatic over the metabolic antigens. The ubiquity of the somatic antigens in nature is a major inconvenience to obtaining specific reactions with these reagents. As is shown in succeeding chapters, cross-reactivity among diverse parasitic infections is a widely spread phenomenon. On the other hand, the presence of somatic antigens in a diagnostic preparation may have some advantages: because they are commonly shared by the diverse developmental stages of a parasite, the corresponding reactions will be positive in the different phases of the clinico-biological course of the infection. Of 21 antigens demonstrated in an extract of adult *S. mansoni,* 14 were shared by the cercariae and 11 by the eggs:

this means that this preparation should be expected to detect the immunological reactivity elicited by the cercariae in the very early infection, by the adults, and by the eggs in the chronic disease. In addition, sharing of somatic antigens permits the use of extracts of stages or parasites that are easy to obtain in the diagnosis of infections whose etiologic agents are difficult to obtain in adequate quantities. In the case of schistosomiasis, adult worms are easier to obtain than migrating schistosomula; in the case of human filariases, the serological diagnosis is often done with extracts of the filaria of the dog heart, *Dirofilaria immitis,* or with other filariae of lower animals.

The striking recent advances in biochemical concepts and methodologies have been used only sparsely in the study of parasitic antigens in the late 1960s and in the 1970s, but there are some indications that this situation is changing now, and important progress is expected in the years ahead.

IMMUNODIAGNOSTIC TESTS IN CLINICAL PARASITOLOGY

Clinical parasitologists have the distinct advantage that they can often detect the presence of the etiological agent in the host by comparatively simple procedures. This is not true in all cases, however; on occasion they must resort to the indirect demonstration of the parasite—because it or its elements are not readily accessible, or because the direct procedures are not efficient enough, or because considerable damage to the host can occur before the parasite becomes directly demonstrable. Besides, some of the techniques for the direct identification of the organisms may be cumbersome when applied to large populations for epidemiological purposes.

The exquisite specificity and sensitivity obtainable with immunological techniques make them remarkably appropriate for identifying an etiological agent by investigation of the peculiar reactions that it elicits in the host. Also, these methods are easily adaptable to populational studies and may suggest the degree of damage produced by the parasite or the state of resistance exhibited by the host. It is not surprising, then, that immunodiagnostic procedures have been utilized in the study of virtually every parasitic infection.

A growing number of books and review articles on immunological methods or on their specific application to parasitology have been published in the last few years: a selected list of them appears at the end of this chapter. In this text, only the most basic aspects of the immunological tests routinely used in clinical parasitology are discussed, so that the reader may gain some understanding of their advantages, limitations, and proper interpretation. The diagnostic tests commonly used in connection with each particular parasitosis are mentioned with the discussion of the corresponding infection. The techniques themselves are not explained in any detail since excellent manuals are already available for the use of the laboratorist.

Attributes of an Immunodiagnostic Test

Three major attributes are demanded of an adequate immunodiagnostic test: sensitivity, specificity, and proper timing.

Sensitivity Sensitivity refers to the ability to detect slight immunological responses, such as those occurring at the beginning of an infection or those remaining after parasitological cure. The more sensitive a technique is, the earlier it will demonstrate the presence of nascent immunity. On occasion, highly sensitive tests create inconvenience for the clinician, since they can detect residual immunity after the infection has passed, and make the distinction between current and remote episodes difficult. In most parasitic infections the host immune responses develop quickly after infection until reaching a peak, stabilize for awhile, and later decrease more or less rapidly depending on whether or not the parasite was totally eliminated from the host. Comparison of the results of a test done twice about 10 days apart will indicate whether the host's immune response is increasing, leveled, or decreasing; from this information, the clinician may attempt an educated guess as to the stage of the infection he is facing.

The evaluation of the strength of cell-mediated immunity is rather imprecise and is mentioned later for each particular case. The assessment of the intensity of the humoral response is commonly done by performing the same test with increasing dilutions of the suspected fluid (normally serum). Conventionally, the dilutions are increased in twofold steps beginning at 1:4 or 1:10. The results are usually reported as the *titer* of the serum, which corresponds to the reciprocal of the highest positive dilution. A serum that showed positive reactions only up to a dilution of 1:512, for example, has a titer of 512. High, medium, or low titers are highly relative terms that depend on the inherent sensitivity of the tests used: titers of 64 are considered high for the complement fixation test whereas titers of 2560 are only moderate for the indirect hemagglutination test. It is important to remember that the titer of a serum measures the activity and not the quantity of its antibodies. Depending on their affinity for the antigen or on their class, similar concentrations of antibodies against the same antigen can exhibit widely divergent activities.

The sensitivity of a test is in direct relationship to the number of molecules of antibody that must react for the test to be read as positive. In some tests, only a few antibody molecules will produce the phenomenon that the observer can detect as a positive reaction; with other tests, a large number of antibody molecules must react before the results become evident to the laboratorician. The former tests, therefore, are more sensitive than the latter ones. Since the way in which the antibody molecules react in each group of tests is characteristic for the group, the sensitivity of each technique, within limits, is also characteristic of each category. Table 2 gives examples of the limits of sensitivity of some common serological tests.

Specificity Specificity refers to the ability of a test to detect the immune responses elicited by one single class of invader (a parasitic species in this case), and not those produced by other invaders. Actually, unlike sensitivity, specificity is a property of the antigens, and not of the test, utilized. Certainly the most sensitive tests may demonstrate weak cross-reactions that would not be revealed by less sensitive techniques, but the selection of a proper antigen should avoid the cross-reactivity completely. It seems that misconceptions about this simple concept have been widespread in the last 15 years, since a number of reports that tried to solve problems of nonspecifity by devising more sensitive immunological methods have been repeatedly published. The basic approach to solve the question of lack of specificity, which is very important in many parasitoses, consists in isolating potent antigens that are exclusive of the parasite under investigation, and using them in tests of the most adequate sensitivity for the purposes in mind. There has not been any particular encouragement to do this kind of research in the last decade or so, but the work of a few established investigators seems to indicate some return to the basic task again.

Timing The timing of an immunodiagnostic test varies according to the objectives of the worker: clinicians usually prefer precocious but evanescent reactions that appear soon after infection and vanish in a few weeks, thus indicating the acute period of the disease with

Table 2. Approximate limits of sensitivity (in μg of antibody N_2/ml of serum) of common immunological tests

Test	Antibody detected
Precipitation, in tube	3–15
Precipitation, interfacial	30–50
Precipitation, in gel	15–30
Agglutination, direct	0.01–0.05
Hemagglutination, indirect (IHAT)	0.001–0.005
Complement fixation (CFT)	0.05
Fluorescent antibody, indirect (IFAT)	0.1–1.0
Radioimmunoassay (RIA)	0.0001
Enzyme-linked immunosorbent assay (ELISA)	0.0002
Transfer of skin immediate hypersensitivity (P-K test)	0.001

which they are most often concerned. Epidemiologists, in addition, also need persistent reactions that represent the prevalence of a parasitosis in a population, whenever the infection was acquired. In a few cases, like in the presence of a pregnant woman with antibodies to *Toxoplasma gondii,* the clinician may want to exclude the existence of an acute infection. The most sensitive diagnostic methods are able to detect incipient immune responses earlier and fading immune reactions longer than the least sensitive tests, but their differences are more quantitative than qualitative. As in the case of specificity, the timing at which an immune response is detected in the host may be governed to a large extent by the adequate selection of antigens: antigens restricted to the invasive stages of the parasite (e.g., cercariae of schistosomes) are bound to elicit early and transient responses, whereas antigens characteristic of later and persistent stages (e.g., eggs of schistosomes) are likely to produce tardy and sustained immunity. Not much work has been done in this area yet, but investigations with *T. spiralis* have demonstrated that antigens that react in the acute or the chronic infection may be obtained from the same developmental stage (Barriga, 1977). Since antibodies of the IgM class are commonly produced earlier and vanish sooner than the corresponding IgG antibodies, the modification of some serological techniques for the exclusive demonstration of IgM antibodies has assisted in the identification of acute or subacute infections on some occasions.

Immunodiagnostic Reactions

All immunodiagnostic reactions begin with the chemical union between the antibodies (or the specific receptors on T cells) and the corresponding antigenic determinants on the antigenic molecule. This binding, called *primary reaction,* is virtually instantaneous and usually invisible by ordinary means. When the conditions are appropriate, the primary reaction may trigger other phenomena that are detectable in vitro (e.g., precipitation, agglutination, lysis) that are called *secondary reactions.* Primary or secondary reactions that occur in vivo may, in turn, initiate further events that manifest themselves as physiological changes (e.g., immediate-type allergy, Arthus reaction, delayed-type allergy); these manifestations have been called *tertiary reactions.*

Most of the routine diagnostic tests are secondary reactions; a few are tertiary, and some primary reactions have been modified to make them visible for practical use. It is important to keep in mind, however, that many times an antigen-antibody reaction will not be detected if the accessory conditions to render it perceptible are not adequate. In the following pages the basic principles and characteristics of the immunodiagnostic reactions commonly used in parasitology are reviewed.

Precipitation Reactions

Precipitation consists of the insolubilization of a previously soluble antigen; it is initiated by the formation of complexes by cross-linking of the antigen with the corresponding antibody molecules, followed by the aggregation and sedimentation of the complexes. This reaction requires the presence of ions and of a divalent or multivalent antigen, and it is strongly dependent on the proportion of antigen to antibody. When antibody is used in excess, each antigen molecule becomes surrounded by all the antibodies that it can bind, so that cross-linking or cross-bridging among various antigen molecules does not occur and large sedimenting aggregates are not formed (Figure 3). Despite the antigen-antibody reaction, a precipitate is not observed and the reaction appears negative; this phenomenon is called *prozone*. Excess of antigen has the same effect, but this time is the antibody molecules that become saturated with as many antigenic molecules as they can bind (two for IgG) (Figure 3); this phenomenon is sometimes called *postzone*. The maximum of precipitation occurs at an antigen : antibody ratio that varies somewhat with different antigens and is known as *optimal proportion*.

Originally the precipitation test was performed by mixing aliquots of the antigenic solution and the corresponding serum in a tube (precipitation in tube), but a modification that minimizes the prozone effect (*interfacial* or *ring* precipitation) was soon developed. In this case, the antigenic solution is carefully laid over the serum in a small tube; both

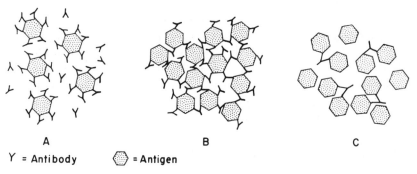

A   B   C

Y = Antibody   ⬡ = Antigen

Figure 3.   Effect of the antigen : antibody ratio on precipitation and agglutination tests. A relative excess of antibody (**A**) produces a *prozone* phenomenon; each molecule of antigen becomes totally surrounded by as many molecules of antibody as it can bind, so that cross-linking and large aggregates are not produced. A relative excess of antigen (**C**) causes a *postzone* phenomenon; in this case, every molecule of antibody is saturated by as many molecules of antigen as it can bind (two when the antibody involved is IgG) and the final effect is the same. An *optimal proportion* of antigen and antibody favors the production of extensive cross-linking and the formation of large, visible aggregates (**B**).

reagents diffuse into each other and form a cloudy disk of precipitation when they reach the optimal proportion. To save reagents, this latter test can be performed in a capillary tube by drawing the antigenic solution into it, followed by the serum.

Precipitation can also take place in semisolid media if they are permeable enough to allow the diffusion of the serum and of the antigenic solution toward each other. These techniques are commonly called *diffusion* or *immunodiffusion* tests, and 1% agar is often used as the medium. The best known of the several varieties of this technique is the *double diffusion* or *Ouchterlony's* test that consists of depositing the antigenic solution and the serum in different wells in the agar, separated by a few millimeters. Both reagents diffuse into the agar and form a band of precipitation in the zone where they found each other and reached the optimal proportion. When complex mixtures of antigens are used, the overall rate of diffusion of several of them may be similar, so that they will precipitate in the same place and the test will be incapable of discriminating among them. This has been partially solved by separating the antigenic proteins according to their isoelectric points, by the passage of an electric current through the medium, before attempting their precipitation. Once the electricity has distributed the antigens along its path, a trough is cut in the agar and is filled with the serum, and the diffusion is allowed to proceed (*immunoelectrophoresis*). This technique permits the separation and identification of a score of antigens (against less than a third of that in the Ouchterlony's test) and is the preferred diagnostic method for hydatidosis for some workers.

The diffusion tests, however, may take several days before providing reliable results and, since the reagents diffuse in all directions, they are wasteful of materials. The *counterimmunoelectrophoresis* (electroimmunodiffusion, immunoelectroosmophoresis, or cross-over electrophoresis) test has been specifically designed to solve these problems. Since many proteins migrate toward the anode (positive pole) when subjected to an electric current in a moderately alkaline buffer, whereas IgG moves toward the cathode under the same conditions, disposition of the serum in a well on the anodic side of the plate and the antigenic solution in a well on the cathodic side and passage of an electric current will impulse the reagents toward each other. The results with this technique may be ready in less than an hour, and all the reagents accumulate in the zone where the reaction occurs. On occasion, *microprecipitation* tests (or precipitation on living organisms) have been utilized with diagnostic purposes, e.g., circumlarval precipitation with *T. spiralis,* circumoval precipitation with eggs of *Schistosoma,* and the Cercarienhüllenreaktion with schistosome cercariae;

this later apparently requires complement in addition to precipitating antibodies.

In a growing number of parasitic infections it has been found that antigens of the parasite are present in the fluids of the host, sometimes before the production of antibodies begins. Also, the concentration of parasite antigens in the host is expected to be a more reliable indicator of the parasitic load than the evaluation of the antibody activity. The demonstration of circulating antigens has been often done by reactions of precipitation, using serum of the patient as the antigenic solution, and potent specific antisera raised in the laboratory as the antibody solution. The results of these tests have not been satisfactory enough for practical diagnosis yet, possibly because a large proportion of the circulating antigen is already bound to the corresponding antibodies so that it does not react in vitro. A *uroprecipitation* test in which the parasite antigen is demonstrated in the concentrated urine of the patient has been used as an assay in trichinellosis and schistosomiasis, but the results have been inconsistent.

The antibody class most efficient in precipitation reactions is IgG. Although IgM and, probably, IgA contribute to a minor extent, for practical purposes the results of these reactions are conventionally attributed to the content of IgG antibodies in the fluid under study. Since IgG antibodies are produced later than IgM antibodies in most cases, precipitating reactions are inherently more slow to appear positive than other techniques that demonstrate IgM antibodies preferentially. Also, the optimal antigen: antibody ratio for precipitation with most antigens requires several molecules of antibody for each molecule of antigen; this, in conjunction with the large number of molecules necessary to produce visible precipitates, makes the precipitation tests wasteful of antibody and, therefore, not very sensitive.

Agglutination Reactions

The agglutination tests are essentially similar to the precipitation tests, but in this case the antigen is particulate rather than soluble (0.2 μ is considered by some authors to be the limit of antigen size for agglutination and for precipitation). Since these antigens will precipitate even in the absence of a reaction with their antibodies, positive results are detected by the pattern of sedimentation rather than by its mere presence.

Direct agglutination consists of aggregating the particles of antigen (e.g., bacteria, erythrocytes, yeasts, protozoa) by cross-linking them with the antibodies in their corresponding antisera. This technique is simple, but cannot be used with parasites larger than protozoa; it requires a permanent source of living organisms or methods of preservation that do not alter the relevant antigens, and it is restricted to detecting only surface antigens. In *passive* or *indirect* agglutination, antigens extracted from the desired source are artificially bound to particles that are subsequently reacted with the suspected serum. The coating of the particles with antigen is traditionally called *sensitization*. A number of different particles have been assayed for these purposes and several of them are in current use (e.g., latex, bentonite, charcoal, cholesterol-lecithin, collodion), but the most popular method seems to be the sensitization of erythrocytes. The name *indirect hemagglutination* has been given to this latter technique; the tests that use other particles are often called *flocculation* tests. Flocculation reactions are actually precipitation reactions with rather strict pro- and postzones and whose precipitates are in clumps rather than amorphous; they occur predominantly with sera of horses immunized with proteins.

The approximate contribution of IgM antibodies to an agglutination reaction may be investigated by running the test with native serum and with mercaptoethanol-treated serum: the drop in titer after reductive treatment is taken as an indication of loss of IgM activity. This technique also abolishes IgA and some IgG activity, so the results must be taken with caution.

The agglutination reactions can be easily used to detect antigen rather than antibody by using the method of *inhibition of agglutination*. In this case, an agglutination test with a known positive serum is set up, aliquots of the diluted serum are mixed with dilutions of the fluid suspected of containing the antigen, and the agglutination test is performed with these aliquots. Inhibition of a reaction that was positive before will indicate that the antibodies in the serum were preoccupied by the antigen and implies the presence of the antigen in the fluid under study.

The agglutination tests can be done in test tubes, capillary tubes, plaques of microtubes, slides, or with the particles adsorbed on cards. The particles are often stained to facilitate the reading of the results.

IgM antibodies are most efficient for producing agglutination reactions, although IgG and IgA antibodies are also quite effective. Since IgM antibodies usually appear early in the process of immunization, the agglutination techniques are normally precocious tests. Visible agglutination may result from the aggregation of relatively few particles, which, in turn, may be brought about by comparatively few antibody molecules; this makes agglutination, particularly hemagglutination, one of the most sensitive conventional methods for the demonstration of antibodies. Unfortunately, these tests, especially hemagglutination, are markedly affected by the prozone phenomenon and show considerable day-to-day variation in their results.

Tests Based on the Action of Complement

There are a number of tests that take advantage of the properties of complement. The oldest, most general, and best known of them is the *complement fixation* test. This technique consists of detecting the consumption (or activation) of the complement, consecutive to the reaction of IgM or IgG antibodies with the corresponding antigen. Normally, the antigen and carefully measured quantities of complement are provided by the laboratorician and the antibodies may or may not be provided by the (heat-inactivated) fluid (serum commonly) under study. After incubation of this mixture, the complement is measured again— if it was used up, it indicates that an antigen-antibody reaction took place and implies the presence in the serum of antibodies corresponding to the antigen; if it was not, it implies the absence of complement-fixing antibodies. Detection of complement is normally done by adding IgM antibody-sensitized red blood cells. Their hemolysis will show that the complement is still present; the lack of hemolysis will indicate that it was consumed during the first reaction.

Complement fixation is a reasonably sensitive technique, very reproducible, little affected by the prozone phenomenon (in contrast with the precipitation and agglutination methods), and adequate to use with soluble or particulate antigens. Since IgM and IgG antibodies are functional in this test, but the former are much more efficacious, the corresponding reactions usually appear early in the infection. These advantages are considerably weakened by important drawbacks, however: the execution of the test requires careful preparation and standardization of the reagents, and the use of numerous controls; some antigenic preparations and some sera adsorb the complement by themselves (they are called *anticomplementary*) so that they cannot be used with this technique; and some animal species do not provide adequate sera (horse antibodies do not produce hemolytic complement complexes, pig serum promotes complement activity and yields unreliable results, dog serum often becomes anticomplementary after heat-inactivation, and avian antibodies do not activate mammalian complement). The existence of simpler sensitive tests of more general use has considerably reduced the popularity of the complement fixation technique in the last years; however, recent standardization of this method by the U.S. Center for Disease Control may stimulate its use again in the future.

Another common test based on the activity of complement is Sabin and Feldman's *methylene blue dye test* for toxoplasmosis. Incubation of the parasites with serum containing anti-*Toxoplasma* antibodies and with an *accessory factor* of human serum lyses the organisms in such a way that they lose the affinity for methylene blue that is characteristic of healthy toxoplasms. Conventionally, the test is positive when 50% or more of the protozoa lose their tinctorial affinity; the titer of a serum is the reciprocal of the highest dilution able to produce this result. There is solid evidence now that the accessory factor is identical to the complement system. The sera of many animals and of numerous people contain a nonspecific, heat-labile anti-*Toxoplasma* substance that prevents their use as a source of accessory factor. This material must be destroyed by heating at 60°C for 30 min before testing for anti-*Toxoplasma* antibodies by the dye test. Some still limited work suggests that the nonspecific anti-*Toxoplasma* activity might be connected with the activation of the complement by the alternative pathway. The dye test is a sensitive and specific technique, but the necessity for maintaining and manipulating living parasites is a deterrent to its use.

The skin test (see "Tests for Immediate-type Hypersensitivity," below) occasionally demonstrates the presence of Type III or intermediate-type hypersensitivity. In this case, the inoculated antigen precipitates in situ with the corresponding circulating IgG antibodies and the complement is activated. The complement-derived compounds produced cause vascular changes and attract neutrophils and mononuclear leucocytes that give origin to a soft skin swelling. This reaction begins in 1–2 hr, peaks in 4–8 hr, and subsides in 12–24 hr.

Tests with Labeled Antibodies

Most of the techniques using labeled antibodies are rather new methods that exhibit an exquisite sensitivity: those in current use or being evaluated now for parasitic diseases are the indirect fluorescent antibody test (IFAT), a radioimmunoassay (RIA) in the form of the radioallergosorbent test (RAST), and the enzyme-linked immunosorbent assay (ELISA).

Indirect Fluorescent Antibody Test   The IFAT consists of revealing the presence of antibody attached to its antigen by means of a fluorescent anti-immunoglobulin. The antigen (e.g., a section of the parasite) is flooded with the serum under investigation and washed carefully to remove the unreacted materials; if antibodies were present in the serum, they will remain bound to the antigen despite the wash. The preparation is then flooded again with a (commercial) solution containing anti-immunoglobulin antibodies labeled with a fluorescent dye, and washed for a second time; if antibodies stayed on the antigen the first time, the anti-immunoglobulin fluorescent antibodies will attach to them and will resist the second wash. The final preparation is examined under a microscope with an ultraviolet light that will excite the fluorescent material and show it as bright, colored areas (Figure 4).

The main advantages of the IFAT are that it does not require previous purification of the antigen, it indicates its location in the parasite, and it can be easily modified to demonstrate diverse classes of antibodies by using the respective anti-immunoglobulin preparation (e.g., anti-IgM, anti-IgG). Its sensitivity is high, but it is limited by the ability of the human eye to discriminate tiny points of fluorescence. The major drawbacks are the relatively elevated cost of the necessary equipment, the confusing presence of natural fluorescence in some tissues or dyes, and the difficulty of establishing the dilution at which the results must be considered negative on a purely subjective basis.

In a modification called the *soluble antigen fluorescent antibody* (SAFA) test, the antigen is adsorbed on a disk and the fluorescence may be read in a fluorometer. This technique permits comparison of different sera and assigning of a constant value to the weakest reaction acceptable as positive.

Radioimmunoassay   RIA has undergone a number of minor modifications that have turned it into a family of closely related tests. Most of them are utilized to detect minute amounts of antigens (hormones or drugs), but a particular technique that measures IgE antibodies to a given antigen (*radioallergosorbent test,* or RAST) is being assayed in parasitology. In this test, the allergen (or an antigenic preparation that contains it) is coupled to a solid matrix that is incubated with the

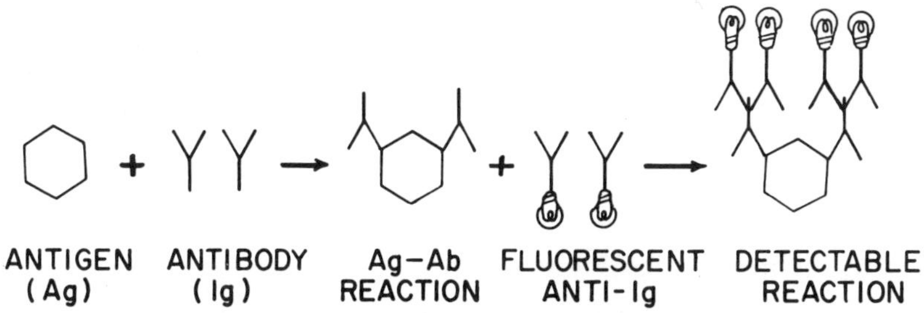

ANTIGEN   ANTIBODY   Ag—Ab   FLUORESCENT   DETECTABLE
(Ag)   (Ig)   REACTION   ANTI-Ig   REACTION

Figure 4.   Mechanism of the indirect fluorescent antibody technique (IFAT). The antigen fixed on a microscope slide is reacted with the suspected serum and subsequently washed; if specific antibodies present in the serum bound the antigen, they will remain attached despite the wash. The slide is then reacted with a solution of anti-immunoglobulin antibodies labeled with a fluorescent dye and washed again; the anti-immunoglobulin antibodies will attach to the antibodies already coating the antigen and will resist the wash. Microscope examination under untraviolet light will excite the fluorescent dye and will show the respective antibodies as bright spots. The observation of fluorescence therefore implies that specific antibodies were present in the suspected serum.

suspected serum for the antibodies to bind the antigen, and subsequently washed to remove the unreacted materials. Now the matrix is treated with (commercial) anti-IgE antibodies labeled with ^{125}I and washed again to eliminate the unreacted protein, and the radioactivity remaining on the matrix is counted. Curves provided with the anti-IgE solution permit expression of the radioactivity in nanograms of IgE antibody. Appropriate modifications in the reagents used will allow detection of nanogram quantities of any other class of antibodies also. The RAST is an extremely sensitive method that, unfortunately, requires equipment and procedures too sophisticated for many laboratories of parasitologic diagnosis in areas of endemia.

Enzyme-linked Immunosorbent Assay ELISA has also experienced some modifications directed to meet particular purposes; the technique used for antibody detection is the *indirect* ELISA. The test begins by coating tubes (usually of polystyrene or polypropylene) with the antigen and incubating this with dilutions of the suspected serum so that the antigen-antibody reaction can take place. After washing to remove unreacted materials, an anti-immunoglobulin preparation labeled with an enzyme (e.g., alkaline phosphatase) is added, so that the enzyme-carrying anti-immunoglobulin will bind the antibodies that remained attached to the antigen. After a new wash, the substrate of the enzyme is added (e.g., *p*-nitrophenyl phosphate) and the whole system is incubated so that the enzyme can act. After an appropriate period, the reaction is stopped and the conversion of the substrate is evaluated qualitatively by simple inspection, or quantitatively by spectrophotometric measurement of the color that has developed. Enzymatic degradation in the system implies the presence of specific antibody in the suspected serum and the activity of the enzyme is proportional to the number of antibody molecules that bound the antigen (Figure 5).

The ELISA approaches the sensitivity of RIA but is less expensive and much simpler to perform. Commercial kits for the diagnosis of some infectious diseases are already on the market.

Test for Immediate-type Hypersensitivity

Three tests are currently in common use in clinical practice to determine the presence of immediate-type allergy or Type I hypersensitivity: the skin test, the histamine release assay, and the RAST. The skin test is much simpler than the other two and provides almost immediate results; it is influenced by the particular sensitivity of the subject to the vasoactive materials released during the reaction, but this might not be a drawback when evaluating a patient. A feared inconvenience of the skin testing is the possibility of triggering a systemic allergic reaction in the individual tested. Although this rarely occurs during the

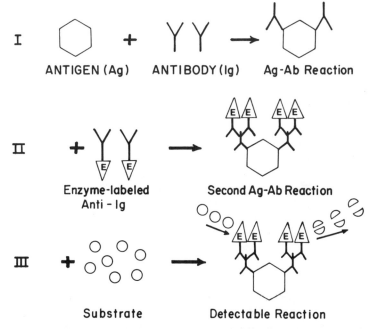

Figure 5.   Mechanism of the indirect enzyme-linked immunosorbent assay (IELISA). The antigen attached to the interior of a tube is reacted with the suspected serum and subsequently washed; if specific antibodies present in the serum combined with the antigen, they will remain bound despite the wash. A solution of anti-immunoglobulin antibodies labeled with an enzyme is then added to the tube and the contents are washed again; these antibodies will react with the antibodies already bound to the antigen and will resist the wash. The substrate of the enzyme is then added and its transformation is investigated after an appropriate period. Enzymatic degradation of the substrate indicates the presence of the enzyme and therefore implies that specific antibodies were present in the suspected serum.

diagnosis of parasitic diseases, it is not unheard of during evaluation of patients sensitive to hymenoptera stings. Repeated skin tests also have a possibility of sensitizing a subject against the antigen used or of boosting a preexisting immune response.

The histamine release assay and the RAST are in vitro methods that cannot produce anaphylaxis or sensitization of the patient, but their general correlations with the patient's clinical history are only 90% and 70%, respectively. The RAST has the advantage that it is done with serum, which can be stored almost indefinitely. Both techniques, however, are expensive and complex to perform.

The *skin test* is done by injecting about 0.05–0.1 ml of the appropriate antigenic solution in the epidermis of the volar surface of the forearm in humans or in the ear of domestic animals (*intraepidermal test*). Some clinicians prefer to prick the skin through a drop of antigenic solution (*prick test*). In allergic patients, it is likely that some local mast cells have IgE antibodies attached to their membranes; the combination of the specific antigen with these antibodies will cause the release of vasoactive substances from the cells. In allergic patients, a circumscribed area of edema (*wheal*) surrounded by a zone of erythema (*flare*) appears around the point of inoculation within 30 min, and usually subsides in a couple of hours. Comparison of the size of this reaction with a control inoculation of the straight vehicle where the antigen is diluted will indicate the intensity of the reaction and whether or not it is specific. The results of this test are highly dependent on the dose of antigen administered. Kagan found that its specificity in hydatid disease diminished considerably when hydatid fluid containing more than 12–15 μg of N_2 per ml was used as antigen; we found the same with hydatid fluid containing more than 4 μg of protein per ml in *Echinococcus granulosus*–infected dogs.

The *histamine release assay* consists of incubating leucocytes of the patient with the relevant allergen. If his basophils are coated with the corresponding IgE antibodies, the antigen-antibody reaction that takes place triggers the release of histamine from the cells, which is subsequently measured by fluorometric techniques. The RAST measures the amount of specific IgE antibody against a given antigen in the serum of the patient (see "Tests with Labeled Antibodies," above). Both techniques may be modified to evaluate blocking IgG antibodies. Because of the complexity of these methods, their cost, and the equipment they require, they are rarely available at large for parasitological diagnosis.

Tests for Delayed-type Hypersensitivity

A number of techniques to demonstrate Type IV hypersensitivity are currently available, but the most popular methods in clinical medicine are the skin test, lymphocyte transformation (or blastogenesis), and the inhibition of leucocyte migration.

Skin Test The skin test for delayed-type hypersensitivity is similar to, and has the same advantages and inconveniences as, that described for immediate-type hypersensitivity above, but in this instance the antigen reacts with effector T lymphocytes previously sensitized with the specific antigen and causes the production of lymphokines. The cell-recruiting activities of these substances produce a local inflammatory infiltration (with predominance of lymphocytes and macrophages) that is expressed externally as an area of induration in the skin. The size of the induration is taken as an indication of the intensity of the cell-mediated immunity to that particular antigen. The reaction commonly is evident after 24 hr and may reach its peak 48–72 hr after the inoculation of the antigen. Similar to the Type III hypersensitivity but contrary to the Type I allergy, this reaction is inhibited by the administration of corticosteroids. The delayed reaction to tuberculin is partially inhibited by the injection of histamine, within 10–40 min, in the same place as the inoculation of tuberculin, probably because the local vasodilation washes out the antigen or the lymphokines. This finding suggests that the local manifestations of Type IV hypersensitivity may be abolished when they are preceded by an immediate-type reaction to the same antigenic preparation.

Lymphocyte Transformation Test The lymphocyte transformation test consists of incubating peripheral lymphocytes of the patient with the desired antigen: the reaction of the antigen with effector T lymphocytes that have been previously sensitized by contact with the same antigen will stimulate the proliferation of these cells and, probably, the production of lymphokines. These lymphokines, in turn, may have a mitogenic effect on other cells and add to the cellular multiplication. The rate of cell division in the tubes with antigen is assessed by measuring the incorporation of ^{3}H-labeled thymidine, and compared with the incorporation in control tubes without antigen. Any important increase of the cellular proliferation in the experimental tubes over the control tubes indicates a mitogenic effect mediated by the antigen, and implies the existence of lymphocytes sensitized against that particular antigen (presumably as a result of an infection) in the patient. The ratio of isotope incorporation between the control and the experimental tubes is usually considered to be a measure of the intensity of the reaction. The lymphocyte transformation test is commonly regarded as an in vitro correlate of cell-mediated immunity rather than an evaluation of the humoral competence, because memory B cells are much less abundant in the circulation than sensitized T cells, and do not have the additional amplifier mechanism represented by the lymphokines. The correlation between this assay and the presence of Type IV hypersensitivity as detected by in vivo procedures is not perfect, however.

As an in vitro method, the lymphocyte transformation test does not represent any immunological risk for the patient, but it has a number of drawbacks as a clinical test: it is expensive, complex, and requires sophisticated paraphernalia; the patient's cells do not tolerate more than a few hours of transportation or storage; the results are not evident before 4 or more days; and the data obtained are often too variable to be of any use in individual cases.

Inhibition of Leucocyte Migration The inhibition of leucocyte migration consists of depressing the normal translation of the neutrophils by incubation of peripheral leucocytes with an antigen. In this assay, peripheral leucocytes (obtained from the buffy coat, for example) are set up in agar or in a microtube in a culture media to which the desired antigen is added. If effector T lymphocytes sensitized against that particular antigen are present in the cell population, their reaction with the added antigen will result in the production of lymphokines. The presence among these of the factor that inhibits neutrophil migration will be evident because these cells will not move away from their original location as will the cells in a control without antigen. As in the previous assay, observation of the effect of the lymphokine implies the existence of T cells sensitized against the specific antigen utilized, and presumes the corresponding infection in the patient. The intensity of the reaction is assessed by determining the area of migration of the neutrophils in the experimental plaques as a percentage of the migration in the control plaques. The test for inhibition of macrophage migration (caused by a different lymphokine) is done in a manner similar to that for leucocytes but the patient provides only the T cells and the macrophages are obtained from the peritoneal cavity of guinea pigs. This technique is somewhat simpler than the lymphocyte transformation, requires less sophisticated equipment, and may provide results within 48 hr. In addition, its correlation with the presence of Type IV hypersensitivity in vivo is excellent.

CONCLUSION

In general, techniques such as the RAST, the histamine release, the lymphocyte transformation and the inhibition of the leucocyte migration have wide and important applicability in biomedical research and in several clinical specialties. However, their routine use in clinical parasitology appears doubtful when one remembers that human parasitoses occur preponderantly in areas of scarce resources and that animal parasitoses are strictly subjected to economical considerations.

SOURCES OF INFORMATION

Barriga, O. O. 1977. Reactivity and specificity of *Trichinella spiralis* fractions in cutaneous and serological tests. J. Clin. Microbiol. 6:274–279.

Bloom, B. R., and David, J. R. (eds.). 1976. In vitro Methods in Cell-Mediated and Tumor Immunity. Academic Press, Inc., New York.

Fife, E. H., Jr. 1972. Current state of serological tests used to detect blood parasite infections. Exp. Parasitol. 31:136–152.

Garvey, J. S., Cremer, N. E., and Sussdorf, D. H. 1977. Methods in Immunology. W. A. Benjamin, Reading, Mass.

Kagan, I. G., and Norman, L. 1976. Serodiagnosis of parasitic diseases. In: N. R. Rose and H. Friedman (eds.), Manual of Clinical Immunology, pp. 382–409. American Society for Microbiology, Washington, D.C.

Kwapinski, J. B. G. 1972. Methodology of Immunochemical and Immunological Research. Wiley-Interscience, New York.

Pan American Health Organization. 1967. Immunological Aspects of Parasitic Infections. P.A.H.O., Washington, D.C.

Rice, C. E. 1968. Comparative serology of domestic animals. Adv. Vet. Sci. 12:105–162.

Smith, C. E. G. (chairman). 1976. Symposium on recent advances in the serology of tropical diseases. Trans. R. Soc. Trop. Med. Hyg. 70:93–113.

Soltys, M. A., and Woo, P. T. K. 1972. Immunological methods in diagnosis of protozoan diseases in man and domestic animals. Z. Tropenmed. Parasitol. 23:172–187.

Torres, P., and Barriga, O. O. 1975. Inter- and intra-specific antigenic relationships among some Ascaroidea. Acta Parasitol. Polon. 23:441–451.

Weir, D. M. (ed.). 1978. Handbook of Experimental Immunology. 3rd Ed. Blackwell Scientific Publishing, Oxford.

Williams, C. A., and Chase, M. W. (eds.). 1967–. Methods in Immunology and Immunochemistry. Vols. I–V. Academic Press, Inc., New York.

Wilson, R. J. M. 1978. Circulating antigens of parasites. In: Immunity in Parasitic Diseases, pp. 87–101. Colloque INSERM-INRA. Editions INSERM, Paris.

World Health Organization. 1975. Parasitic antigens. Bull. W.H.O. 52:237–249.

Chapter 3

Immune
Reactions to Parasitic Protozoa

THE PROTOZOA PREVAILING IN THE LUMENS

ENTAMOEBA HISTOLYTICA

The infection by *Entamoeba histolytica,* acquired by ingestion of cysts eliminated with the stools of infected hosts, is restricted almost exclusively to humans. Natural, asymptomatic infections are occasionally found in lower primates, dogs, and rats. Findings of the parasite in cats and pigs are exceptional. Experimentally, the parasite can be established in monkeys, dogs, cats, guinea pigs, rats, mice, and hamsters. Mattern and Keister (1977) have recently developed experimental models of hepatic and of cerebral amebiasis in hamsters and in newborn mice, respectively. A natural outbreak of symptomatic amebiasis in monkeys, producing intestinal and hepatic lesions, has been reported recently (Amyx et al., 1978), which in time may provide a better model to study the human infection.

Natural Resistance

Little is known about the mechanisms of susceptibility to *E. histolytica.* The available evidence suggests that the protozoan requires fairly strict physicochemical constants that are present only in certain mammalian species. Several experiments have demonstrated that the colonization of the intestine of rodents and the producton of histological lesions necessitate the association of bacterial flora. Strains of *Entamoeba* that did not affect the tissues of rats became invasive, however, when transferred to splenectomized rats. This finding indicates that the pathogenicity of the parasite may be partially controlled by the immune system of the host. The same conclusion is supported by the recent observation that rats immunodepressed with drugs or irradiation developed more severe pathology than the untreated controls on infection

41

with *E. histolytica*; their susceptibility to the infection remained unchanged, nevertheless. Apparently the resistance to the infection has a nonimmunological basis, but the protection against the disease includes the activity of the immune system.

Acquired Immunity

Recent reviews on acquired immunity to *E. histolytica* have been written by Balamuth and Siddiqui (1970), by Kagan (1973), and by Krupp and Jung (1976).

The amebic infection in humans, especially when tissue invasion occurs, gives origin to a variety of antibodies and elicits a cell-mediated response. IgM, IgG, and IgE have been already identified among the former, but IgA was not found in the serum in the same study (Harris, et al., 1978).

Despite the presence of specific immune responses, epidemiological observations indicate that humans are completely susceptible to amebiasis. The occurrence of repeated attacks of amebic dysentery in the same patient suggests that lasting protection does not develop in these cases. Hepatic amebiasis, on the contrary, rarely affects the same patient more than once. We do not know whether this indicates that the parenteral location of the parasite elicits a more effective immune resistance or the statistical chances of acquiring hepatic amebiasis twice are too low.

Patients followed for years in Colombia and in South Africa have shown that symptomatic intestinal amebiasis and hepatic amebiasis can coincide with titers, very elevated sometimes, of specific antibodies against the protozoa. On some occasions, these patients suffered reinfections despite their strong humoral response. A recent report communicated that patients with invasive amebiasis (intestinal or hepatic) exhibited skin tests predominantly of the immediate type whereas patients with noninvasive infections had skin reactions predominantly of the delayed type. This correlation may be a coincidence or may suggest that cell-mediated immunity opposes the invasive capacity of the parasite.

The lack of solid evidence of protective immunity in humans has stimulated some research in laboratory models. Observations in dogs have demonstrated that the infection terminates spontaneously, which may indicate the production of protective immunity. It is not known whether the same occurs in man, although a few cases of spontaneous cure of hepatic amebiasis are on record. With regard to immune resistance to reinfections, Krupp and Jung (1976) have reviewed most of the available evidence: in a study of 29 dogs that were infected with *E. histolytica*, treated, and challenged again, 24 of them remained totally refractory to the reinfection for periods of $2\frac{1}{2}$ to $9\frac{1}{2}$ months. Blood transfusions from these dogs to normal dogs transferred some degree of resistance. In another study, repeated subcutaneous inoculations of *E. histolytica* extracts in rats reduced their subsequent susceptibility to 63%, from the 92% found for noninoculated rats.

A different experiment with hamsters gave interesting results. The experimental animals received inoculations of parasite extract in Freund's complete adjuvant and the controls received only the adjuvant. All of the hamsters were subsequently infected with the protozoan directly in the liver, and the presence of specific antibodies and hepatic lesions was investigated 10 days later. Among the experimental hamsters, 96% had antibodies and 29% developed liver lesions; among the controls, only 43% had antibodies, but 92% presented lesions. This report suggests that a strong antigenic stimulation may prevent a large part of the pathology. The resistance does not have to be necessarily related to the humoral response, however, since Freund's complete adjuvant is a notorious stimulant of cell-mediated immunity (which was not investigated in this work). In addition, another study with hamsters found that antibody titers correlated fairly well with the size of the lesions. At this time is not known, however, whether larger lesions represent a stronger antigenic stimulation or the antibodies play a role in the production of the pathology.

Recent studies of transformation of peripheral lymphocytes in individuals with hepatic amebiasis, individuals with dysenteric amebiasis, asymptomatic carriers, and subjects with no infection or history of infection revealed that only those with liver lesions responded to the *E. histolytica* extract (Harris and Bray, 1976). Since most of the peripheral lymphocytes are T cells, this reaction probably indicates cell-mediated immunity, which is also supported by the fact that lymphocyte transformation was independent of the presence of precipitating antibodies in the same patients. These results, however, are at variance with the reported absence of delayed skin reactions in patients with invasive amebiasis.

Recent work on vaccination of guinea pigs with complete extract of the parasite or with three Sephadex fractions, in Freund's complete adjuvant, showed that a 650,000-dalton fraction induced complete resistance to a challenge, whereas the extract produced only 70% resistance and the other fractions were less effective.

In summary, the capacity of *E. histolytica* to induce resistance to the current infection or to a reinfection has been documented in lower animals, but the relevance of these findings to humans, the natural and most susceptible host, remains to be determined. Observations in humans rather negate the production of immune refractoriness, but some association (inhibitory or enhancing) may exist between the immune response and the pathology. Considerable work in this area is still necessary; possibly recent findings of spontaneous amebic disease in spider monkeys will provide a more valid model for the human infection.

Immunodiagnosis

Although *E. histolytica* is detectable in the stools of virtually all infected individuals, its finding and identification require considerable expertise and patience, and do little to help decide whether the parasite has invaded the tissues or remains as a lumen dweller. The adequate immunodiagnosis of amebiasis is expected to facilitate the identification of the infection, make extensive surveys practicable, and assist in the recognition of the invasive forms of the infection. Some caution is necessary in the interpretation of reports generated prior to 1960, since *E. histolytica* was cultured in association with bacteria until about that date and many of the antigenic preparations utilized previously may have been contaminated with extraneous materials. Around 1960, most investigators began to use cultures with *Trypanosoma cruzi,* which greatly facilitated the separation of the organisms, and axenic media were developed later. These media prevent contamination but also constitute an artificial environment quite different from the normal intestinal habitat of the parasite.

Recent information on the immunodiagnosis of amebiasis has been published by Elsdon-Dew (1976). Valuable information is also presented in some references listed in Chapter 2.

The complement fixation test (CFT) was the first technique assayed for the serological diagnosis of amebiasis, but at present it has been superseded by more sensitive methods. In an early evaluation, this test identified only 63% of 363 cases of intestinal amebiasis, but it was positive in 28 of 29 cases of hepatic infection.

The indirect hemagglutination test (IHAT) is considerably more sensitive: in one series, it was positive in 91% of 314 patients with hepatic amebiasis, in 84% of 514 with amebic dysentery, in 9% of 191 healthy carriers, and in 2% of 658 noninfected individuals. The sensitivity of the IHAT sometimes conspires against its usefulness, since it can detect specific antibodies in a large proportion of healthy carriers and in some recovered patients with residual antibodies. Curiously, it has been reported that children under 5 years of age with acute dysentery tend to give negative tests. The U.S. Center for Disease Control reports as positive only titers of 128 or above. Both the CFT and the IHAT have been used to detect specific antibodies in the stools of patients (*coproantibodies*) with good results.

Agglutination tests with the antigen adsorbed on latex or bentonite particles are somewhat less sensitive than the IHAT, but their overall effectiveness coincides with that of the latter.

A number of precipitation tests have also been used in amebiasis. The double diffusion in gel was positive in 92% of 622 cases of amebic hepatitis, 72% of 595 cases of amebic dysentery, 55% of 19 asymptomatic cyst carriers, and 10% of 198 noninfected people. Immunoelectrophoresis and counterimmunoelectrophoresis (CIEP) are at least as sensitive as the gel double diffusion, but the precipitation in capillary tubes is less sensitive. In most cases, the precipitation tests are very sensitive in detecting invasive amebiasis, specially of the liver, and react very little with the serum of healthy carriers; for these characteristics and for the relative rapidity of the results, some workers think that CIEP is the test of choice to identify hepatic amebiasis (Farid et al., 1977). The antibodies detected by these tests are different from those reactive in the IHAT and they are present as often in infected children as in infected adults.

The indirect fluorescent antibody test is as sensitive as the IHAT in the hepatic forms of amebiasis, but less so in the dysenteric forms.

Skin reactions of the immediate and the delayed type are common in amebiasis: the former may appear after a week of overt disease, whereas the latter may take up to $1\frac{1}{2}$ months of symptomatic infection before becoming positive. Some investigators have found some association between invasive amebiasis and immediate-type skin reactions, but this correspondence has failed in other studies. Harris and Bray (1976), on the contrary, found that only the peripheral lymphocytes of 6 patients of hepatic amebiasis underwent transformation with specific antigen, whereas 9 dysenteric patients did not respond. In general, it appears that the sensitivity of the cutaneous reactions to detect invasive amebiasis and to differentiate it from the noninvasive infection is not as good as those of alternative techniques.

Other tests, such as enzyme-linked immunosorbent assay, radioimmunoassay, the immobilization test, and the direct agglutination test, are in the process of evaluation or are used mostly in experimental research.

In general, most of the immunological techniques are so sensitive in detecting antibodies in hepatic amebiasis that some clinicians exclude the disease in the presence of a negative serological test. The immunological techniques are less sensitive, but still give a very high proportion of positive results in infections with important invasion of the intestinal tissues. The percentage of positivity in healthy cyst carriers, however, is quite variable with any test. This characteristic may depend on the persistance of antibodies from past infections (which remain detectable in the serum for a year or more) or on limited invasion of the tissues that may trigger an immune response, despite being insufficient to cause clinical symptoms. A variable proportion of presumably healthy carriers may actually have asymptomatic but invasive infections that elicit antibody formation. Noninfected individuals with positive serology may have had similar episodes in the past. It is possible that the controversies that occasionally arise on the value of different techniques for the immunodiagnosis of amebiasis spring from the variety of antigenic preparations in common use. Only recently have systematic investigations on the antigens of *E. histolytica* been initiated. Reportedly, the main antigens in the CFT and in the IHAT are glycoproteins of 220,000 and 650,000 daltons.

Krupp (1977) studied sera of patients of three continents by immunoelectrophoresis with extracts of the parasite and found up to 14 bands of precipitation. The number of bands and the IHAT titers were directly related only in some patients; similarly, other workers have been unable to differentiate among diverse strains of the parasite by hemagglutination. It is possible that those 14 antigens are the antigenic core of *E. histolytica,* shared by all strains.

Antigenic comparison of *E. histolytica* with other members of the *Entamoeba* genus has revealed some cross-reactivity with *E. coli* and with *E. invadens* of snakes, but not with *E. moshkovskii.* Only avirulent strains of *E. histolytica* cross-reacted with *E. hartmanni.* Certain antigenic differences between virulent and avirulent strains of *E. histolytica* have been reported, but their relationship to the pathogenic capacity remains to be investigated.

At present, the immunological techniques are unable to replace the microscope methods in the diagnosis of *E. histolytica* infection, but they constitute a valuable tool to help decide whether a given infection is invasive, or to diagnose the hepatic forms of the infection. The current availability of commercial kits for latex agglutination tests, double diffusion tests, and counterimmunoelectrophoresis will probably produce additional information in clinical situations in the near future.

Immunopathology

The relationship between antibody levels and extension of pathology found in infected hamsters and the correspondence between manifestations of cell-mediated immunity and hepatic lesions reported in some human patients have made some researchers wonder whether the host's immunity plays some role in the production of damage. It is not known whether the tissue lesions are the result of the immune response or represent only a stronger antigenic stimulation: in both cases a direct relationship between lesions and increased immunity should be expected. The recent report (Faubert et al., 1978) of the finding of antiliver antibodies by CFT in patients of hepatic amebiasis and in rabbits immunized with *E. histolytica* extracts favors the former alternative, but still much research in this area is needed.

Immunological and histopathological findings in patients with diverse forms of amebiasis by Harris and Bray (1976) revealed that only the hepatic forms exhibited manifestations of cell-mediated immunity, although no evidence of it was found in the histological study of the lesions. They proposed that the parasite in the intestinal habitat would suppress the cellular immunity, whereas the hepatic parasites would only inhibit its local manifestations. Alternatively, it is possible that intestinal stimulation by *E. histolytica* is a poor inducer of cell-mediated responses and that the absence of lymphocytes in amebic lesions of the liver is a consequence of local physicochemical conditions rather than of an immunological suppression. Recent reports have communicated that both branches of the immune system are depressed in amebiasis patients with fatal outcome; we do not know in this case, either, whether this corresponds to an instance of immunologically mediated inhibition or simply to a profound alteration of the physiology of terminal patients.

THE COCCIDIA

Until a few years ago, the coccidia of man and domestic animals traditionally included species of the genera *Eimeria* and *Isospora*. The last years have seen tremendous advances in the biology of the sporozoa that have unveiled new systematic relationships among these parasites. Among the intestinal coccidia of the carnivores alone, six species of *Isospora*, eight of *Sarcocystis*, two of *Besnoitia*, one of *Toxoplasma* and one of *Hammondia* (Levine, 1977) are currently recognized.

In this section, immunological aspects of the intestinal infection by *Eimeria* and *Isospora* are reviewed, and the few known features of the intestinal phase of the other coccidial genera of medical importance are mentioned. The immunology to the parenteral stages of *Toxoplasma*, *Hammondia*, *Sarcocystis*, and *Besnoitia* are discussed among the tissue protozoa. Avian and mammalian coccidiosis are considered separately since their respective ecologies in captive animals recommend diverse approaches to the control of these infections.

Eimeria and *Isospora* infect their hosts through the ingestion of oocysts passed with the feces of infected individuals. The mature, infective oocyst carries in its interior a number of sporozoites. Once ingested, the sporozoites are released, and they penetrate the intestinal tissues and divide to form a schizont (conglomerate of small parasites individually called merozoites). The merozoites are freed by rupture of the host cell and invade new cells to repeat the process of multiplication. After a short number of generations that is characteristic for each species, the invasive merozoites differentiate into male and female gametocytes inside the intestinal cells, fertilization takes place, and the new generation of oocysts develops and appears in the feces (Figure 6).

As we know now, the genus *Eimeria* is exclusively restricted to

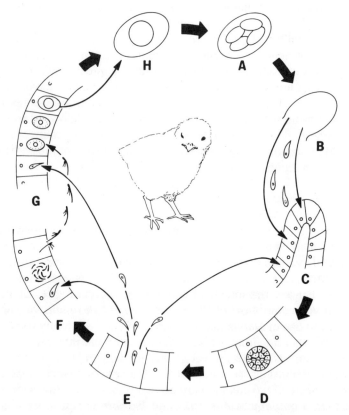

Figure 6.   Life cycle of an eimeriid coccidium, *Eimeria tenella*. Oocysts sporulated in the external environment (**A**) release their sporozoites (**B**) when ingested by an appropriate host. The sporozoites enter the intestinal mucosa (**C**) and multiply in its cells (**D**) by schizogony to form numerous merozoites. The merozoites leave by rupturing the cell (**E**) and invade new cells (**C**) to repeat the process of proliferation. After a few generations, the invasive merozoites become masculine (**F**) and feminine (**G**) gamonts that undergo fecundation and develop into oocysts. The unsporulated oocysts (**H**) are released by rupture of the host cell and pass out with the feces to sporulate in the external environment.

this enteral phase, but *Isospora* of carnivores may multiply to a limited extent in the extraenteral tissues of rodents that act as intermediate hosts. Nevertheless, rodents appear to have a role of only marginal importance in the diffusion of isosporosis of carnivores in nature. The existence of intermediate hosts, on the contrary, is very important, if not essential, to the life cycle of the other coccidia named above.

Because of the biology of the coccidia, the enteral infection is self-limiting (but see "Mammalian Coccidiosis," below) and the parasites are eliminated spontaneously as the oocysts pass out. As a general rule, the intestinal parasitism by coccidia is very specific and affects only one species or a group of closely related species of host.

Avian Coccidiosis

Natural Resistance   Natural resistance to avian coccidiosis has been recently surveyed by Cuckler (1970). Field observations have demonstrated that not all the birds of a flock are affected equally by coccidiosis; some of them suffer a violent disease, and others develop milder symptoms. Studies of genetic selection have allowed establishment of strains of chickens with different susceptibility to the infection. In one experiment, seven selected strains of White Leghorn chickens were infected with 100,000 oocysts of *Eimeria tenella*; the least susceptible strain exhibited 38% mortality whereas 91% of the most susceptible birds died.

Apparently, this resistance depends on multiple genes that have an additive effect. It is not known, however, whether this phenomenon is connected with a greater and earlier ability of the resistant animals to respond immunologically to the infection or with nonspecific mechanisms.

It is also a common observation that coccidiosis is normally a disease of young birds and that the adults usually present a considerable resistance to the infection and to the pathology. Chickens raised in the absence of coccidia infection until they were 6 months old, however, were as susceptible to the infection and to the disease as young birds. It seems, therefore, that the elevated resistance of adult birds under natural conditions is the result of previous experiences with the parasite rather than a consequence of their age. Similar experiments with turkeys, however, indicated that this host develops true age resistance.

Acquired Immunity   Recent reviews on immunity to coccidial infections have been written by Rose (1972, 1976) and by Ogilvie and Rose (1978).

The intimate contact between the parasite and the host tissues in coccidiosis is expected to produce a respectable degree of immune response. In fact, antibodies of the IgG and IgM classes (with ample predominance of the former) have been found in the circulation of infected birds, and IgG and IgA antibodies have been recently reported to be present in the intestinal secretions of similar birds. The first circulating antibodies are detectable during the second week of infection, peak in 1 or 2 months, and decline until disappearing after 3 to 7 months. Reinfections do not produce evidence of a secondary response of the seric antibodies, possibly because the short permanency of the parasite and the preexistence of antibodies in the digestive tube in these cases prevent the passage of enough antigen to the systemic circulation.

Cell-mediated immunity in coccidia-infected birds has been demonstrated by skin tests and by tests of inhibition of macrophage migration.

Infection with a reduced number of oocysts (50 to 100) induces a total protection against reinfections after a few days with some species of coccidia (*Eimeria maxima*), but with other species (*E. tenella* and *E. necatrix*), a large number of oocysts are required to produce only partial resistance after 2 or more weeks. This partial immunity, however, is effective enough to prevent the appearance of clinical symptoms and becomes absolute with subsequent infections. Unfortunately, the immunizing capacity has no relationship with the pathogenicity of the parasite. The highly immunogenic species *E. maxima* is only moderately pathogenic, whereas *E. tenella* causes serious mortality.

In the absence of reinfections, the immune resistance wanes after a few months, but repeated infections will maintain it for the life of the animals. Experimentally, it has been demonstrated that small repeated infections produce stronger resistance than the same number of oocysts administered in a single dose. From this, it seems evident that the immune protection against coccidiosis develops secondary responses on restimulation.

A few attempts to verify transfer of immune resistance between generations through the egg have given conflicting results on different occasions.

Acquired immunity in coccidiosis is highly specific, to the extent that it distinguishes among different strains of the same species; the resistance against one strain is still expressed against other strains, but with less efficiency than against the homologous variety. A well-known exception is the ability of *E. tenella* to produce some degree of cross-protection against *E. necatrix*.

The general lack of cross-protection in coccidiosis indicates that these parasites do not share functional antigens and that the effect of the protective immunity is exerted on the parasite itself rather than on its immediate environment, as happens with many nematode infections. Sharing of functional antigens among developmental stages of the same species must exist, however, since the induction of immunity on occasions is produced by stages different from those that are affected by it (see below).

In several areas of immunity to coccidiosis, there is a frank divergence between laboratory findings and field observations; the ample prevalence of *E. maxima* in chicken farms, for example, is in open conflict with the efficacy of this species to produce immune protection as found in the laboratory. As a matter of fact, absolute resistance to coccidiosis is uncommon in nature; most birds develop some degree of immunity that protects them from the clinical disease, but act as healthy carriers of the infection. Evidently, there must exist a complex of ecological factors that modulates the efficiency of the immunity under natural conditions. The virus of Marek's disease, for example, is known to interfere with the acquisition of resistance to the infection and to the disease by coccidia.

Mechanisms of Acquired Resistance   Experimental infections with intermediate stages of the life cycle of coccidia have identified those forms that preferentially stimulate protective responses and those that are most affected by them. In the cases of *E. tenella* and *E. necatrix*, the second generation of schizonts (the most pathogenic stage for these species) produces a strong protection against reinfections, whereas the gametocytes are less effective. In the case of *E. maxima,* most of the protection is also induced by the second generation of schizonts, but the most pathogenic stage in this species is the gametocytes. Thus, there appears to be no consistent correlation between immunogenicity and pathogenicity. It is interesting to note that any treatment that is able to destroy the parasite before it reaches the immunity-inducing stage is bound to interfere with the production of resistance to subsequent infections.

In hosts exhibiting complete protection, this is expressed by a reduction in the number of sporozoites that penetrate the intestinal tissue, by inhibition of their multiplication once inside the cells, and by phagocytosis and subsequent digestion of the parasitic elements. Transfer of sporozoites from resistant birds to susceptible birds has demonstrated that the parasites lose their ability to proliferate in a nonimmune host only after they have been subjected to the activity of the immune system in a resistant host for 24 or more hours. The same phenomenon occurs when the experience is repeated with merozoites. Evidently, the immune response requires some time to affect the parasite irreversibly; it is not known whether this gap is necessary to attract the immune elements that will damage the parasite to its location or to slowly produce the structural or physiological changes that will kill it. In birds only partially immunoprotected, some sporozoites will multiply to the stage of schizont, but these forms look abnormal and finally disappear.

Apparently, not all coccidia are affected by the immunity at the early stages of infection, because recent reports have indicated that the stages destroyed in birds resistant to *E. maxima* are the gametocytes.

In many instances, it has been demonstrated that there is an inverse correlation between the number of oocysts administered to a susceptible host and the total number of oocysts resulting from the infection. Chicks infected with 50 oocysts of *E. tenella*, for example, produced 80,000 new oocysts whereas similar animals receiving 40,000 oocysts produced only 1200 new oocysts. This phenomenon has been called "crowding effect," and a number of hypotheses have been offered to explain it. In the case of the coccidia, at least, it is possible that the partial inhibition of the reproduction is mediated by immune phenomena that, conceivably, could be stronger in proportion to the amount of antigenic stimulation, within certain limits.

Specific antiserum causes agglutination, complement-mediated lysis, and reduction of the infectivity of sporozoites and merozoites of the corresponding species in vitro. In vivo, however, no relationship has been observed between levels of seric antibodies and resistance, and this latter often persists beyond the period in which antibodies are detectable in the serum. Many attempts to transfer resistance by inoculation of serum of immune birds have failed in the past; some success has been obtained recently, but inconsistently. Comparison of the course of the disease in normal chickens and in chickens with suppressed humoral responses has also failed to show any important protection mediated by antibodies. Since serum antibodies deleterious to the parasites in vitro do not appear to protect the host, it seems fair to assume that these antibodies do not reach the normal location of the coccidia in the intestine or that the parasite uses some host mechanism (not available in vitro) to avoid their activity.

Rose (1972) has obtained some recent evidence that cecal secretions of birds recovered from coccidiosis can transmit resistance to susceptible birds.

The role of cell-mediated immunity in protection has been studied by transfers of lymphoid cells and by specific suppression of this branch of the immune response in birds that were subsequently challenged. In all cases, some evidence of the participation of cellular immunity in acquired resistance has been obtained, but this still does not explain all the strong protection produced by natural infections. Peritoneal macrophages of immunized birds are more active in ingesting the parasites, but do not kill them.

Most specialists feel that some form of cooperation between the humoral and the cellular branches of immunity must exist in coccidiosis. An alternative possibility that has been little explored yet is that the protective immunity in this infection is a phenomenon mainly localized in the intestine. If this is the case, which appears likely, systemic immune responses will not necessarily correlate strictly with protection.

Artificial Production of Resistance   Because of the economic importance of avian coccidiosis, numerous assays to produce resistance artificially have been attempted. Injections of extracts of oocysts or schizonts have caused the formation of precipitating antibodies similar to those found in natural infections (Horton-Smith et al., 1963). The birds so treated were totally refractory to infection by intravenous inoculation of sporozoites, but were completely susceptible to oral infection with mature oocysts. These experiments support the hypothesis that circulating antibodies do not reach the parasite in its intestinal habitat.

The use of light infections followed by drug treatment has been sufficiently effective in eliciting resistance to stimulate the commercial production of a vaccine in the U.S. (Coccivac) that contains all the species virulent for chicken and is administered in the water. This method, however, has two major drawbacks: it contaminates the environment, and it relies on the prolonged administration of small doses of coccidiostats, which facilitates the selection of drug-resistant parasites.

Production of resistance by infection with attenuated strains (especially by irradiation) that cause an asymptomatic infection has been moderately successful. It is not known, however, whether the attenuation treatment actually reduces the virulence of the parasite or simply kills some of the oocysts and results in lighter infections.

Researchers have recently succeeded in adapting a strain of *E. tenella* to grow in chicken embryo (strain TA), and a second strain (WIS-F) has been selected through multiple passages in chicken. Both strains are considerably less pathogenic than the field strains, but conserve their capacity to elicit immune protection. Apparently, they are stable and do not revert to their original pathogenicity. A strain of *Eimeria acervulina* adapted to chicken embryo has been demonstrated to be too unstable for its use in massive programs of immunoprophylaxis.

Immunodiagnosis   Coccidia infections stimulate the production of a variety of antibodies in birds: tests of precipitation, agglutination, lysis, immobilization, neutralization, complement fixation, fluorescent antibodies, and delayed skin reactivity have been utilized to identify the infection on various occasions. Since patency appears earlier than the corresponding immune response and is easy to verify, and because the lack of relationship between systemic immune reactions and protection precludes establishment of epidemiological conclusions based on immunological tests, little interest has existed in developing immunodiagnostic techniques for practical use.

Immunopathology   Several years ago, Augustin and Ridges (1963) reported the finding of precipitating antibodies against the parasite and against normal host tissues in turkeys infected with *Eimeria meleagrimitis*. They could not determine whether antigenic sharing existed between parasite and host or host materials became autoantigenic as a result of alterations caused by the parasitism. It appears that these interesting observations have not been researched further.

Mammalian Coccidiosis

With the exception of cattle and rabbits, coccidial infections of the alimentary tract of humans and domestic animals are usually infrequent or the production of the corresponding disease is sporadic. For this reason, the studies on immunity to digestive coccidiosis of mammals have been more limited than those on avian coccidiosis and our knowledge of it is correspondingly more deficient.

Apart from humans and the domestic carnivores, the digestive tracts of all the hosts of importance to medicine are affected almost exclusively by species of the genus *Eimeria*. Humans may harbor species of *Isospora* and *Sarcocystis* in their intestines (the old *Isospora hominis* is actually *Sarcocystis hominis* and *Sarcocystis suihominis*), and the digestive canal of the domestic carnivores may be invaded by species of *Isospora, Sarcocystis, Hammondia, Toxoplasma,* and *Besnoitia*. These last four genera also possess extraenteral phases that are considered under the tissue-dwelling protozoa (below).

Most of our current information on the immunity to digestive coccidiosis of mammals is based on work with species of *Eimeria*. With regard to *Isospora*, the studies in dogs are sparse, and they are practically nonexistent in humans and cats. Also, recent discoveries on the systematics and biology of coccidia (Frenkel, 1977; Levine, 1977) make the interpretation of earlier findings particularly uncertain. For example, the species of dog coccidia reported until recently as *Isospora bigemina* may have been *Hammondia heydorni* or any of four species of *Sarcocystis*. The species commonly identified as *Isospora rivolta* in the dog may have been *Isospora ohioensis* or *Sarcocystis bertrami*. Research on the immunology of the enteral phase of *Toxoplasma* in cats has been limited, and the investigations on the intestinal cycles of *Hammondia, Sarcocystis,* and *Besnoitia* are just beginning.

There exists the unspoken consensus among most parasitologists that the intestinal coccidiosis of mammals is essentially comparable to avian coccidiosis. The similarity of the biology of the coccidia that affect both groups of vertebrates and some experimental findings support this presumption in general terms, but a few differences in the behavior of the parasites of mammals and of birds and in the immune reactivity of both hosts warn against sweeping generalizations.

Natural Resistance   As in the birds, the coccidia in mammals are very strictly host specific, but some notable exceptions exist. Although *Eimeria separata* of rats normally do not develop in the mouse, they can complete their life cycle in mice with a determined genetic constitution. Species of *Isospora* were originally believed to be less host specific than species of *Eimeria*; it was then thought that both genera were strongly specific, but lately Hendricks (1977) has been able to infect six species of primates belonging to two different families, four carnivores of four different families and one marsupial with the monkey coccidium *Isospora arctopitheci*. These examples may be more exceptions than rules, but it is interesting to consider the possibility that mammalian coccidia might be less host specific than the corresponding parasites of birds.

Studies with the rat coccidium *Eimeria miyairii* have demonstrated that, within the same host species, the susceptibility to the infection varies with the genetic structure of the host. As with birds, the reasons for this are unknown.

Also in mammals, coccidiosis is a disease particularly prevalent in young animals. Experiments with the rabbit coccidia *Eimeria intestinalis, Eimeria magna,* and *Eimeria perforans* suggested that this relative resistance of the adults is more related to the age of the host than to its previous experiences with the parasite. It is not known whether the same is true for other mammals.

Acquired Immunity   Coccidial infections induce the production of a variety of antibodies in mammals. Studies in rabbits infected with *Eimeria stiedae* revealed complement-fixing antibodies between the 10th and the 20th days of infection; they reached a peak 20 days later and began to decline in a month, but were still detectable 165 days after the infection. No secondary responses were detected on reinfection.

Cell-mediated immunity has been verified in rabbits, pigs, and cattle by delayed skin tests. In cattle infected with *Eimeria bovis,* it has been found that their peripheral lymphocytes respond with transformation to the extracts of oocysts of the homologous parasite, and to extracts of oocysts of *E. stiedae* to a more limited degree.

As in birds, mammalian coccidiosis follows a course toward spontaneous cure, leaving resistance to reinfections. In mammals, however, the protection is rarely complete. This characteristic may be due to ecological as well as to purely immunological reasons. Ecologically, mammals are normally subjected to infections that are less intense and less frequent than those in birds, so the antigenic stimulation must be weaker; immunologically, the coccidia of mammals may be less immunogenic for their hosts than the coccidia of birds. Cattle, for example, require at least three prior infections before tolerating a dose of 20 million oocysts of *E. bovis* without developing symptoms. In contrast, a single moderate infection with *E. stiedae* in rabbits produces resistance to the disease and a single massive infection makes them resistant to a subsequent infection. The stronger immunogenicity of *E. stiedae* has been attributed to the persistence of the antigenic material that is trapped in the liver of infected animals. It is possible, however, that immunogenicity is a characteristic inherent in the parasitic species: 20,000 oocysts of *Eimeria polita* produce resistance that lasts for 3 months in pigs, for example, whereas 200 oocysts of *Eimeria scabra* cause a resistance that is verifiable for 4 months.

It has been traditionally held that coccidial infections are self-limiting because of the biology of the parasite. Numerous observations in birds and in mammals support this idea, but a few known exceptions pose a disturbing problem: patencies of 1 to 2 months are not rare in humans, and infections lasting for more than a year are on record; periods of patency of $2\frac{1}{2}$ to 4 months have been recorded in dogs apparently kept from reinfections; elimination of oocysts in the feces of cattle has been followed during the entire winter; passage of *Toxoplasma* oocysts has been restored in cats by superinfection with other coccidia. Also, our work (Barriga and Arnoni, 1979) has demonstrated that the length of *E. stiedae* patency varies in proportion to the infective dose. For all these reasons, it is possible that the elimination of the parasite is actually due to an immune response (at least in mammals) and that, in the absence of this response, the infection progresses until resistance develops or the host dies. If this is the case, only the hosts that controlled the infection will stay alive for study, biasing the results in their favor. In our experiments, rabbits infected with 10,000 oocysts of *E. stiedae* persisted at the peak of patency at least until day 50 postinfection, but the same parameter could not be recorded in animals infected with 100,000 oocysts—80% of them had died by day 40. I believe that this is a fundamental problem in coccidiosis that deserves study.

A phenomenon difficult to explain is the "winter coccidiosis" of cows that affects calves after weaning. In locations such as Montana, winter temperatures are too low for the oocysts to mature, so exogenous infection must be insignificant. Marquart (1976) has presented circumstantial evidence that intermediate stages of the parasite remain inhibited in the host and reinitiate their development in coincidence with bad weather and changes of alimentation. Since it is well known that environmental and nutritional factors interfere with the expression of the immunity (Hudson et al., 1974), Marquart's hypothesis may be extended to assume that the inhibition of the parasitic development is caused by an immunological mechanism that is repressed during the winter.

Once established, the duration of the immune resistance is variable: different reports indicate that it lasts for 3–4 months in swine, 2–7 months in cattle (in direct proportion to the age of the host and to the size of the infective dose) and up to 2 years against *E. stiedae* in rabbits.

Because of the particular importance of *Toxoplasma gondii* of the intestine of cats in public health, it may be convenient to review briefly what is known about its immunology. The cat may acquire the intestinal infection by ingestion of mature oocysts from the feces of an infected cat or by consumption of tissues of infected intermediate hosts. Patency begins 20 or more days after the ingestion of oocysts, 5–10 days after the ingestion of active tissue parasites (tachyzoites), and 3–5 days after the ingestion of resting tissue forms (bradyzoites). It is not known at present whether these different periods of maturation represent an inherent characteristic of the diverse forms of the parasite or they are influenced by differences in the host's immune reactivity to the various stages. In any case, ingestion of tissue parasites appears to be the most common or effective manner of infection, since most cats develop anti-*Toxoplasma* antibodies around the age they begin hunting. Experimental administration of resting tissue forms infects almost all cats, whereas administration of the other stages produces infection in less than 50% of the cats. Again, it is not known whether this different infectivity depends on the parasitic stage in itself or on the reaction that it is able to elicit in the host.

At any rate, cats develop a strong immunity to enteric toxoplasmosis; a primary infection remains patent for only 1 or 2 weeks, rarely longer, and subsequent challenges do not reach patency or oocysts are passed only for a brief period. The parasites are not eliminated from the intestine, however, since restoration of oocyst passage has been achieved in cats recovered from an infection by administration of oocysts of *Isospora* or of corticosteroids. The length of the resistance is not known, but it probably persists as long as the parasites are present.

Hammondia in the cat also sheds oocysts for 1–3 weeks after a primary infection and terminates patency spontaneously. The parasite remains, however, and new episodes of patency may occur from time to time.

As judged by experimental infections with *Sarcocystis* or *Besnoitia*, these genera produce a weak protective immunity that suppresses patency after a few weeks or months, but is unable to inhibit the development of the parasites of a later challenge.

Mechanisms of Acquired Resistance   Immune protection to coccidial infection is considered to be as specific in mammals as it is in birds: rabbits previously infected with *E. perforans* are totally susceptible to *E. stiedae*. These two species, however, are located in different habitats (intestine and biliary passages, respectively) so that the corresponding immune reactions might not have a chance to operate on the heterologous parasite if they are of exclusively local expression (as seems likely). On the other hand, the lymphocytes of cattle infected with *E. bovis* undergo some degree of blastic transformation when stimulated with extracts of *E. stiedae* of rabbits. It might be interesting to experimentally review the assumed specificity of anticoccidial immunity in mammals.

No information is available about the stages of the parasite that elicit immune protection or about the forms preferentially affected by it. Studies with *E. stiedae* have shown that the sporozoites penetrate the intestinal epithelium of immune rabbits but do not appear in the biliary ducts; it is believed that they are destroyed in the mesenteric lymph nodes, possibly by a process similar to that found in birds.

Immunization of rabbits with extracts of coccidia has produced the same systems of precipitating antibodies detected in animals recovered from an infection, but no resistance to subsequent infections was detected. Similarly, transfer of serum from immune animals to susceptible animals in several host species has not modified the susceptibility of the latter to challenges. As in birds, circulating antibodies that are lethal to the parasites in vitro appear not to affect them in vivo. A recent report by Tsang and Lee (1975), however, has shown that intramuscular injections of anti–*Isospora felis* serum in puppies has preventive and curative effects when administered up to 8 days after infection with 30,000 oocysts of the homologous parasite.

Transfer of lymphoid cells from resistant to susceptible pigs did not transmit resistance, but transfer of large numbers of lymph node or thoracic duct cells from rats resistant to *Eimeria nieschulzi* into syngeneic recipients reduced the number of oocysts produced by the latter on a homologous infection. The experiment with pigs may not be valid since the transferred cells were probably rejected because of genetic differences. The experiment with rats indicates that cell-mediated immunity plays some role in protection, but the production of only partial resistance by this procedure and the large numbers of thoracic duct cells (6×10^8) that had to be inoculated to obtain 50% reduction in oocyst passage suggest that other mechanisms must also operate in natural infections. A few recent reports claimed that injections of transfer factor obtained from cattle infected with E. bovis will prevent the infection of susceptible cattle with E. bovis and the development of Eimeria ferrisi in mice. This work disagrees with the strong specificity of the immune response believed to exist in coccidial infections (which might not be the case—see above), and requires further confirmation.

As in birds, the current evidence does not permit assigning the major protective role in coccidiosis to humoral or cell-mediated immunity, although the latter seems to have some importance. It is possible again, that immune resistance to coccidial infection is a predominantly local phenomenon with little systemic expression (see "Nematodes of the Digestive Tract," in Chapter 4).

Immunopathology It is commonly assumed that the damage caused by the intestinal coccidia is due to direct destruction of host cells during the proliferation of the parasites. It has been found recently that the parasitism also produces a transient inhibition of the production of intestinal enzymes in birds. Our work (Barriga and Arnoni, 1979) with E. stiedae in rabbits has demonstrated that damage to hepatocytes and decreased weight gain have already occurred by the 8th day of infection, when the parasites are quietly multiplying in the ductal epithelium. This suggests that the pathogenic effects of the parasitism are indirect rather than derived from direct destruction of host cells by the organism; the relationship of the pathogenicity to immune phenomena has not been studied yet.

We also found a direct correlation among size of the infective dose, length of patency, and mortality rate. We believe that these correspondences indicate that large doses of parasites depress the immunological reactivity of the host and prevent it from inhibiting the multiplication of the organism, which eventually results in death.

OTHER PROTOZOA OF THE ALIMENTARY LUMEN

Our knowledge of the immunology of other intestinal protozoa is considerably more limited than the examples already discussed. Studies with Giardia intestinalis of humans are just beginning (possibly because of the long-held belief in the Anglo-Saxon scientific community that this protozoan was not pathogenic), and experimentation on intestinal protozoa of domestic animals is sporadic at best.

Some of the studies of immunity to *G. intestinalis* have been recently reviewed by Kulda and Nohynkova (1978). The observation that outbreaks of giardiasis subside by themselves and the spontaneous termination of the infection in less than 4 months in 12 of 14 human volunteers suggest that the infection produces an immune response effective enough to eliminate the presence of the parasite from the intestine. Indirect immunofluorescence tests revealed circulating antibodies in 32 (89%) of 36 patients and in none of the healthy controls. Curiously enough, 10 (28%) of 36 patients with malabsorption syndromes, but free of *Giardia*, were also positive. Osipova and Shahabutdinov reported in the IVth International Congress of Parasitology in Warsaw that the peripheral lymphocytes of patients with giardiasis released a migration-inhibitory lymphokine on stimulation with extracts of the protozoan.

Recent studies have demonstrated that the sera of infected people and infected hamsters have the ability to immobilize and to lyse *Giardia* in vitro. At least the lytic effect must be mediated by complement, because it disappears with heating or aging of the sera. The significance of these phenomena in vivo has not been assessed yet. Other reports have indicated an association between *Giardia* infection and reduced production of IgA at the intestinal level, but no correlations with the seric levels of IgE have been observed.

Some workers have found that some of the intestinal alterations commonly attributed to immunodeficiencies may have been actually caused by *Giardia* infections (Eidelman, 1976). The association between the parasite and the immunodeficiency syndrome, however, is far from being clear and deserves further research.

Histomonas meleagridis of chickens, turkeys, and other birds infects chickens and turkeys very readily, but its pathogenicity for the former is much less than for the latter. Studies with gnotobiotic turkeys have demonstrated that the pathogenicity of the parasite depends on its association with bacteria, a fact that most immunological studies have not considered. The conflicting reports on the results of vaccination against histomoniasis may have resulted from neglected immune responses to the associated flora. In support of this possibility, it has been reported that the parasite still persisted in the ceca of birds that had recovered from the acute condition and were resistant to a new episode of disease.

Turkeys are susceptible to the infection and the disease at any age, although the symptoms tend to be more chronic in older birds. Precipitating antibodies appear in the circulation of chickens and turkeys toward the end of the second week of infection, but their significance in protection is unknown. The occurrence of outbreaks of disease when preventive treatment is withdrawn suggests that protection against reinfection does not develop under natural circumstances. A few attempts at vaccination by administration of pathogenic parasites or by termination of a pathogenic infection with drug treatment have produced resistance against subsequent infection with *H. meleagridis* inoculated intracecally, but not against infection carried by the cecal nematode *Heterakis gallinarum*, which is the common form of transmission in nature. It is possible that the mechanism of protection in these cases includes the production of secretory IgA antibodies, and that the protozoa may elude their activity by being carried directly into the intestinal mucosa by the larval worms.

Although the progress of immunology in the last years offers a number of experimental possibilities that were not available to earlier researchers, the lack ot axenic techniques for the culture of the parasite and the reduction of the practical importance of immunology with the new methods of turkey production have not encouraged further work.

TRICHOMONAS VAGINALIS

Urogenital trichomoniasis by *Trichomonas vaginalis*, an essentially venereal infection, affects exclusively humans in nature, although experimental infections have been obtained in monkeys, guinea pigs, hamsters, rats, and mice.

Natural Resistance

The rather frequent observation that some individuals may maintain regular sexual contacts with infected subjects without acquiring the infection has brought up the point of how susceptible humans actually are to *T. vaginalis* infection. Determination of "noninfection" is difficult however, since a proportion of individuals harboring the parasite may be asymptomatic and the demonstration of the protozoan in their genital secretions is sometimes uncertain. In addition, lack of susceptibility in sexually active adults may indicate acquired resistance rather than natural insusceptibility. The general opinion of the specialists is that humans do not exhibit natural resistance to urogenital trichomoniasis.

The refractoriness of lower animals seems to be due to physiological peculiarities not related to immunity: female monkeys will accept experimental intravaginal infections for short periods without showing symptoms and will present a minimum number of parasites in the interestral period; artificial infections in guinea pigs will also invade the uterus and cause abundant symptoms, with possible precancerous transformation of epithelial cells; repeated intravaginal inoculations in hamsters will establish *T. vaginalis*, but superinfection with a nonpathogenic trichomonad will terminate the original infection; rats and mice will accept lasting vaginal infections, but only if they are maintained in permanent estrus artificially. The direct relationship between sexual hormones and persistence of the protozoa appears to hold true in humans also since trichomoniasis is rare in prepubertal and in menopausal women.

The "mouse assay," which consists of provoking abscesses by subcutaneous, intraperitoneal, or intramuscular inoculation of the parasites, was originally developed for drug evaluation, but it has also been utilized occasionally in immunological investigations. It is not clear whether abnormal locations in abnormal hosts should yield results relevant to the immunology of the natural infection in humans, however.

The serum of uninfected humans and most domestic animals has been demonstrated to possess natural antibodies that agglutinate and lyse *T. vaginalis* in vitro. In a study with sera of teenagers negative to parasitological examination, most of the sera were agglutinating at titers of 40 or 80. Lysis usually occurs at titers up to 32, and no differences have been found among healthy or infected subjects. Heat inactivation of the serum (56°C for 30 min) eliminates most of the lytic action (which indicates that it is complement-mediated), but does not affect the agglutination titers. Whether the complement-fixing antibodies and the agglutinating antibodies are identical or not remains to be determined. In any case, the natural antibodies appear to be species specific, because they are not removed from the serum by adsorption with other species of trichomonads. The relationship of these antibodies to natural resistance in unknown at present.

Acquired Immunity

Despite the wide prevalence of human urogenital trichomoniasis, research in the immunology of the infection is sparse at best, and a considerable portion of it has been done in Poland and the U.S.S.R., not infrequently published in their national languages. Detailed reviews have been published by Honigberg (1970, 1978a).

Patients of trichomoniasis exhibit a variety of specific circulating antibodies and produce cell-mediated responses, as judged by the occurrence of delayed skin reactions. The presence of these manifestations of immunity indicates that the parasite is immunogenic enough to produce systemic responses even when restricted to its urogenital location. The refractoriness to the infection by *T. vaginalis* in numerous individuals that maintain regular sexual contacts with infected subjects suggests that the parasite confers resistance to reinfections. The infrequency of spontaneous cures, on the other hand, is an argument against the development of any important degree of protective immunity. Some investigators think that all people are equally susceptible to a primary infection but that there are variations in their ability to develop immune resistance. This may be true, but it does not seem to be solid proof of the theory.

The participation of circulating antibodies in protection has been studied by inoculating mice with serum of infected patients, with serum of noninfected individuals that cohabitated with infected subjects, and with serum of parasitologically healthy people without a history of infection. The sera of the first two groups produced considerable protection against the lesions caused by a subsequent injection of parasites, whereas the serum of the last group had no effect. When the infected individuals were successfully treated, their sera diminished in protective ability until it disappeared in 6 months. The protection afforded by these sera had no relationship with their titers of specific agglutinating or complement-fixing antibodies. Evidently the protective antibodies are not the same as those detected by the serological tests. This experiment indicates that *T. vaginalis* infections produce transient protective antibodies, although they do not appear to be active in the patient. Possibly the circulating protective antibodies do not reach the normal location of the parasite or the parasite has mechanisms to elude their action in the urogenital tract.

It has been recently reported that patients of trichomoniasis have elevated levels of IgE. It is now known yet whether this immunoglobulin corresponds to anti–*T. vaginalis* antibodies or it has a protective action on the infection.

After many unsuccessful attempts, it has recently become possible to demonstrate the presence of IgG and IgA in the cervical secretions of women with trichomoniasis. In another study (Ackers et al., 1975), IgA antibodies were found in the vaginal secretions of 76% of 29 infected women and 42% of 19 apparently uninfected ones. No relationship was detected between the presence of antibodies and the symptomatology, but there was some indication that the women that had antibodies carried fewer parasites. Despite this lack of correspondence, the notion has been offered that the existence of antibodies in the cervical secretions may be the reason why *T. vaginalis* does not invade the uterus.

Although cell-mediated immunity against *T. vaginalis* occurs, its role in protection has not been investigated so far.

In summary, there is solid evidence that repeated or prolonged contact with *T. vaginalis* elicits production of local and systemic immune responses, but there is no proof that they exert a deleterious action on the parasites lodged in the urogenital mucosae.

Artificial Production of Resistance

Almost all the work in this area has been directed to the verification of the production of protective immunity or of its mechanisms in animal models.

Intramuscular inoculation of living parasites protected over 80% of mice against the tissue damage produced by later injections. The resistance lasted, to a decreasing degree, for 15 weeks, but it did not correlate with the titers of circulating antibodies developed. The functional antigens involved in this resistance must have been excretions/ secretions since dead parasites did not elicit any protection. Later work showed that intramuscular inoculations protected well against intraperitoneal challenges but weakly against subcutaneous infections. Similarly, an intraperitoneal primary infection protected better against intramuscular than against subcutaneous secondary infections. The subcutaneous inoculation of parasites, in turn, did not produce resistance to subcutaneous or intraperitoneal challenges, but elicited adequate protection against intramuscular reinfections.

These results are difficult to interpret and may reflect the longer persistence of parasites inoculated intramuscularly (which must result in maintained antigenic stimulation) and the abundance of macrophages in the peritoneal cavity (which presumes better processing of the antigen and possibly an increased efficiency of the effector mechanisms of the immunity). At any rate, they emphasize the importance of the parasitic habitat in the final effect of the immunity and discourage careless extrapolation from animal models to the human infection.

Other authors have communicated that the intraperitoneal inoculation of dead *T. vaginalis* induces an important degree of protection against homologous challenges and a lesser but still significant resistance against reinfection with other trichomonads.

A report by Aburel et al. (1963) appears to have been the only attempt to induce resistance artificially in people. These authors inoculated heat-killed trichomonas in the vaginal mucosae of 100 infected women refractory to the chemotherapeutic treatment: 40 of them eliminated the infection subsequently and 49 remained infected but became asymptomatic. Three patients refractory to all treatments, including vaccination, that were inoculated with anti–*T. vaginalis* serum in the cervix and vagina showed great clinical improvement, and one eliminated the parasite. These studies do not seem to have been confirmed, but the recent finding of antitrichomonas antibodies in genital secretions provides theoretical support for them. As in other parasitisms of the mucosal surfaces, protective immunity in human trichomoniasis may be a predominantly local phenomenon.

Immunodiagnosis

The diagnosis of human genital trichomoniasis commonly is secured by finding the parasite under microscope examination. This method, however, is too time consuming for extended surveys and its efficiency is not satisfactory: in a series of 600 infected women diagnosed by culturing the parasite from their genital secretions, the routine Papanicolaou-stained smear showed 64% of the infections, wet mounts detected 71% of the infections, and the complement-fixation test identified 75% of the infections. Cultivation is currently the most efficacious procedure when performed properly, but it may take several days before arriving at the definite diagnosis. Although little used in practice yet, appropriate immunodiagnostic techniques should be able to provide quick and reliable results. Honigberg (1978a) has written an update in this area.

Numerous assays with the complement-fixation test have given amply divergent results: reported sensitivity has varied from 30% to 86% among various authors and false positive reactions have reached up to 16.5%. In a recent series in which an antigen derived from parasite samples of 20 different origins was utilized, this procedure diagnosed 57% of 275 female patients, 40% of 155 male patients, and 8% of 440 persons free of parasites. Limited observations indicate that the antibodies appear in the second week of infection and persist for a year after the cure.

The earlier results published about the direct agglutination test are difficult to interpret because they usually did not distinguish between natural and acquired agglutinating antibodies. In a large series of exams in which only reactions at dilutions of 1:320 or higher were considered positive, 249 of 256 infected women and 116 of 130 infected men were diagnosed. The test was frequently positive in noninfected subjects that cohabited with patients, but it was negative in people free of parasites and without history of infection. The respective titers disappear within 6 months of the parasitological cure in men and within 9–12 months in women.

The indirect hemagglutination test with a polysaccharide antigen was positive at an average titer of 46 in 183 of 186 infected women, at a titer of 29 in 171 of 194 infected men, and at a titer of 9 in 384 of 619 people selected at random. Further examination of a sample of this latter group revealed 39% infection in it; the sera of the infected individuals were positive at titers of 20 or higher while the persons free of parasites yielded negative tests.

In a limited study with formolized parasites, the indirect fluorescent antibody technique was positive in 19 infected subjects and negative in 9 noninfected individuals.

Intraepidermal inoculations of parasites or of semipurified extracts have produced immediate-type skin reactions that have been regarded as nonspecific by the respective investigators. Similar inoculations of phenolic extracts produced delayed skin reactions in 35 of 43 patients and in 6 of 59 noninfected persons; 6 of the patients and 7 controls presented reactions that the investigators did not interpret. Delayed skin reactivity to a homogenate of parasites was also found in 21 infected women but not in 40 noninfected people. These results and the simplicity of the cutaneous tests make it advisable to repeat these investigations with purified fractions of the parasites.

The results of the immunodiagnostic procedures appear to be strongly influenced by the strain of parasites utilized in the tests; in 99 patients the direct agglutination test was positive with homologous parasites in 97, with the least reactive strain in 56, and with all four strains tested in only 26. The complement-fixation technique with four strains of *T. vaginalis* was positive in 170 of 171 female patients and in 82 of 83 male patients when a mixture of the strains was used as antigen, but the least reactive strain alone diagnosed only 140 infections among the women and 60 among the men.

In most cases, the titers of antibodies do not have a relationship with the severity of the symptoms, although recent reports from the U.S.S.R. claim that antibody levels in acute infections are higher but return to normal more quickly after treatment than in chronic infections.

Numerous researchers have found positive serologic reactions in subjects free of parasites. Although this may be an expression of the inspecificity of the tests, some evidence suggests that many of these cases may correspond to occult infections, to repeated contacts with the parasite without permanent colonization, or to residual antibodies of past infections. Positive reactions in noninfected individuals are more frequent in adult populations than in pubertal or prepubertal populations, which suggests a relationship between the risk of infection and the serologic positivity.

No attempts have been made yet to develop practical immunodiagnostic procedures with the genital secretions of infected individuals.

Immunopathology

Several studies have demonstrated common and specific antigens among diverse species of *Trichomonas*. At least some of the common antigens must elicit resistance, because the "mouse assay" has demonstrated cross-protection among *T. vaginalis*, *Trichomonas gallinae*, and *Tritrichomonas foetus*. Antigenic comparison of samples of *T. vaginalis* of different origins has permitted identification of at least four strains. These studies, which unfortunately appear to have been discontinued, were particularly interesting because, apart from their different reactivity in immunological tests, the diverse strains coexisted in common geographic areas, which is difficult to explain from the purely epidemiological standpoint.

Kua-Eyre and Honigberg (1976) recently studied by immunoelectrophoresis five types of *T. vaginalis* that differed in pathogenicity for women and for mice and found some inverse correlations between pathogenicity and number of antigens. The significance of this correlation is not known yet, but it is easy to hypothesize that more antigenic parasites would exert a stronger stimulatory effect on the host's immunity, which may result in better neutralization of their ability to cause damage.

Recent reports have indicated that patients of trichomoniasis have elevated levels of IgE in their circulation. This finding suggests that the local swelling and pruritus (this latter often relieved by antihistaminics) that accompany symptomatic trichomoniasis may have an allergic etiology.

Cappuccinelli (1975) has communicated that intraperitoneal inoculation of *T. vaginalis* in mice reduces the ability of the animals to produce immune reactions to subsequent challenges with sheep erythrocytes or with oxazolone. It has not been determined yet whether this is a specific property of the parasite or a simple manifestation of the immunoregulatory mechanisms that must operate in all immunological reactions in vivo. That the phenomenon may also occur in natural infections is suggested by a recent report from the U.S.S.R. that trichomoniasis causes a marked activation of preexisting genital mycoplasma infections.

TRITRICHOMONAS FOETUS

Tritrichomonas foetus is a parasite of the genital system of cattle that is transmitted exclusively by sexual contact in nature; in domesticated animals, however, there is some opportunity for transfer by fomites. The parasite has occasionally been identified in horses and in deer, and may be identical to *Tritrichomonas suis* of the nasal cavity and digestive tract of pigs. Morphological, metabolic, and immunological studies have demonstrated that the cattle and swine parasites are very closely related if not the same (Levine, 1973). Some laboratory animals can be infected experimentally.

Unfortunately, most of the investigations on immunity to bovine trichomoniasis have been done prior to 1960 and have not benefited from the recent advances in immunology. The widespread use of artificial insemination in cattle production in the last decades has reduced greatly the prevalence and, consequently, the economic impact of the infection, therefore discouraging the funding and research in this area. A detailed review of the information available at that time was published by Morgan in 1946, and recent reviews have been produced by Honigberg (1970, 1978b).

Natural Resistance

Very little information is available about the natural resistance of cattle to trichomoniasis, but from the numerous reports of successful experimental infections, it appears that cows are uniformly susceptible to the parasite on first contact. Artificial establishment of the infection in the bull has been considerably less successful; perhaps a degree of maceration of the penil mucosa during coitus favors the infection in nature. Studies in the late 1930s also showed that old cows that had not had previous experience with the protozoan did not develop age resistance.

Rats and mice are refractory to the genital infection, but the "mouse assay" has been used subcutaneously, intraperitoneally, and intrascrotally (see *"Trichomonas vaginalis,"* above). Persistent genital infections can be obtained in hamsters, rabbits, and guinea pigs. In these latter, some authors have found that production of estrus or administration of corticosteroids aid the permanency of the parasites. It has also been reported that *T. foetus* in the guinea pig causes inflammation of the genital tract, can produce abortion, and provokes histological lesions that resemble carcinoma.

A large number of animal species, including cattle and members of the five systematic classes of vertebrates, have demonstrated possession of natural antibodies in their sera that agglutinate the parasite at dilutions from 1:2 (in the poikilotherms) to 1:1024 (in horses). The sera of noninfected cattle most often show agglutinating activity up to dilutions of 1:48 or 1:96. Natural antibodies that lyse *T. foetus* in the presence of complement have also been found in numerous species.

Natural agglutinins do not exist in newborn calves, but they are acquired with the colostrum. These passive antibodies decline until disappearing in 1 or 2 months, and the young animal forms its own natural agglutinins from the second or third month of life. The adult levels are reached by the fourth month of age. As in the case of *T. vaginalis*, these antibodies are specific for *T. foetus* and cannot be adsorbed from the serum with other species of *Tritrichomonas* or *Trichomonas*.

It has been proposed that the natural anti–*T. foetus* antibodies are produced by occult infections or by common antigens amply distributed in nature or that they correspond to a genetic predetermination of the vertebrates. It seems difficult to accept the existence of occult infections in such a range of animals with such a high prevalence, and the specificity of the antibodies negates the idea of widespread common antigens. The explanation of the existence of specific anti–*T. foetus* antibodies (and of antibodies against other trichomonads) in vertebrates actually appears to be a challenge to the ingenuity of immunologists.

Acquired Immunity

Following a primary infection with *T. foetus*, bulls usually retain the infection for years without producing a detectable immune response. Circulating antibodies are sporadically found, but only in cases of severe infections that reach the testicular tissues. A recent communication from Australia, however, stated that, of 7 bulls infected with the Brisbane strain, 6 eliminated the parasite spontaneously within 2 weeks, whereas the remaining animal continued to be infected for at least 59 weeks. It is possible, therefore, that strains exist with a diverse ability to survive in the male host (because of metabolic or antigenic characteristics).

The infection in the cow, on the contrary, normally persists for shorter periods (1 to 23 weeks). The spontaneous elimination of the parasite in this case is believed to be an immunologically mediated event, because the recovered animals become resistant to reinfection for almost a year. The duration of the infection, however, seems to be largely influenced by the physiological conditions of the genital tract. The parasite disappears before or a little after parturition when the infected cow proceeds through a normal pregnancy; it is eliminated about a week after abortion when this occurs; it persists for up to 4 months in cases with uterine edema; and it disappears along with the pus and exudates in cases of pyometra. It is not known at present whether these local modifications act directly on the parasite or on the relevant manifestations of the immunity, although the latter seems more likely. Some authors have insisted that light infections, often difficult to demonstrate, may persist well beyond these periods.

Immune resistance is expressed by complete refractoriness or decreased susceptibility to colonization by the parasite, by retardation of its multiplication, or by a more precocious elimination of the infection. A primary inoculation of parasites of the Brisbane strain in cows produced infections lasting for 6 to 12 weeks, whereas the parasites of a second infection were recovered only for 2 weeks. Similarly, intravaginal inoculation of *T. foetus* in rabbits produced an infection in 83% of them that lasted 23 days on average; a second inoculation took in only 17% of the animals and lasted only 2 days.

Unlike bulls, cows often develop antibodies in the serum and in the genital secretions as a consequence of the infection. Similar antibodies have been detected in experimental infections of laboratory animals. In natural infections, antibodies are formed inconsistently and at low titers; only in severe infections, especially when abortion took place, are they present in most of the affected animals and at rather high levels. Antibody formation may be induced by deposit of living or dead parasites or their extracts in the uterus, or by injection of the same materials by extragenital routes (subcutaneously, intramuscularly, intravenously, or intraperitoneally), but not by exclusively intravaginal inoculation. It appears evident that the humoral response depends on the intensity of the parasitic invasion, and that the corresponding antigens are absorbed from the uterus but not from the vagina.

Production of tissue-sensitizing antibodies has been demonstrated by immediate-type skin tests in infected cattle and by allergic symptoms of the uterus, as a consequence of the infection, in cows previously inoculated intrauterinely with *T. foetus* antigen or antiserum.

The only evidence of cell-mediated immunity in trichomoniasis has been the production of specific delayed-type skin reactions in rabbits immunized with the parasite. Similarly, lymphocytes of immunized mice are able to destroy the parasites on contact, which may also be an expression of cell-mediated immunity. There are no compelling reasons, however, to assume that the same phenomenon should occur in cattle with natural infections restricted to the genital tract.

Mechanisms of Acquired Resistance

The older literature registers several attempts to induce protection by paragenital injection of tritrichomonas or of serum of resistant animals. These methods were often reported to produce some protection against challenges with cultured parasites, but rarely against the infection transmitted by coitus. Since many variables not related to immunity (derived from the cultivation techniques, the actual procedures of infection, and the diagnosis of the presence of the parasite) were involved in these experiments, it is difficult to determine whether the friction of the mucosae during coitus could overcome an incipient protection or whether the results reflected procedural difficulties.

More recent studies in cows and rabbits have been unable to demonstrate correlation between titers of circulating antibodies and resistance to the infection, or between levels of antibodies in the circulation and in the genital secretions. The consensus of the investigators at present is that circulating antibodies do not participate in any important way in protection against bovine trichomoniasis.

Agglutinating antibodies are frequently demonstrable in the uterine and vaginal secretions of infected cows, beginning about 6 weeks after the infection. Some differences in the activity and persistence of the antibodies recovered from the uterus and those retrieved from the vagina suggest that they may not be identical. In any case, strict inverse correlation has been demonstrated between the presence of antibodies and of parasites in genital secretions, which argues very strongly in favor of an antiparasitic action of these antibodies.

It is not clear, however, how the agglutinating antibodies dispose of the parasites. Phagocytosis of the trichomonas has been observed in infected cattle, in immunized rabbits and mice, and in tissue cultures. This may be an indication of the usual mechanism of elimination of the parasites, but confirmation and exploration of other possibilities is still necessary.

The presumably allergic changes of the uterine mucosa on reinfection may also play some role in protection, by increasing the permeability of the membranes to other immune elements or by altering the parasitic habitat.

Immunodiagnosis

Circulating antibodies to *T. foetus* have been demonstrated in infected cattle by skin tests, complement fixation, precipitation, agglutination, immobilization, and hemagglutination. In most cases the results of serological techniques have been erratic and only slightly sensitive, and the antibodies have disappeared within 6 months of the infection. The cutaneous reactivity is inhibited when large amounts of specific antigen are absorbed (as a result of natural infection or experimental inoculation), when corticosteroids are administered, and subsequent to parturition.

Acquired agglutinins may be distinguished from natural agglutinins by reacting the serum with diverse strains of *T. foetus*; whereas acquired agglutinins react more strongly with the homologous strain, the natural antibodies exhibit the same reactive titer for all the varieties. Also, serum containing only natural antibodies does not sensitize homologous skin, and its titers remain unchanged by the injection of specific antigen.

A "mucoagglutination test" with genital secretions is more reliable than the serological procedures. This test is positive in only about 60%

of infected cattle, which precludes its use on an individual basis, but it may be a valuable tool to identify the presence of the parasite in a herd.

Immunopathology

Strains of *T. foetus* that exhibit different pathogenicity have been identified by the "mouse assay," and Kulda and Honigberg (1969) found that parasites that differed in virulence, or their extracts, had various effect on the cells of tissue cultures. Although several strains have also been determined by antigenic analysis, it appears that no relationship has been established yet between the presence of determined antigens and the ability of the parasite to cause damage.

The accurate identification of nonpathogenic strains may facilitate vaccination against virulent parasites, since the limited evidence available indicates complete cross-protection among diverse strains. The findings by Kulda and Honigberg may even open the possibility of isolating the materials responsible for damage to the host and devising immunoprophylactic procedures with them.

The occurrence of symptoms of allergy in the uterus of immune cows by intrauterine instillation of *T. foetus* extract prompted some investigators to propose an allergic etiology for the trichomoniasis abortion. This notion has not been properly investigated yet.

THE PROTOZOA PREVAILING IN THE TISSUES

TOXOPLASMA GONDII AND RELATED ORGANISMS

Toxoplasma gondii is a coccidian parasite that infects the intestine of its definitive hosts, the domestic cat and other felids. More than 200 species of mammals and birds (including humans and the cat) may be utilized as intermediate hosts. Clinical disease is known to occur at least in humans, domestic carnivores, ruminants, swine, poultry and fur animals (chinchillas, foxes, and rabbits). Recent studies have demonstrated that *T. gondii* is closely related to species belonging to the genera *Hammondia*, *Besnoitia*, and *Sarcocystis*. Recent reviews that include immunological as well as biological aspects of *Toxoplasma* and associated organisms have been written by Dubey (1977) and Jacobs (1973).

The enteral infection in the felids rightfully belongs with the intestinal coccidia, so it is considered, along with some comments on its associated species, in the respective section (see "The Coccidia,"

above). It must be borne in mind, nevertheless, that the full understanding of the immunology of toxoplasmosis includes the immunological events that occur in the definitive host. It is advisable to read
both sections together.

The intermediate hosts may become infected by ingesting oocysts
passed by an infected cat and matured in the external environment or
by consumption of infected raw tissues of other intermediate hosts.
Mammals may also acquire the infection by transplacental passage of
the parasite from an infected mother (Figure 7).

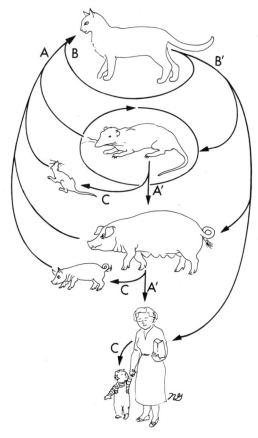

Figure 7. Major mechanisms of transmission of *Toxoplasma gondii* in nature. The definitive host (cat) can acquire the infection by consumption of tissues of infected intermediate hosts (**A**) or by ingestion of oocysts sporulated in the external environment (**B**).
The intermediate hosts, in turn, can acquire the infection by consumption of the tissues
of other infected intermediate hosts (**A′**), by ingestion of sporulated oocysts (**B′**), or
congenitally (**C**). Other modes of infection are possible but less important. Cannibalism,
as seen in rat communities, contributes to the persistence of the infection in a population.

The infective forms of the parasite enter nucleated cells of the new host and multiply rapidly inside them. These quickly dividing stages are called *tachyzoites*, trophozoites, or proliferative forms, and they characterize the acute or active phase of the infection. The proliferation of the tachyzoites destroys the host cell, and the released parasites invade new nearby cells or spread in the circulation either free or by parasitizing migratory leucocytes. Most parasites are captured by macrophages or penetrate the cells of lymph nodes, liver, or lungs, but a few may spread throughout the body.

The cycles of tachyzoite reproduction continue for 1 or 2 weeks until the parasites that invade new cells develop into smaller stages that divide slowly in cysts formed inside the host cell. These forms are called *bradyzoites*, merozoites, or cystic forms, and characterize the chronic or inactive phase of the infection.

Natural Resistance

The natural resistance to *T. gondii* infection varies markedly among the diverse animal species: guinea pigs, rats, monkeys, and possibly humans are rather resistant and rarely develop a symptomatic infection; mice, hamsters, and rabbits, on the other hand, are very susceptible and frequently die of the infection. Studies with seven strains of mice have suggested that the natural resistance has a genetic basis. The influence of several factors on the susceptibility to toxoplasmosis has been reviewed lately by Kozar (1970) and Frenkel (1973).

Acquired Immunity

The acquisition of immune resistance to reinfections with *T. gondii* is usually investigated by recording the reduction of the mortality or morbidity, or by the prolongation of the survival time to a challenge infection in animals recovered from a previous infection. In this way, it has been possible to demonstrate that the original infection produces a strong but not absolute immunity against subsequent invasions by the parasite. The immunity is commonly expressed as a supression of the symptoms caused by the new infection and a marked reduction in the number of parasites that survive in the host. Massive reinfections or particularly virulent strains of parasites, however, may overcome the resistance and initiate a new episode of clinical disease.

The immunity is not able to terminate the original infection, nevertheless, and examinations of tissues of infected animals and people have revealed that the parasites remain in the host for long periods, maybe for life. Although the resistance to reinfections seems to decline with time, it appears to persist to some degree as long as parasites are still present in the body. Artificial termination of the primary infection is promptly followed by the return of susceptibility to the parasite. Evidently, immunity to *T. gondii* is of the premunition type.

Experimental studies with rodents have shown that the immunity of the mother does not prevent transplacental transmission to the litter. Clinical observations in pregnant women indicate that the same is not true for people: women infect the fetus only during the acute phase of the infection and the subsequent immunity impedes further transmission in future pregnancies. Studies in sheep and pigs also demonstrated that congenital transmission occurs only once, and solely during the active infection of the mother.

The actual mechanism of congenital infection is still unknown, but experiments in mice suggest that it is related to the virulence and tropism of the parasite. The mere presence of the infection in the placenta is not enough to assure the infection of the fetus.

There is ample consensus among researchers that antibodies are effective only on tachyzoites making their way from one host cell to another, whereas the bradyzoites are protected from the activity of the humoral response in their intracellular cysts. This does not imply, however, that parasites are absent from the circulation. On the contrary, parasitemia is the rule in the acute phase, and parasites have been transferred with the blood of chronically infected animals and people on several occasions. Probably the parasites of the chronic phase are parasitizing white blood cells rather than free in the circulation.

The assumed lethal effect of antibodies on free *Toxoplasma* and the coincidence of the appearance of circulating antibodies with the passage to the chronic phase of infection have led many researchers to believe that the formation of bradyzoites is induced by the host immune response. This notion is supported by the well-known facts that toxoplasmosis is more severe in immunoincompetent individuals and that reversion of the chronic to the acute phase may be achieved by administration of immunosuppressor drugs. Furthermore, the parasites appear to persist longer and to cause more damage in locations beyond the normal reach of the immune system, such as eye, central nervous system, and fetus. Despite all this circumstantial evidence, Stahl et al. (1965) observed that bradyzoites in immunodepressed mice developed earlier and in greater numbers than in normal mice. It is possible that, with independence of the effect of the immune response on the maintenance of the chronic phase of infection, the opportunity of the first passage from tachyzoites to bradyzoites may be an inherent characteristic of the biology of the parasite.

Studies in man and lower animals have shown that *T. gondii* infection induces the production of antibodies of the classes IgM, IgG, and IgA. IgM antibodies are the first to appear, stay at high levels for

2 months, and usually wane after 3 to 5 months of infection; occasionally they reappear in the course of the chronic infection and persist for a few months. These episodes may be the expression of localized reactivations of the infection that are not detected clinically. Since IgM does not cross the placenta and is produced early in the ontogeny, detection of specific anti–*T. gondii* IgM antibodies in the newborn indicates that the antibodies were produced by the child itself and presumes that in utero infection took place. High titers of IgM antibodies associated with rising levels of IgG antibodies in an adult may suggest recent infection.

A heat-labile anti-*Toxoplasma* "hostile" factor found in the plasma of numerous animal species (notably ruminants) has been tentatively identified as a natural IgA antibody.

Recent investigations in humans and lower animals have demonstrated that *T. gondii* stimulates cell-mediated immunity that is detectable by skin tests, by inhibition of the macrophage and leucocyte migration, and by lymphocyte transformation with specific antigens. In humans, at least, the production of cell-mediated responses is rather slow; its first manifestations are detected about 2 months after infection and frequently later.

Immune responses in general are long-lived in toxoplasmosis, but the manifestations of cellular immunity persist longer than the expressions of humoral immunity, which coincides with the prolonged presence of acquired resistance to reinfections. Possibly the scanty presence of bradyzoites in the tissues of the host is enough to maintain the cellular but not the humoral immunity. Recent studies with immunofluorescence have shown reactions around bradyzoite cysts, which suggest that antigen leaks outside the parasitized cells.

Mechanisms of Acquired Resistance

Antibodies do not appear to play a critical role in the protection against *T. gondii* reinfections. A number of experiments directed to transfer resistance from immune to susceptible animals by inoculation of serum have failed or have achieved only an insignificant degree of protection. Similarly, vaccination with dead organisms, which produced a humoral response comparable to that elicited by natural infections, did not induce any important protective immunity. In vitro studies, however, have shown that anti–*T. gondii* antibodies bind the organisms and facilitate their ingestion and digestion by macrophages. Antibody-coated parasites are less able to invade host cells and their external and internal membranes are lysed in the presence of complement.

It is not known why these same mechanisms do not operate efficiently in vivo to prevent reinfection. Possibly the parasite is able to utilize some yet unidentified system provided by the host (and therefore absent in the in vitro systems) to elude the effect of the humoral response or is simply protected by its intracellular location.

Cell-mediated immunity has been shown to play a much more important role in immune protection. Resistance has been transferred from immune to normal hamsters by inoculation of spleen and lymph node cells. Curiously, there appears to be a dissociation between cutaneous delayed hypersensitivity and cell-mediated protection; cutaneous reactions and antibodies were detected in the recipients by the fifth day of transfer, but protective immunity did not appear before the third week.

Treatment of mice with antilymphocyte serum, which is an effective depressor of the T cell activity, greatly increased the mortality resulting from a primary *T. gondii* infection and notably reduced the resistance to a reinfection (although it did not affect the preexisting antibody levels). Finally, "nude" mice, which lack thymuses and therefore are unable to mount a cell-mediated response, do not develop immunity to *T. gondii*; they do, however, when transfused with thymus from normal mice, but not when inoculated with bone marrow cells or with high-titered anti–*T. gondii* serum.

Macrophages of normal animals are readily invaded by *T. gondii*, which divides undisturbed in them. Macrophages of immune animals, on the other hand, show an increased ability to engulf and destroy the parasites. Several in vitro experiments have demonstrated that the activation of macrophages in immune animals depends on the production of lymphokines by T cells, after stimulation with specific *T. gondii* antigens. Transfer of immunity with populations of macrophages from resistant animals has failed until a proportion of lymphocytes from the same origin has been added to the inoculum.

The stimulation of T cells by *T. gondii* antigens and the corresponding production of macrophage-activating lymphokines is an antigen-specific event, but the ensuing phagocytic effector activity is nonspecific. Macrophages activated in the course of a *T. gondii* infection show increased phagocytosis against viruses, bacteria, fungi, other protozoa, and tumor cells (McLeod and Remington, 1977). The utilization of this characteristic as a therapeutic tool in humans (in the fashion of "malaria versus syphilis" of the past) has been proposed occasionally, but the risks seem too high to justify it.

Recent work has demonstrated that intracellular *Toxoplasma* exerts an effect on the membranes of the phagocytic vacuole in such a way that the fusion with the membranes of the lysosomes does not occur. The parasite is then spared the action of the lysosomal hydrolytic enzymes that could destroy it otherwise. The effect of the parasite on the membranes must be an active mechanism since it does not occur when the parasite is dead. It appears that the activation of the macrophages by lymphokines develops in these cells the capacity to defeat the "barrier" established by the parasite or to initiate the attack before the barrier becomes effective.

All the evidence accumulated up to now indicates that cell-mediated immunity is the main factor of acquired resistance in toxoplasmosis (Remington and Krahenbuhl, 1976). It is possible, however, that humoral immunity plays a far more important role in vivo than it has been credited for so far. In most cases, a primary infection by *T. gondii* is controlled by the body during the second or third week of invasion; at this time production of specific antibodies is at the logarithmic phase and cell-mediated immunity will be still undetectable for another month or so. It is not improbable that the humoral response restricts the accelerated proliferation of the parasites during those critical first 2 months in which cellular immunity has not reached an effective level yet. Research in this area seems to be lacking at present.

Artificial Production of Resistance

Numerous attempts to induce protective immunity by experimental inoculation of dead parasites or by injection of soluble or particulated extracts of *T. gondii* have been met with a variable degree of success. Although it has been possible to prolong the survival or to diminish the mortality of laboratory animals challenged with virulent parasites on occasion, the protection produced by these methods has always been much weaker than the protection afforded by the actual infection. Vaccination with living parasites of reduced virulence, on the other hand, has produced an appreciable degree of resistance to a later challenge with fully virulent organisms. The inherent risk of using living parasites for immunization of humans or domestic animals (see Chapter 8) advises against this procedure, however.

Apparently the attainment of an effective level of protective immunity requires the sustained production of antigens compatible only with the prolonged presence of metabolizing parasites. Recently Seah and Hucal (1975) reported that inoculation of mice with *Toxoplasma* irradiated with 10,000 roentgens produced 100% refractoriness to reinfection during the 3 ensuing weeks. Since an appropriate dose of radioactivity will prevent the reproduction of the parasites without killing them, this method may assure the necessary extended production of antigens, avoiding the risks of potentially dangerous organisms replicating in the host. Further research in this direction is being actively pursued currently.

Immunodiagnosis

The ambiguity of the symptoms of toxoplasmosis and the difficulties of demonstrating the presence of the parasite often force the clinician to rely on indirect means of identifying the infection. The most widely accepted methods for the immunological diagnosis of *T. gondii* infections are the methylene blue dye test (MBDT), the complement-fixation test (CFT), the indirect hemagglutination test (IHAT), and the indirect fluorescent antibody test (IFAT) (Jacobs, 1976).

The MBDT is based on the loss of the affinity of the parasite for methylene blue stain as a result of a partial lysis of the parasitic membranes by specific antibodies plus an "accessory factor" of the serum. Since the sera of many animals and of some people possess a heat-labile component that nonspecifically lyses the membranes of *T. gondii*, it is convenient to heat all sera at 60°C for 30 min before attempting to perform the test. The antibodies active in this test appear during the second week of infection in humans and rise to very high titers (up to 16,000) in the following 2 or 3 weeks. They stay high until convalescence and then decrease to remain at low levels for years. The MBDT is sensitive and highly specific, but requires maintenance and manipulation of living parasites, which is expensive and dangerous for non-specialized laboratories.

Complement-fixing antibodies are commonly detected at the beginning of the fourth week of infection, rise to a peak, and diminish after a few months. Very low titers may be found up to 2 or more years after the infection. Because of the rather transient nature of these antibodies, many clinicians tend to associate high titers in the CFT (32 or more) with recent infection. This test is satisfactorily specific, but its sensitivity depends to a great extent on the antigen utilized. The rather involved technique of the CFT, its unpredictable results with sera of many domestic animals, the lack of proper standardization, and the existence of other adequate tests have tended to diminish the routine use of this test lately.

Hemagglutinating antibodies normally appear during the third week of infection; titers greater than 256 are often considered indicative of recent infection. The titers of the IHAT have correlated very well with the titers of the MBDT in most studies, except in some cases of congenital infection in which IHATs have been negative for months in the presence of high-titered MBDT. Special precautions must be observed when sheep erythrocytes are used in IHAT with serum of domestic animals, since many of these have natural agglutinins against sheep red blood cells. The efficacy and comparative simplicity of this test and the availability of commercial kits have made the IHAT a routine procedure for the diagnosis of toxoplasmosis at present.

The IFAT is very consistent and often correlates very well with the MBDT and with the IHAT; as in the other two, a titer greater than 256 is frequently regarded as representative of recent infection. The IFAT has given occasional cross-reactions with antigens of *Sarcocystis* and of *Besnoitia* and with sera containing antibodies against nuclear material. This test may be easily modified to demonstrate IgM antibodies in newborn children suspected of having congenital toxoplasmosis. A method that permits the simultaneous investigation of IgM antibodies against *T. gondii* and *Trypanosoma cruzi* is currently available. The absence of anti–*T. gondii* antibodies in a newborn child does not rule out the infection, since about 50% of congenitally infected children do not show a detectable immune response. The reasons for the lack of reactivity in this group are not known, but the immaturity of the immune system of the child, a feedback mechanism induced by antibodies transferred from the mother, or the production of a specific immunotolerance by massive doses of antigens early in life might account for it.

A few serological tests such as precipitation in liquid or in gel, direct agglutination, and agglutination with latex have been proposed at different times, but have failed to compete successfully with the established tests. A commercial kit for direct agglutination has become available lately, which may help popularize this test.

Recent advances in serologic methodology are in current assay in toxoplasmosis: the use of counterimmunoelectrophoresis has detected antibodies in infected mice on the third day of infection, whereas the MBDT became positive only on the seventh day; enzyme-linked immunosorbent assay was used in parallel with the MBDT and the IHAT and demonstrated a good correlation with both exams; a radioimmunoassay was performed in parallel with the IHAT and also showed good concordance. None of these tests is in current use yet, but the experimental results obtained so far are encouraging.

Differentiation of Acute from Chronic Infections Since *Toxoplasma* infection is very prevalent in man and domestic animals, clinicians often face the problem of deciding whether a given serologic titer represents a current acute episode or the remnants of an old infection. The accepted serological criteria for identifying a *T. gondii* infection as acute is the appearance of serological titers in previously negative patients or a three-fold increase of serological titers in a 2-week period.

Karim and Ludlam (1975) have recently reviewed the evolution of specific antibodies in glandular and ocular toxoplasmosis, as detected by six serological tests. In the glandular syndrome, the MBDT, the CFT, the IFAT, and the direct agglutination test were all already positive at the onset of the symptoms, reached their highest titers (approximately 3000 to 8000) 3 to 4 months later and were still positive (over 256) after 18 months. The IHAT was positive at a low titer at the onset, peaked about a year later, and was still very high after 18 months. The IFAT, restricted to the investigation of IgM antibodies, was very near its peak at the onset (titer 256), increased slightly over the next 2 months and fell rapidly (to about 32) after 6 months (Figure 8). Although the titers among individual patients were quite variable, this work reports the averages of 160 clinical cases, so its conclusions must be strongly representative of what actually happens in the infection

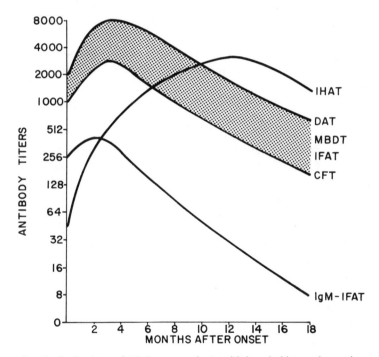

Figure 8.   Antibody titers of 160 human patients with lymphoid toxoplasmosis, as detected with different tests. The direct agglutination test (**DAT**), the methylene blue dye test (**MBDT**), the indirect fluorescent antibody technique (**IFAT**), and the complement-fixation test (**CFT**) were all positive at high titers at the onset of the symptoms, reached their highest titers about 3 months later and remained at fairly elevated titers for another 15 months. Titers with the indirect hemagglutination test (**IHAT**) were low at the onset, but increased slowly to reach a maximum after a year. Titers with the IFAT modified to detect only IgM antibodies (**IgM-IFAT**) were near their peak at the onset; the peak was reached shortly afterward, and the antibody activity fell rapidly thereafter. (Based on data by Karim and Ludlam, 1975.)

with lymphoid manifestations. According to the results, the IFAT for IgM antibodies should be the most reliable test to identify the acute infection, and high IHAT titers would be more characteristic of the chronic infection.

In 77 cases of *Toxoplasma* choroidoretinitis, Karim and Ludlam found high titers with the IHAT (1000 on average), low titers with the IgM-IFAT (about 5), and medium titers (140 to 500) with the other techniques. Although the titers also varied widely in these patients, these results support the notion that ocular toxoplasmosis is a late sequel of the infection.

Magliulo et al. (1976) recently devised an immunocytoadherence test that appears to be positive only in the acute phase of the infection. Preliminary studies in rats detected an immune response before the appearance of hemagglutinating antibodies and the reactivity diminished rapidly 1 month after the infection. Assays with the sera of 10 patients tended to corroborate these findings.

A delayed cutaneous test ("toxoplasmina") becomes positive a few months after the infection and remains so for life. It has had limited use to verify old infections in epidemiologic studies, in cases of evaluation of gravid women (to rule out acute infections of possible congenital transmission), and in cases of ocular toxoplasmosis.

Since ocular toxoplasmosis often is a late sequel of the infection, serological titers are very low and lend little assistance to the clinician who wants to identify this affliction. Desmonts proposed the calculation of the "aqueous humor antibody coefficient" by multiplying the ratio of antibody titer in the aqueous humor to that in the serum, by the ratio of globulin concentration in the serum to that in the aqueous humor. The diagnosis of ocular toxoplasmosis is suggested with coefficients from 2 to 7 and it is probable with coefficients of 8 or greater.

A number of methods to verify cell-mediated immunity in toxoplasmosis are available, but, since the cellular immunity appears so late in the course of the infection and it is not a safe indication of absolute resistance to reinfections, there has been no particular interest in their clinical applications.

Antigens

The study of the antigens of *T. gondii* has been greatly impaired by the lack of an adequate system to grow the parasite in the absence of other cells. The present antigenic preparations are obtained form parasites recovered from parasitized cells of peritoneal exudate of infected mice, from cell cultures of chicken embryos, or from cells of tissue cultures. In every case, contamination with materials of the host cells is impossible to avoid.

Some logical conclusions have been drawn from the results of serological tests: the different kinetics of the titers yielded by the same serum studied by the MBDT and by the CFT suggests that the respective antibodies are elicited in the host or detected in the test by different components of the parasite. Similarly, the adsorption of the hemagglutinating antibodies from a serum does not affect the titers detected by the MBDT, which also suggests that different antigens are operating in each case.

When rabbits are immunized with the supernatant fluid obtained by centrifugation of lysed *T. gondii*, most of the antibodies produced are hemagglutinins; immunization with the sediment, on the other hand, elicits antibodies particularly reactive in the MBDT. Based on these findings and on the fact that the antibodies reactive in the MBDT precede the antibodies detected by the IHAT for about a week in most infections, it has been speculated that the antibodies active in the MBDT are stimulated by the external membranes of the protozoa, whereas the hemagglutinins are induced by internal components of the parasite released after its lysis. If this hypothesis is true, the MBDT should be inherently more precocious than the IHAT.

The antibodies detected in the IFAT may be induced by nucleoproteins of the parasites, since this exam has yielded positive results with the sera of 14 of 18 patients who had antibodies against nuclear material. Experimentally, it has been demonstrated that the antigen that is reactive in the delayed cutaneous test is also a nucleoprotein.

Studies directed to determine antigenic relationships have found that *Hammondia*-infected animals show cross-resistance and serological cross-reactivity to *Toxoplasma*. Some reduced cross-reactivity with IFAT has been found between *Sarcocystis* and *Toxoplasma*. None of this seems to interfere with the routine diagnostic tests.

Immunopathology

There is no solid evidence that the immunity to *T. gondii* has a role in the production of pathology. In 1961 Frenkel suggested that the ocular lesions in toxoplasmosis resulted from a localized reaction of hypersensitivity. O'Connor (1970), however, was unable to substantiate this hypothesis and concluded that the damage was most probably due to the multiplication of the parasite in the cells of the eye.

An attempt to desensitize patients of toxoplasmosis by repeated injections of toxoplasmina yielded some degree of clinical improvement in half the patients. These results may indicate the participation of allergic components as well as of psychosomatic factors in the production of symptoms.

Recently it has been found that mice infected with *T. gondii* had deposits of antigen-antibody in their renal glomeruli. It is not known yet whether the same phenomenon occurs in humans and other animals or these findings are actually related to the pathology of the infection.

HAMMONDIA

Hammondia is a new genus designated in 1974 by Frenkel to accommodate a coccidium very similar to *T. gondii* but that required an obligated cycle through a predator-definitive host and a prey-intermediate host. Only two species are known at present: *H. hammondi* of the intestine of the cat and the muscles of several rodents and *H. heydorni* of the intestines of dogs and the muscles of cattle. Little is known at present of the immunology of *Hammondia* infections, but some preliminary studies have shown serological cross-reactivity and cross-protection between *Hammondia* and *Toxoplasma* in the intermediate host. Further studies of the cattle infection are necessary to determine whether vaccination with *Hammondia* against *Toxoplasma* infection is feasible.

BESNOITIA

The coccidia in this genus infect the intestine of the cat in the few cases in which the definitive host is known, and infect the connective tissues of the intermediate hosts (ruminants, horses, rodents). Six species are currently known, all with fairly strict specificity for their obligatory intermediate hosts. *B. besnoiti* is pathogenic for cattle and *B. bennetti* for horses.

Limited immunological studies have revealed a remarkable similarity to the immunology of *Toxoplasma* infections. The infection produces a resistance of the premunition type, which is transferable with spleen and lymph node cells. As in toxoplasmosis, *Besnoitia* infections induce an increment of the phagocytosis that is effective against unrelated organisms also. No cross-protection or serological cross-reactivity has been found between *Besnoitia* and *Toxoplasma*, but some common antigenicity with *Sarcocystis* has been detected with IFAT.

SARCOCYSTIS

The endothelial and muscle forms of *Sarcocystis* in the intermediate hosts have been known for more than a century, but the intestinal phase in the definitive host was discovered only in 1972 (Markus, 1978). The species in humans and domestic animals recognized nowadays are *S. cruzi* (definitive host dog, intermediate host cattle), *S. hirsuta* (cat, cattle), *S. hominis*, formerly *Isospora hominis* (humans, cattle), *S. oviscanis* (dog, sheep), *S. tenella* (cat, sheep), *S. miescheriana* (dog, swine), *S. porcifelis* (cat, swine), *S. suihominis*, formerly *I. hominis* (humans, swine), *S. bertrami* (dog, horse), *S. fayeri* (dog, horse), and *S. lindemanni* (definitive host unknown, intermediate host humans). *S. cruzi* and *S. oviscanis* are important pathogens of cattle and sheep, respectively.

Calves infected with *Sarcocystis* develop antibodies reactive in the IFAT between 28 and 35 days of infection and hemagglutinating antibodies between 30 and 45 days of infection. IFAT titers were between 40 and 640 depending on the time elapsed since infection; IHAT titers reached up to 39,000 by the 90th day of infection. In the IHAT, titers greater than 162 are considered to be specific for the infection, and titers over 1458 are regarded as indicative of acute infection. Infection of old cows produced a sharp increase in hemagglutinating titers, suggestive of a secondary response. Complement-fixing antibodies appeared in infected sheep between 15 and 47 days of infection.

Random samples of human sera studied by IFAT with *Sarcocystis* from bovine muscle yielded titers up to 1024, although they were negative by CFT with sheep *Sarcocystis*. Since humans are the definitive host of cattle *Sarcocystis* but not of the sheep parasite, these titers may represent instances of infections of the human intestine with *S. hominis*, which are fairly common, rather than extraenteral infections with *S. lindemanni*, which are rare. Apparently the infection of the intermediate host with *Sarcocystis* does not produce protective immunity: mice can be readily reinfected with *S. muris*, and sheep infected with *S. tenella* accept *S. oviscanis* infections 4 weeks later.

Some cross-reactivity has been found between *Sarcocystis* and *Toxoplasma* with the IFAT, but not with the MBDT or the IHAT. The different species of the genus *Sarcocystis* possess common antigens and antigens peculiar to each species.

THE LEISHMANIAS

The leishmanias are flagellates transmitted by flies of the genus *Phlebotomus* (sandflies) or closely related genera that affect primarily humans, canids, rodents, and lizards. The forms in the gut of the biological vector and in cultures are elongated, have an anterior flagellum, and are referred to as *promastigotes*. The parasites in the vertebrate invade macrophages and occasionally other cells; they are small, round bodies without external flagella, and are called *amastigotes* or *micromastigotes*.

With the exception of *Leishmania enriettii* of guinea pigs, the common species of leishmanias are morphologically indistinguishable, which has created serious problems for their taxonomic classification. Originally, they were classed according to the clinical syndromes produced in man and their geographic distribution. It was soon evident that these criteria were insufficient to identify parasites recovered from lower animals, and assays to separate them by biochemical and immunological characteristics were attempted later. Unfortunately, none of these attributes has provided clear-cut results so far. The present classification is a changing combination of all these properties that has grown more and more confusing with time. Many authors are returning to the three original species, considering them now as complexes that enclose several related species. A lucid account of the current species and of their relations has been recently given by Zuckerman and Lainson (1977). This centralizing classification is adopted here, in part for simplicity and in part because much of the research in immunology of leishmaniasis has been reported under the name of the classic species and it would be an impossible task to try to assign the corresponding findings to the newer species.

A number of reviews on, or that include, the immunology of *Leishmania* infections have been published in the last years. The papers by Maekelt (1972), Zuckerman (1975), Bray (1976), and Preston and Dumonde (1976) are excellent introductions to the pertinent literature.

Three fundamental types of leishmanial disease are recognized in humans:

1. Cutaneous leishmaniasis, which affects only the skin. The classic form (CCL) is the most common presentation, but a diffuse form (DCL) is not infrequent, and cases of leishmaniasis recidiva (LR) appear occasionally. The causative agent is *L. tropica* and associated species (*L. major*, *L. aethiopica*, *L. mexicana amazonensis*, *L. mexicana pifanoi*, and *L. peruviana*).
2. Mucocutaneous leishmaniasis, which originally infects only the skin but later metastasizes to nasopharingeal or auricular tissues. The etiological agent is *L. braziliensis* and related forms (*L. mexicana mexicana*).
3. Visceral leishmaniasis, which produces a transient skin lesion followed by invasion of the macrophages of the visceral organs. It is caused by *L. donovani* and its varieties (*L. infantum* and *L. chagasi*).

Natural Resistance

It has been traditionally assumed that humans do not exhibit natural resistance to *Leishmania* infection in any significant degree. Recent findings, however, challenge this belief. The high prevalence of some species of *Leishmania* in reservoir animals does not correlate with a much lower prevalence in the coexisting human population, and this difference is not always explained by the habits of the vector. Also, it has been found that a high proportion of individuals in endemic areas have positive immunological reactions to leishmanial antigens, despite the fact that they have never had a history of leishmaniasis; it is suspected that these are cases of brief contact with the parasite, in which it was promptly rejected. Whether the rejection is due to natural resistance or to a quickly acquired immunity is open to investigation, but evidently the contact is enough to elicit an immune response.

In lower animals, some species of leishmanias affect a single host species [e.g., *L. hertigi* of porcupines and *L. enriettii* of laboratory (*Cavia porcellus*) but not of wild (*Cavia aperea*) guinea pigs], whereas others are found in diverse animals (e.g., *L. braziliensis panamensis* of humans, lower primates, rodents, procyonids, and sloths). This suggests that the host specificity may be partially dictated by characteristics inherent in the parasite.

On the other hand, the constitution of the host must also play an important role, since guinea pigs are refractory to the infection with human leishmanias, selected strains of mice can be infected, and hamsters develop a fulminating disease. Studies with mice have demonstrated that their susceptibility to *L. tropica* and *L. donovani* depends on the genetic composition of the particular animals.

It has been recently reported that the serum of rodents, and to a lesser degree human serum, possesses natural IgM antibodies able to agglutinate and lyse in vitro the promastigotes from cultures of human leishmanias. It is possible that these antibodies also operate in vivo to eliminate part of the parasites inoculated by the vector.

Acquired Immunity

The study of the immunology of *Leishmania* has demonstrated that leishmaniases are spectral infections whose manifestations may vary from nil to lethal, depending on the activity and efficiency of the immune system in each case. All forms of leishmaniasis begin with the introduction of the parasite in the vertebrate and its multiplication in the macrophages and histocytes of the skin of the host.

Classic Cutaneous Leishmaniasis   In classic cutaneous leishmaniasis (CCL) the initial lesion develops into an ulcer that heals over a period that varies from 3 to 12 months. The resolution of the ulcer leaves a strong immune protection (which usually persists for life) specific for the species of parasite involved. In the course of this infection, it is sporadically possible to detect the presence of specific antibodies. Tests of gel diffusion, complement fixation, indirect hemagglutination,

immunofluorescence, Prausnitz-Kustner, and skin reactions of the immediate and intermediate type have been reported to be positive in
patients. The presence of antibodies, however, is inconstant and their
titers are usually very low. Undoubtedly, the humoral response in cutaneous leishmaniasis is minimal.

Cell-mediated immunity has been demonstrated by delayed skin
reactions, by antigen-specific transformation of lymphocytes, and by
inhibition of the macrophage or leukocyte migration. These manifestations of immunity appear rather late in the course of the disease and
progress with it, reaching their peak in coincidence with the time of
cure. Unlike the humoral response, they occur with great constancy
in all patients. New infections only take before the primary ulcer heals,
and during the phase in which cell-mediated immunity is just developing; if the healing or the immunity are well established, the reinfections are abortive. Similarly, surgical removal of the ulcer before definitive cure prevents the full development of cell-mediated immunity
and resistance. The correlation of the systemic and local manifestations
of cell-mediated immunity with the cure of the lesions and with the
production of immune resistance to reinfections has been demonstrated
so consistently that few researchers doubt that a cause-effect relationship exists between them.

As opposed to the classic form, there are two other manifestations
of the cutaneous disease that show little tendency to spontaneous resolution: the diffuse and the recidiva varieties.

Diffuse Cutaneous Leishmaniasis   In diffuse cutaneous leishmaniasis (DCL) the original lesion is a nodule that extends or sends cutaneous metastases through the systemic route, although the parasite
does not invade the organs. The lesions are huge masses of heavily
parasitized macrophages in the thickness of normal tissues.

Specific antibodies in this variety are found as inconstantly and
at titers as low as in the classic form, but the levels of IgM and IgG
in the serum are normal, which suggests preservation of the usual
mechanisms of immunoglobulin production. The specific cell-mediated
reactivity against the parasite, on the other hand, is absent, although
the manifestations of cell-mediated immunity against other antigens
(e.g., lepromina, tuberculin) remain normal. When the lesions begin
to heal, spontaneously or by treatment, the cell-mediated reactivity to
leishmanial antigens returns and lymphocytes invade the lesions. It
seems evident that the diffuse form corresponds to a state of specific
anergy, in which the host is unable to mount a cell-mediated response
against the invader.

Because the DCL has a rather well defined geographic distribution,
it has been hypothesized that its causative agent may be a strain of *L*.

tropica that lacks the necessary antigens to stimulate cellular immunity, or that possesses substances that induce tolerance to the specific antigen. On the other hand, inoculation in volunteers of parasites obtained from patients with the diffuse form has provoked the classic cutaneous disease, which suggests that the clinical manifestations of the infection are dictated by peculiarities of the host rather than of the parasite. Ultimately, it is possible that DCL is the result of the interaction between parasitic strains with particular characteristics that affect hosts with predisposition to defects of the immune response.

Whatever the primary cause of the depression of specific cell-mediated immunity may be, the great number of parasites existing in the lesions may contribute to it by production of immunological tolerance because of an excess of antigen. This type of phenomenon has been induced in laboratory animals by injections of large amounts of leishmanial antigen. The fact that chemotherapeutic destruction of a proportion of the parasites of a patient often induces the return of cell-mediated immunity and the resolution of the disease lends support to this hypothesis.

The logical therapeutic approach to DCL consists of stimulating the inhibited cell-mediated immunity. Recent studies have demonstrated that the injection of nonspecific stimulants of cellular immunity (e.g., BCG) effects reductions in the severity of the infection and the mortality of mice experimentally infected with *L. tropica.*

Leishmaniasis Recidiva   In leishmaniasis recidiva (LR), the original ulcer may heal but tubercules of indolent growth appear in its periphery; they contain few parasites but a large number of lymphocytes. As in the other varieties, this form produces a minimal humoral response. The manifestations of cell-mediated immunity, on the contrary, are grossly augmented. In contrast to DCL, the pathogenesis of the lesions in LR is an excessive reactivity of the cellular immunity to the parasite or its products.

As in the diffuse form, there are arguments that favor the role of strains of parasites with particular characteristics and others that indicate predisposition of the host in the production of this clinical variety of leishmaniasis. Again, it may be that LR only occurs when there is association of certain potentialities of the invading parasite with certain peculiarities of the host's immune system.

The rational treatment of LR consists of reducing the exaggerated cellular reactivity of the host and removing the antigenic stimulation represented by the parasite in the tissues.

Mucocutaneous Leishmaniasis   Mucocutaneous leishmaniasis also begins with a cutaneous ulcer but, during the course of this primary lesion or, more frequently, after it has healed, destructive and persistent metastases appear in the nasopharyngeal mucosa and in its associated cartilages. *L. mexicana mexicana* causes metastatic lesions only in the ears in 40% of the cases; in the remaining occasions it behaves as a CCL.

Humoral and cell-mediated responses are somewhat more intense than in the cutaneous leishmaniases, but the antibodies characteristically develop in conjunction with the appearance of the mucosal metastases. After healing of the skin lesion, the patients are refractory to intradermal inoculations of the homologous parasite, but the mucosal lesions continue progressing. The lack of defensive reaction at the location of the metastasis has been attributed to the protection offered by the cartilage, which is notoriously indifferent to the effects of the immune response.

Visceral Leishmaniasis   In visceral leishmaniasis, the cutaneous lesion is transient and is rapidly followed by parasitic invasion of the reticuloendothelial tissue of the organs. As a result of the parasitism, a marked hyperplasia of the lymphomacrophagic system occurs, which is responsible for the hepatomegaly, splenomegaly, and hypergammaglobulinemia characteristic of the infection.

The elevated gammaglobulins correspond essentially to the IgG class, with a smaller proportion of IgM. A part of them are specific antibodies against the parasite, but the remainder constitute a nonspecific response, probably caused by the stimulation of the lymphoid cells by the parasite. Recent examination of 10 patients showed that 7 had elevated numbers of B lymphocytes in the circulation (Rezai et al., 1978).

In contrast to the humoral response, the patients of visceral leishmaniasis do not show any evidence of cell-mediated immunity against the parasite. In the study by Rezai et al., 9 of 10 patients had reduced numbers of peripheral T lymphocytes. In those cases in which the infection is cured, spontaneously or by treatment, the levels of nonspecific and antiparasite immunoglobulins decline, and a specific cell-mediated immunity that leaves a strong protection develops.

It therefore appears that an exaggerated production of immunoglobulins (and B cells?) associated with a depression of the cell-mediated immunity (and T cells?) occurs in visceral leishmaniasis, and that this situation reverses at the moment of cure. Recent findings of the mutually exclusive effect of the humoral and cell-mediated responses in many instances suggest that the same phenomenon may be operating in this case. At this moment there is not enough evidence available to give preference to either the stimulation of immunoglobulin production or the inhibition of cell-mediated immunity as the primary event.

An increasing number of cases of asymptomatic or purely cutaneous infections by *L. donovani* are being reported currently. It is possible that the original cutaneous lesion in these cases elicits a protective immunity strong enough to prevent the visceral invasion by the parasite, and even to produce the rapid cure of the skin damage.

Mechanisms of Acquired Resistance

Observations in human patients indicate very strongly that protection against leishmaniasis is associated with manifestations of cell-mediated immunity and not with antibody production. This conclusion is widely supported by animal experiments that have been able to transfer resistance with lymphoid cells but not with serum of immune animals.

In vitro, leishmanias invade and multiply inside macrophages of normal animals without showing any evidence of inhibition. Macrophages of animals with active infections, on the contrary, have the ability to retard the proliferation of the parasites. Some researchers have found that this capacity is enhanced by the prior incubation of the leishmanias in the serum of recovered animals, but others have not been able to demonstrate this effect. No evidence has been found, however, that antibodies play a protective role in the human infection.

Lymphocytes activated specifically with leishmanial antigens, or nonspecifically, are able to produce lymphokines that do not exert a direct action on the parasites in vitro but destroy parasitized macrophages. More than defensive, this activity may contribute to the pathology in vivo, since it destroys tissue without eliminating the parasite. Other reports have communicated that the supernatant fluid of cultures of activated lymphocytes (presumably containing lymphokines) stimulates the macrophages to kill the intracellular parasites. Most likely, more than one lymphokine operates in the different systems tested so far.

Although considerable research is still necessary, the general opinion of the specialists is that destruction of leishmania is achieved by macrophages that have been previously activated by lymphokines.

Antibodies appear not to have any significant participation in the resistance to leishmaniasis, but a novel theory has been offered recently. According to this hypothesis, antibodies cytophilic for macrophages would be able to bind and immobilize the parasites on the surface of these cells, preventing their entry, and therefore their subsequent proliferation. Experimental confirmation is still necessary.

Artificial Production of Resistance

Since the course of the CCL is relatively benign except in the cases of disfiguring lesions in the face, this variety has been utilized in immunoprophylactic procedures. Vaccination against CCL, common in the U.S.S.R. and in Israel, is done by inoculation of virulent strains of the parasite in unexposed areas of the body. In order to achieve effective protection, the lesion must be allowed to continue its natural course until healing. Although the resulting scar may be as severe as those produced by the natural infection, the wise selection of the site of inoculation greatly reduces the impact that the natural disease might have in the future life of the subject.

All studies with inoculation of dead parasites in humans have failed to elicit significant protection. Some hope still remains, however, because it has been found that inoculation of extracts of *L. enriettii* containing subcellular particles produces protective immunity in guinea pigs, whereas soluble extracts were not effective. Experiments in monkeys have demonstrated that it is possible to induce protection against *L. braziliensis* infections by inoculation of *L. mexicana mexicana*; this latter may cause persistent ulcers of the auricular cartilages of humans, but does not invade the nasopharynx or the viscera. Subsequent studies have offered little encouragement, however. A dermotropic strain of *L. donovani* recovered from a squirrel was used on one occasion in an attempt to vaccinate human populations; although the pilot experiment gave encouraging results, a further trial on a larger basis did not show the production of significant protection (Manson-Bahr, 1971).

In general, the protection induced by leishmanias is very specific for the infecting species. Some known exceptions, however, encourage further research in this area, especially now that a number of new "species" of diverse pathogenicity are being distinguished. Russian authors have described two varieties of *L. tropica:* the "major" variety that causes moist ulcers of short duration (3–6 months) principally in the legs, and the "minor" variety that produces dry lesions that evolve over long periods (a year or more) and affect predominantly the face. Prior infections with *L.t. major* protect against reinfections with *L.t. minor*, whereas the reverse is not true.

Leishmania appears to be a genus in the middle of an active process of speciation; the ample overlap of biochemical and immunological characteristics among the different species and with parasites of other genera (*T. cruzi*, for instance) suggests so. This sharing of attributes is particularly troublesome for the proper characterization of the various parasites, but it provides some hope that varieties with different pathogenicity but that share functional antigens may be found.

Immunodiagnosis

Serological diagnosis is utilizable only in cases of visceral leishmaniasis, mucocutaneous leishmaniasis with nasopharyngeal lesions, and cutaneous leishmaniasis with compromise of the lymphatic vessels.

The presence of antibodies is too inconstant and their titers are too low for practical purposes in the other forms of the disease. The most common tests currently used are complement fixation, indirect hemagglutination, and indirect immunofluorescence. Other tests, such as precipitation in fluid or semisolid media, direct agglutination, and the cultivation of the parasite in dilutions of immune serum (Noguchi-Adler test), are used with experimental rather than practical purposes.

In most cases, the respective antigens are prepared with promastigotes from cultures. Since some studies have shown antigenic differences between promastigotes from cultures and micromastigotes from vertebrate hosts, it has been assumed, and occasionally verified, that the latter must provide better antigens for diagnosis. At least one recent report, however, did not find important differences between the serological results obtained with antigens of both origins. At any rate, the question deserves more investigation.

The complement-fixation test is the oldest and probably still the most widely used technique for the serological diagnosis of leishmaniasis. However, sera from people with visceral leishmaniasis and of dog carriers of the visceral and the cutaneous forms are frequently anticomplementary, and the test is occasionally positive in cases of tuberculosis, leprosy, actinomycosis, and other chronic diseases.

The indirect hemagglutination test is sensitive and reasonably specific: cross-reaction among diverse species of *Leishmania* is the rule and reactivity with sera of Chagas' disease patients is found occasionally. Reactions at low titers with sera of patients of tuberculosis or leprosy may also occur. At least the cross-reactivity with heterologous species of *Leishmania* and with *Trypanosium cruzi* can be removed by previous adsorption of the suspicious serum with the corresponding parasites.

The indirect fluorescent antibody technique used to be done with fixed promastigotes, but current assays with micromastigotes in tissue section or in tissue culture suggest that these preparations improve its sensitivity and specificity (Shaw and Lainson, 1977). This test is also specific for the genus *Leishmania* rather than for one or another of its species, and may react with the sera of patients of Chagas' disease and of malaria. The cross-reactivity can be removed by adsorption with the proper parasites, although recent reports claim that the use of tissue culture micromastigotes prevents reactions with anti–*T. cruzi* sera.

The specific anti–*Leishmania* antibodies disappear soon after cure, so that the presence and titer of antibodies is an adequate criterion to evaluate the result of individual treatments or of campaigns of control or eradication in a population. The presence of antileishmanial antibodies in apparently healthy populations suggests the existence of occult infection.

The leishmanina or Montenegro test is a delayed-type skin reaction, specific for the genus *Leishmania*, that becomes positive in infected individuals considerably earlier than the advent of protective immunity. The reaction usually persists for years or for life. This test is normally negative in cases of visceral or diffuse cutaneous leishmaniases; when it becomes positive, it indicates the return of the cell-mediated response and usually heralds the beginning of the cure. False-positive reactions to leishmania are not rare; they are often attributed to previous asymptomatic infections by human parasites or to earlier abortive infections by leishmanias of reptiles.

Immunopathology

It has been verified recently that the typical anemia of visceral leishmaniasis is not due to the parasitism of the bone marrow, as believed before, since this tissue incorporates iron normally in anemic patients. The half-life of the erythrocytes, on the contrary, is considerably reduced, mainly as a consequence of an active erythrophagocytosis. This phenomenon, associated with the characteristic hypergammaglobulinemia, suggests that an autoimmune mechanism may be responsible for the anemia.

All the attempts to demonstrate antibodies bound to the erythrocytes of the patients have failed so far, but it has been found that they are coated by serum proteins other than immunoglobulins. The identification of this material and its significance in the genesis of the anemia must await further research.

In vitro evidence suggests that the tissue destruction in leishmaniases is at least partially due to destruction of parasitized macrophages by lymphocytes.

Recent work has reported the finding of antigen-antibody complexes in the kidneys of patients of visceral leishmaniasis. This and the communication by Rezai et al. (1978) that complement levels were depressed in 2 patients who also exhibited proteinuria suggest that renal damage in visceral leishmaniasis may be caused by immunological mechanisms.

Studies in lower animals have demonstrated that leishmanial infections reduce their ability to mount humoral and cell-mediated responses to nonrelated antigens. This topic is discussed in Chapter 7.

THE PROTOZOA PREVAILING IN THE BLOOD

THE BABESIAS OR PIROPLASMAS

Babesia are protozoa related to the coccidia that multiply in the erythrocytes of numerous species of vertebrates and are transmitted by ticks. The life cycle in the arthropod intermediate host is not completely known, but phases of sexual reproduction are postulated. In many species, the parasite infects the eggs of the vector in such a manner that the infection is passed from generation to generation of ticks. Once the vector injects the parasites in the vertebrate during a blood meal, *Babesia* enters the erythrocytes and multiplies in asynchronous cycles. A preerythrocytic phase of reproduction has been suggested by some authors, but there is no solid proof of this hypothesis.

The species of *Babesia* of major importance in the Americas are *B. bigemina* and *B. argentina* (syn. *B. bovis*) in cattle, *B. caballi* and *B. equi* in horses, and *B. canis* in dogs. *B. microti* and *B. rodhaini* of rodents have been used as laboratory models of the infection. *B. microti* has become particularly important to medicine of late by the finding of over a dozen cases of spontaneous symptomatic infection in people, most of which occurred in Nantucket Island, Massachusetts. Babesias are emerging as zoonoses the importance of which still remains to be determined.

Natural Resistance

Until recently it was believed that babesias were strictly host-specific parasites that affected only a single species of host, or species that were very closely related. Recent findings of human infections by *B. microti* and several instances of successful transmission across phylogenetic barriers (*B. canis* in mice, *B. bigemina* in horses) suggest that their host specificity may not be so strict after all.

Nonspecific phagocytosis must play an important role in natural resistance, since primary infections are easier to establish in insusceptible hosts when these are splenectomized previously. Chapman and Ward (1977) have demonstrated that penetration of babesia in the erythrocytes is achieved subsequent to the activation of complement directly by the parasite. Probably some component of complement binds the red cell and acts as a receptor for the protozoan. It is possible that the differential susceptibility of diverse mammalian species for the same species of *Babesia* may be due to variations in the composition of their complement or to the particular sensitivity of this to be activated by the parasite.

There are no systematic studies on the influence of the host's genetic constitution on its susceptibility to the infection, but it has been observed that *Bos indicus* is more resistant to *B. argentina* than *B. taurus*. This difference, however, might be attributable to the fact that *Bos indicus* is also more resistant to attack by the vector tick. Shorter feeding periods of the arthropods may result in the inoculation of fewer parasites, so that the host will develop an effective immunity before the parasitism reaches symptomatic levels.

Epidemiologic and experimental observations indicate that calves and colts are considerably more resistant to the infection and to the disease than the corresponding adults; puppies and young rats or mice, on the contrary, are more susceptible than the adult animals. The resistance of calves extends until they are 6–7 months old and becomes almost insignificant 3–6 months later; in colts, it is still considerable until 2 years of age. A few recent studies, however, have been unable to demonstrate differences in the severity of the disease between adult cattle and calves under 3 months of age (Mahoney, 1972).

The demonstration that calves of resistant mothers were less susceptible than calves of nonresistant mothers fostered the idea that the protection of the youngsters was the result of colostral antibodies transmitted from the mothers. It seems difficult to accept, however, that antibodies acquired passively during the first day of life may persist in the recipients for periods in excess of 6 months. Besides, this hypothesis does not explain why the age resistance occurs in herbivores, which transfer immunity to their offspring exclusively through the colostrum, and not in carnivores and rodents, which transmit it by the transplacental and the colostral routes. The alternative that colostral antibodies would protect the young animals only while they develop their own antibodies does not agree with the fact that the youngsters become more susceptible during their fourth trimester of life. Apparently, the explanation of the mechanisms of age resistance in babesiosis must await further investigations.

Acquired Immunity

Reviews on the immunology of *Babesia* infections have been produced by Riek (1968), Ristic (1970), Mahoney (1972, 1977), and Aragon (1976). It is rather surprising that our knowledge of the resistance to a disease as important economically as babesiosis is in many areas remains so full of gaps as it actually is. Ironically, it appears that the countries most affected by it, and therefore most interested in its control, are precisely those that can least afford the expenses to prepare scientists and to support the pertinent research.

Babesiosis elicits the production of a variety of antibodies in the vertebrate host, as judged by the various serological tests that appear positive in infected animals. Studies in mice have demonstrated the presence of seric IgM and IgG antibodies during the entire course of the infection, but examination of calves infected with *B. argentina* showed that most of the antibodies corresponded to the IgM class 3 weeks after the infection and to the IgG class 6 months later.

Besides a few marginal attempts that are mentioned in relation to the mechanisms of acquired resistance, cell-mediated immunity does not seem to have been studied in babesiosis.

A primary infection with *Babesia* produces an increasing parasitemia, the severity of which depends on the dose and virulence of the parasites and on the age of the host. If this survives the acute syndrome, the number of parasites in the circulation begins to decrease until becoming undetectable, but leaving a considerable resistance against reinfections. Despite the resistance and the absence of protozoa in the peripheral blood, the parasites appear to persist for prolonged periods in the host. At least in the cases of *B. bigemina* and *B. canis*, new waves of parasitemia, low and generally asymptomatic but detectable, are observed periodically after the primary infection.

It was commonly believed that the resistance to babesiosis was exclusively of the premunition type—that is, it persisted only while some remnant of parasites from the immunizing episode remained in the body. Recent work has found that protection may still be present even years after the parasite has disappeared from the circulation, as judged by the inability of the blood of the experimental animals to transmit the infection to susceptible animals. These experiences have been repeated in cattle, horses, and dogs with similar results. Sterile resistance (in the absence of parasites), however, appears to differ qualitatively from premunition; whereas the former is effective only against the homologous strain of parasites, the latter is active against all the strains of the homologous species.

The duration of the infection (and therefore of the premunition) is thus of critical importance for animals that will move to other ecological zones where they are likely to find strains of *Babesias* different from those at their area of origin. In horses, infections by *B. caballi* that lasted $3\frac{1}{2}$ years and by *B. equi* that persisted for $5\frac{1}{2}$ years have been reported. In dogs, most of the infections terminate spontaneously in about a year. Most researchers estimate that cattle infections by *B. bigemina* and *B. argentina* last for 10–12 months, but shorter and longer periods have been reported (Riek, 1968).

The few reports available mention that resistance disappears soon after clinical cure in dogs and persists for about a year in horses. In the case of cattle, communications on persistence of resistance that ranges from a few months to several years are on record. Undoubtedly, it must be a series of not yet identified factors that are able to modulate the duration of the infection and the residual resistance. The fact that the parasitemia can be increased by administration of corticosteroids suggests that any element that interferes with the immunity could affect the course of babesiosis.

The resistance produced by natural infections is strong but not absolute, and it is expressed mainly as a suppression of the symptomatology and a reduction of the parasitemia in reinfected animals.

A study of the individual, clinically healthy animals in a cattle herd in an endemic area showed that the intensity of the parasitemia augmented progressively in the animals until 2 years of age and then it diminished markedly (Mahoney, 1972). The author explained this variation by hypothesizing that the parasitemia of young animals is the addition of the relapses of the original infection plus the asymptomatic superinfections that the developing resistance could not control totally; as the animals grow older, the resistance becomes stronger so that many of the superinfections are eradicated before reaching patency. Clinical observations support this theory: symptomatic bovine babesiosis is severe in cattle of around a year of age and then diminishes in importance until becoming rare in animals older than 4 years. Animals from areas free of *Babesia*, however, are highly susceptible at any time after age resistance wanes.

The common consensus nowadays is that once the new animals lose their age resistance (if present), they become highly susceptible to babesiosis, but they turn more and more resistant as they suffer infections and superinfections, until the protection is virtually absolute.

Mechanisms of Acquired Resistance

The participation of antibodies in the resistance to *Babesia* infections has been the subject of a number of publications of difficult interpretation. Experimental observations have demonstrated that detectable antibodies in the serum of recovered animals disappear sooner than their resistance to reinfections, and that the titer of anti–*Babesia* antibodies in a serum does not correlate with its ability to transfer protection to susceptible hosts. It is possible, however, that the antibodies measured by in vitro tests are not the same antibodies that affect the parasite in vivo. Transmission of resistance from immune mothers to their offspring, presumably through colostral antibodies, has been taken as proof that antibodies have a protective effect against the infection. Colostrum, however, contains lymphocytes and eventually macrophages that could be adopted, at least temporarily, by newborn animals. Also later reports indicate that cell-mediated immunity may be transferred by colostrum, at least in laboratory animals. Colostral transmission of protection, therefore, does not constitute absolute evidence of antibody participation.

In rodents and cattle, it has been possible to transfer resistance from immune to susceptible animals by injections of serum. This procedure is normally considered a conclusive proof of antibody activity, but, since the serum of *Babesia*-infected animals contains antigens that elicit protection (see "Antigens," below), the doubt still remains of whether the resistance of the recipients is due to passively transferred antibodies or to an active immunization with parasitic materials inoculated along with the serum.

At any rate, the resistance afforded by injection of immune serum has certain peculiar characteristics: it is effective only against the homologous strain; it does not operate when the serum is incubated with parasites in vitro; it only delays patency but does not reduce the parasitemia when the serum and the infected parasites are injected at the same time; and it increases the parasitemia if the serum is administered when the recipients are patent. Most of these effects can now be explained on the basis of the antigenic variation of *Babesia* (see "Antigens," below).

Inoculation of dogs and rats with a soluble antigen present in the serum of infected animals produced protection from reinfections and the formation of antibodies that agglutinated and lysed parasitized erythrocytes in vitro. The presence of equivalent antibodies in natural infections has not been verified yet.

It has been observed in infections of the respective hosts by *B. microti*, *B. canis*, and *B. argentina* that macrophages actively phagocytize, and presumably destroy, parasitized as well as normal erythrocytes. Phagocytosis begins very early in the infection, but intensifies in conjunction with the appearance of antibodies. It is believed that the first phase corresponds to nonspecific phagocytosis, whereas the second may be mediated by antibodies, since the most active organs (spleen, liver, bone marrow) show some invasion by plasma cells. The recent finding that activation of complement is required for erythrocytic invasion by *Babesia* to occur makes one wonder whether the first wave of phagocytosis is really nonspecific or macrophages actually recognize complement factors coating the red cells.

Experiments in rats have demonstrated correlation between phagocytosis and the presence of antierythrocyte agglutinins in infected animals. Since the same agglutinins occur in rats repeatedly bled, it is not clear whether they are formed as a response to the parasite or are an effect of the concurrent anemia. Similarly, it has been found, by in vitro tests at 37°C, that the serum of infected rats contains an opsonin that facilitates the phagocytosis of normal erythrocytes. The significance of this finding in the course of the natural infection is dubious, however, since the serum must be adsorbed with normal erythrocytes at 25°C to demonstrate the opsonizing activity. Along the same lines, protection against *Babesia* has been obtained in rodents by injection of antierythrocyte antibodies, presumably with opsonizing capacity. Similar inoculations have not been successful in cattle, however. Based on the above evidence and results of similar experiments, some authors feel that phagocytosis of infected (and normal) erythrocytes, possibly mediated by antibodies, may be an important mechanism of defense against babesias.

An experiment with rats transferred resistance by inoculation of splenocytes of immune rats into susceptible animals. Unfortunately, it is impossible to decide whether this protection was mediated by antibodies or by lymphokines, since the spleen possesses B cells, T cells, and macrophages. At any rate, the relevance of the spleen to anti–*Babesia* protection is well documented by the increase in susceptibility to the infection and to the disease that follows its surgical removal.

Clark et al. (1977) have suggested that resistance against babesia would depend on nonspecific soluble mediators, because the injection of *Corynebacterium parvum* (*Propionibacterium acnes*) or BCG protected mice against ulterior infections with *Babesia*, although virtually no antibodies against the parasite were detectable. It is well known that these agents are potent stimulants of cell-mediated immunity and of phagocytosis: the protective effect found in this experiment appears not to be connected to the former phenomenon, however, because it was reproduced in athymic mice, which are incapable of producing a cell-mediated response. The authors believe that death of the parasites was intraerythrocytic, which also rules out direct phagocytosis as the responsible mechanism.

In summary, the available evidence suggests that antibodies and macrophages play an important role in protection against babesioses, but the extent and mechanisms of their participation, as well as the intervention of cell-mediated immunity, require further research.

Artificial Production of Resistance

The first attempts to produce resistance artificially against *Babesia* were made more than 80 years ago in Queensland, Australia, by inoculating small amounts of blood of asymptomatic carriers into young animals. After a mild disease, the inoculated animals developed enough resistance to suffer further natural infections without showing symptoms. A decade later, antiparasitic drugs were introduced to control the severity of the disease in the recipients. This treatment did not interfere in any important manner with the production of immunity if it was instituted after the parasitemia had reached detectable levels, but allowed immunization of adult, fully susceptible animals from *Babesia*-free areas with little danger of causing serious disease. With insignificant variations, the same procedure is used nowadays under the name of *premunization*. Australian investigators have recently improved this procedure by passing *B. argentina* through splenectomized cattle prior to immunization; this technique reportedly reduces the virulence of this species for the vertebrate and eliminates its infectivity for the vector (Callow, 1977), so that disease in, and natural transmission from, the vaccinated animals are less likely.

Premunization is reasonably effective in producing protection against the symptomatic disease, but requires careful monitoring of the course of the infection when field parasites are used, permits the transmission of other diseases (bovine leukosis, for instance), and may incite the formation of antibodies against blood groups. In the latter case, the antibodies are transmitted with the colostrum and may produce hemolytic disease in newborn calves.

A recent modification of this procedure is the injection of slow-releasing anti–*Babesia* drugs in association with artificial or natural infections. The slow passage of the drug to the circulation would prevent the parasitism from climbing to dangerous levels while allowing enough antigenic stimulation to produce protective immunity. The assays already done have been encouraging, but the drugs have been toxic on occasions.

Studies with other protozoa have demonstrated that irradiation of the parasites inhibits their reproduction without markedly affecting their general metabolism, or kills them, in relation to the radioactive dose utilized. In the case of *Babesia*, the reported experiments suggest that the inhibitory and the lethal doses are rather close and that they vary widely from one species to the next. Despite this, it has been communicated (Mahoney et al., 1973) that infection of cattle with *B. argentina* irradiated with 20,000 or with 50,000 rads reduced the proliferation of the parasite in the host in such a way that the resulting infection produced mild symptoms but left a strong resistance. Higher doses of irradiation completely stopped the proliferation of the protozoa, which behaved as dead organisms for the purposes of immunoprophylaxis. More research in this area may develop schemes of practical application in the future.

The inoculation of dead parasites, as infected erythrocytes or as parasite extracts (contaminated with red blood cell stromas), has produced a satisfactory degree of protection in rodents and in cattle. Its practical utilization is greatly limited, however, because the resistance is expressed only against the homologous strain and does not last more than 3 months.

The injection of soluble antigen from the plasma of *B. argentina*–infected cattle elicited only slight protection, but against several strains, in susceptible cattle. Conversely, similar attempts in rats, dogs, and, to a lesser extent, horses, caused the production of an important degree of resistance, but only toward the homologous strain. The reasons for this difference (procedural, host dependent, or parasite related) are unknown at present, nor has it yet been determined whether the overall protective effect is induced by parasite antigens or by erythrocytic antigens that have been modified by the parasitism.

Immunodiagnosis

Several techniques have been utilized in the diagnosis of *Babesia* infections. The complement-fixation test (CFT) is quite specific and it yields only weak cross-reactions between *B. bigemina* and *B. argentina*, mainly during the acute infection. In cattle, it becomes positive 7–21 days after infection, peaks in a couple of weeks, and declines to undetectable levels in 3 months for *B. bigemina* and in 6 months for *B. argentina*. In horses, the antibodies appear 11–15 days after the parasitemia becomes evident and persist for the entire course of the infection. In dogs, the first antibodies are detectable at 11–34 days of infection.

Comparison of the CFT with the indirect fluorescent antibody technique (IFAT) showed that the latter detected antibodies in cattle 2–4 weeks before the former and that cross-reactivity between *B. bigemina* and *B. argentina* was approximately equivalent in both tests. In another study, IFAT antibodies persisted for 18–24 months in *B. bigemina* infections and for 13 months in *B. argentina* infections. This method is very sensitive (97–98%) and yields very few false-positive reactions (3–4%).

The indirect hemagglutination test appears to be very specific and sensitive. In a trial, it detected antibodies in cattle up to 4 years after the infection, when 21 of the 22 animals were parasitologically negative.

The agglutination tests, either in a capillary tube or with sensitized latex particles, appear to compare favorably with the CFT (they detect antibodies for a longer period), but up to 12.5% false-positive reactions have been recorded. Because of the simplicity of these tests, they will be very appropriate for field use once better specificity is obtained.

Other tests, particularly precipitation and direct agglutination, have not been properly evaluated for their use in practice yet.

Antigens

Formidable difficulties in separating the parasites from erythrocytic materials have greatly impaired the study of *Babesia* antigens and their use in immunodiagnostic and immunoprophylactic procedures. Most of the assays attempted so far, directed to the isolation of antigens, have revealed varying degrees of contamination with substances of the host. Apparently the association of parasite to erythrocyte is too intimate to achieve separation by the usual methods; a recent study showed that, after a lengthy purification process, some *Babesia* antigens were still firmly attached to host fibrinogen.

Over a decade ago it was observed that the erythrocytes infected with *Babesia* possessed on their surface antigens that were specific only for the strain that caused that particular infection. Later studies showed that, in every relapse of the parasitemia, the antigens present on the red cells were characteristic of that relapse and different from the antigens detected in other relapses. Experiments in rats have demonstrated that immunization with the parasites of a determined relapse protects better against reinfections with the parasites from the same relapse than from a different relapse. Mathematical simulation of these antigenic changes suggested that as many as 100 different antigens may be expressed by *B. argentina*. It is believed that this antigenic variation contributes to the survival of the parasites in the host by keeping them a step ahead of the immune reaction intended to destroy them (see Chapter 7). When the parasite returns to the vector tick, it reverts to a basic antigen that may also be of more than one single type.

Inoculation of serum of infected animals into normal animals has demonstrated that the serum of the former contains substances not present in the body of normal animals. Three types of these "soluble antigens" have been found in infections by *B. argentina*: an autoantigen, formed by complexes of haptoglobin with the hemoglobin released by the parasitism, that induces the formation of precipitating antibodies; a blood group antigen, originated from destroyed red cells, that causes the production of hemolysins when injected into normal animals; and a probably parasitic antigen. The first two antigens do not appear to have any participation in protection, but the inoculation of the third one in susceptible animals provokes antibody formation and partial strain-specific protection against subsequent infections. In a case in which the variable parasitic antigens in the serum of the donor were identical to the antigens of the *Babesia* used in the challenge, the resistance was complete. This finding supports the notion that only the parasites that exhibit variable antigens different from those to which the host has already responded immunologically will survive and persist in the organism of the vertebrate. As is shown in Chapter 7, it is quite possible that the soluble antigens play an important role in the survival of parasites in their hosts.

Immunopathology

The two major events in the pathology of babesiosis are anemia and the occurrence of a number of tissue phenomena attributable to the production of bradykinin and other vasoactive amines. The anemia corresponds rather well with the development of the parasites in infections by *B. bigemina* and by *B. argentina,* but it exceeds the peak of the parasitemia in the cases of *B. canis* and *B. rodhaini*. These latter observations suggest that a mechanism of anemia other than the direct destruction of the erythrocytes by the parasites may be operative in infected dogs and rodents. Inoculation of soluble antigen present in the serum of infected animals produces antibodies that agglutinate and lyse in vitro the parasitized red cells of dogs and rats, and macrophages of infected dogs and rats actively phagocytize parasitized as well as normal erythrocytes. It is not too difficult to assume that the circulating antigen may induce the formation of specific antibodies that will react with erythrocytes that contain parasites as well as with normal erythrocytes that have adsorbed soluble antigen. Since these are agglutinating antibodies, it is likely that they may also have opsonizing activity. Infected and normal red cells would be removed from the circulation in this way, producing an anemia beyond the degree expected by the pure parasitism. Furthermore, since normal erythrocytes coated with antigen would be rapidly agglutinated and phagocytized, their presence in the peripheral circulation should be difficult to verify. As is often proved, a logical deduction is not always true in biology, and this hypothesis requires experimental study.

Production of bradykinin and other vasoactive amines may occur as a result of immune responses involving the production of IgE antibodies or complement-activating immune complexes. The former have not been identified in babesiosis yet, but there is little doubt of the occurrence of the latter. Circulating antigen-antibody complexes have been confirmed in cattle infected with *B. argentina*, and vasoactive amines have been recovered from the urine of rodents with *B. rodhaini*. An objection often voiced against the participation of vasoactive amines in the pathogenicity of babesiosis is that the corresponding symptoms can occur very early in massive infections, before production of antibodies could reach even a detectable level. The recent demonstration that the parasite activates the complement directly and the (possibly related) observation that inoculation of parasite extracts produces bradykinin in vivo explain the early production of vasoactive amines and the consequent symptomatology, without the necessity for a conventional immune response.

The tendency of *B. argentina* to accumulate in the visceral capillaries has no current explanation, but the participation of hemagglutinating antibodies has not been excluded.

THE HUMAN PLASMODIA

In nature, humans are affected by four species of the genus *Plasmodium*: *P. falciparum, P. vivax, P. ovale,* and *P. malariae.* The species *P. knowlesi* and *P. cynomolgi* of monkeys and *P. berghei* of rodents have been abundantly used as laboratory models. It has recently become possible to infect the South American owl monkey, *Aotus trivirgatus,* with *P. falciparum, P. vivax,* or *P. malariae.* This is expected to provide an experimental system more similar to the human infection than those utilized previously.

Malaria begins with the inoculation of sporozoites by the mosquito intermediate host during its blood meal. In a few hours, the sporozoites invade the hepatocytes and start multiplying by schizogony to form several thousand merozoites, which will in turn invade new liver cells. During the second week of infection, the hepatic merozoites pass into the blood stream, enter the erythrocytes, and continue their cycles of reproduction in the red cells. The cycles of invasion, multiplication, and destruction of the blood cells proceed indefinitely, but eventually, some merozoites differentiate into sexual forms (macro- and microgamonts, or macro- and microgametocytes). These forms are ingested by a mosquito during feeding and multiply sexually in it to produce a large number of sporozoites that migrate to the salivary glands to initiate a new cycle of transmission.

In most human malarias, relapses may occur after the original infection has been controlled by the body mechanisms. Those relapses originated from exoerythrocytic parasites that remained in the liver (or elsewhere) are commonly called *recurrences* and appear to be the usual form of relapses in *ovale, vivax* and *malariae* malaria. The relapses from persistent blood parasites are called *recrudescences* and seem to be the only type existing in *falciparum* malaria.

Natural Resistance

This subject has been recently reviewed by Miller and Carter (1976). In general, the plasmodia exhibit quite a strict host specificity, but the extraerythrocytic parasites appear to be less restricted in this regard than the blood forms, since multiplication in the hepatocytes may be completed in abnormal hosts that do not support the erythrocytic parasitism (e.g., *P. vivax* in chimpanzees). Suppression of preerythrocytic parasitism, nevertheless, may be important among the mechanisms of natural resistance: only less than 1% of the sporozoites of *P. berghei* inoculated in the mouse (non-natural host) become proliferative hepatic forms, whereas about 50% of them invade and multiply in the liver cells of the tree rat (natural host).

Recent work has verified that the invasion of the red cells by the merozoites requires the presence of well-defined receptors on the host cell. In vitro studies of invasion of human or monkey erythrocytes by *P. knowlesi* demonstrated that the receptor for this parasite corresponds to the glycoprotein that determines the Duffy blood group. Subjects with Duffy-negative red cells lack the corresponding glycoprotein and therefore are not susceptible to the attack by the parasite. The close association between Duffy negative individuals in a population and their resistance to *vivax* malaria suggests that the receptor for *P. vivax* is also the Duffy glycoprotein or a closely related component. It is currently believed that the preference that some plasmodia exhibit for young erythrocytes (*P. vivax, P. ovale, P. berghei*) or for old red cells (*P. malariae*) also depends on the presence of adequate receptors on erythrocytes of different ages.

Even if the initial invasion of the host cell is successful, some inherited biochemical peculiarities of the erythrocytes may oppose the unrestricted proliferation of the parasites. It has been observed in West Africa that the prevalence of *P. falciparum* in children with sickle cell anemia is as high as that in children with normal erythrocytes, but the former rarely present the elevated parasitemias and the severe symptoms that are common in the latter. S hemoglobin, characteristic of sickle cell anemia patients, differs in one amino acid of the beta chain from the normal hemoglobin A_1; that relatively minor variation seems to be enough to inhibit the metabolism of the parasite in the corresponding cells. Epidemiological observations appear to indicate that the proliferation of the parasites is also impaired in people with hemoglobin C or E (which also have one amino acid different from hemoglobin A_1 in the beta chain) or with beta-thalassemia (in which the beta chains of hemoglobin A_1 are replaced by gamma chains).

Certain histocompatibility antigens of humans and the congenital deficiency of glucose-6-phosphate dehydrogenase have also been shown to possess a high epidemiological correlation with the natural resistance to the malarial disease.

The intimate mechanisms responsible for the restriction of the proliferation of the parasites in these cases are still being debated and some of these epidemiological correlations are not universally accepted. Despite this, most authors think that malarial infections, through the centuries, constituted an evolutionary pressure that helped configure the current genetic characteristics of the populations living in holoendemic areas. The existence of sickle cell anemia almost exclusively in Negroes, for example, is interpreted as an early selection toward resistance to malaria in a race that originated in malarial zones.

Nonspecific phagocytosis must also play a role in natural resistance, as judged by the increase in susceptibility of abnormal hosts subsequent to splenectomy (the human plasmodia in the chimpanzee, for example). Because of the rapidity with which this mechanism operates in primary infections, it is highly unlikely that it corresponds to a manifestation of acquired immunity.

Certain nutritional carencies [paraaminobenzoic acid (PABA), for example], nonidentified factors of the plasma, and the genetic constitution of some mouse strains also protect against the invasion and proliferation of the parasites.

Acquired Immunity

Recent reviews on immunity to malaria have been published by Garnham (1970), the World Health Organization (1974c), Brown (1976), and Zuckerman (1977).

The introduction of sporozoites in the host organism and the invasion of the hepatocytes do not elicit the production of any observable phenomenon during the natural infection, although there is experimental evidence that some resistance against these forms actually occurs (see below). In humans, the extraerythrocytic forms of *P. falciparum* persist for only one generation of multiplication (about a week), whereas the forms of *P. vivax* and *P. ovale* remain for up to 5 years and those of *P. malariae* for up to 30 years. It is not known whether these variations in longevity depend exclusively on the biology of the parasite or whether the host's immune response may play some role in it.

Once the erythrocytic parasitism begins, the infection provokes an intense stimulation of the lymphomacrophagic system that is expressed fundamentally in an increment of the phagocytic activity and an increased production of immunoglobulins. Phagocytosis may be augmented up to 200 times over its normal level, and it is exerted against the parasite as well as against parasitized and normal erythrocytes and against foreign particulate elements of exogenous origin. Most of the newly synthesized immunoglobulins do not have a known antibody activity; only a small proportion exhibits antibody specificity against the parasite, and only a part of this shows the ability to afford protection against the infection.

The first immunoglobulins produced usually belong to the IgM class, contain few specific and still fewer protective antibodies, and are evanescent; in human chronic malaria they have been detected up to 2 years after the original infection. IgG immunoglobulin production rapidly follows the formation of IgM; the IgGs contain most of the specific and protective antibodies, persist for a long time after the original infection (up to 20 years in human chronic malaria), and increase even more after reinfections. Up to now, it has not been possible to establish consistent relations among the levels of IgA, IgE, and IgD in malaric infections.

Evidence of acquired resistance begins to appear at the beginning of the second week of parasitemia and is manifested as a reduction in the reproduction rate and in the number of parasites in the blood. After a variable period, the immunity decreases the parasitemia to unde-tectable levels. The relationship between parasitemia and symptoma-tology is not perfect, however, and the immune response can occa-sionally diminish the clinical manifestations of the infection even in the presence of considerable numbers of parasites in the blood.

In most human cases, new waves of parasitemia that produce new episodes of symptoms, usually less severe than the original ones, occur weeks or months after the initial attack. In the case of *P. falciparum,* these relapses are believed to be recrudescences caused by infected erythrocytes that remained in the deep vessels of the organs; in the cases of the other human plasmodia, they are thought to be recurrences due to new invasions of the erythrocytes by parasites persisting in the liver (and elsewhere?). The possibility of additional recrudescences is not totally excluded in these latter cases, however. Splenectomy or administration of immunosuppressors causes the production of new waves of parasitemia accompanied with severe symptoms.

Antimalarial resistance in nature is specific for the infecting strain and appears to be intense enough to effectively control the proliferation of the parasites, although it does not destroy them completely. Ex-perimental work has indicated that recrudescences and recurrences are mostly (but not exclusively) due to the appearance on the parasites of antigens different from those that were present in the original popu-lation. The new population of plasmodia is thus invulnerable to the mechanisms of specific defense already produced by the host, and will remain so until the host reacts to the new antigens.

For a long time, it was believed that acquired resistance to malaria was exclusively of the premunition type and that, as such, it waned soon after the parasites were eliminated totally from the host. Never-theless, the existence of sterile immunity has been verified recently in rodents, and it has been produced in monkeys by injections of plas-modia with adjuvants. The presence of sterile resistance in humans has not been investigated experimentally yet, but there are no indications that it exists.

In areas of malarial holoendemia, acute malaria is fundamentally a disease of children. Few children under 3 months of age contract the illness, however, which is thought to be due to some passive acquired resistance transmitted from immune mothers, and to some natural re-sistance afforded by their diet of milk that is relatively poor in PABA. From this age on and for some years, children suffer repeated and severe attacks that become milder and milder with time until turning into frequent but low parasitemias, with benign or no symptoms, in the immune adult.

Mechanisms of Acquired Resistance

No evidence of acquired resistance to the natural inoculation of spo-rozoites is observed in humans under normal conditions (but its for-tuitous detection would be surprising indeed). Experiments of vacci-nation with irradiated sporozoites, on the other hand, have had partial success in humans (see below). Studies with *P. berghei* in rodents have provided enough evidence that this stage of the parasite can elicit pro-tection: sporozoite infection of rats maintained under administration of chloroquine or with a milk diet poor in PABA (both prevent the erythrocytic but not the hepatic parasitism) caused a marked reduction in the number of blood forms produced by a subsequent challenge; repeated inoculations of irradiated sporozoites in mice protected more than 90% of them against a subsequent infection.

The antisporozoite resistance is strictly species and stage (but not strain) specific (in humans) and must depend on metabolic antigens since it does not develop following injection of dead parasites. Its actual mechanisms appear to be complex, since there is evidence that anti-bodies, macrophages, and possibly effector T cells play partial roles in it. Participation of the humoral response has been verified by the rapid clearance of parasites (but not resistance to the infection) in mice in-jected with serum of immune animals, and by the reduction of the infectivity of the sporozoites incubated in immune serum. Two types of antisporozoite antibodies have been identified: a precipitating an-tibody that produces a circumsporozoite reaction, usually at one end of the parasite; and a neutralizing antibody that affects the sporozoite infectivity on incubation. Both antibodies are active in the absence of complement, but are not identical.

The impossibility of transfering protection against the infection by administration of serum of resistant animals and the successful vac-cination of mice whose humoral response had been suppressed with anti-mu serum suggest that factors other than antibody may participate in the resistance to sporozoites. Since antibodies alone do not explain the resistance, and this has been demonstrated to be thymus dependent, some authors believe that cell-mediated immunity may have an im-portant participation in this phenomenon. Nevertheless, attempts to transfer protection with cells have failed until now. The observation that the liver is invaded by macrophages, lymphocytes, and plasma cells after the first generation of hepatic schizonts breaks open also indicates response to released antigens, although not necessarily of the cellular type. Histopathological studies have also demonstrated in-creased phagocytosis of sporozoites in resistant animals, but pure macrophagic activity must not be essential, because splenectomy after vaccination does not impair acquired resistance.

In circumstances in which antibodies, cell-mediated immunity, or macrophages cannot explain the acquired resistance against sporozoites, it may be interesting to mention that the administration of three interferon inducers has been shown to protect mice against challenges with *P. berghei* sporozoites. It might be that nonspecific factors play an important role in this phenomenon.

At any rate, acquired resistance to the hepatic parasitism does not help to explain the existence of recurrences in most human malarias. Since it is conceivable that the immune mechanisms are unable to reach the parasites while they are inside the host cells, some investigators have proposed that hepatic plasmodia might undergo prolonged periods of "hypobiosis," not dissimilar to those in some nematodes and some insects, inside the cells. Others have speculated that the liver forms might experience antigenic variations, as do the blood parasites, that would make them invulnerable to the mechanisms of specific defense existing at the moment of their release from the host cell. A similar hypothesis has been offered to explain the sporadic (as opposed to constant) invasion of the blood red cells: only those hepatic merozoites that acquired variable antigens complementary to the receptors on the erythrocytes of the host would be able to initiate the blood cycle. Because of technical difficulties, none of these theories has been verified yet. At any rate, the hepatic parasites do not appear to elicit an effective protection in human malaria, because individuals subjected to chemoprophylaxis against the blood forms in malarial areas are fully susceptible when this is discontinued.

The production of effective immunity (and of symptomatology) in natural conditions is induced by the presence of asexual forms of the parasites in the erythrocytes. The transmission of antimalarial immunity to the offspring, transplacentally in the primates and transplacentally and transmammarily in the rodents, has suggested that protection against plasmodium infection was mediated by antibodies. It is known that colostrum contains lymphocytes and that immunocompetent cells or soluble antigens may cross the placenta, so maternal transfer of immunity cannot be considered an incontrovertible proof of antibody activity any longer. Nevertheless, the rapid presentation and the rates of declination of the protection in the offspring suggest that, effectively, it is an antibody-mediated event. On the other hand, numerous recent experiments have documented that the antibodies present in individuals recovered from malarial episodes confer specific resistance against the infection in susceptible animals. In the case of human malaria, it has been possible to prevent the infection of *Aotus* monkeys by injection of serum of recovered patients.

Studies in vitro have confirmed the previous suspicion that the antibodies were effective against the released merozoites but not against the intracellular parasites. When infected erythrocytes were cultured in the presence of immune serum, the intracellular parasites incorporated normally labeled amino acids added to the culture medium, but the resulting merozoites were agglutinated and could not invade new cells. This inhibitory action was associated mainly with the IgG fraction of the serum and occurred even in the absence of complement.

Experiments in monkeys have shown that the serum obtained after a single infection does not transfer specific immunity, although it has antibodies against the homologous parasite. The serum collected after several infections, on the other hand, inhibits in vitro the invasion of the erythrocytes by merozoites and protects susceptible monkeys against infection by the homologous strain. These results suggest that a single infection produces antibodies only against the original antigenic variant, while repeated infections induce the formation of antibodies effective against an ample array of antigenic variants. These experiments coincide with clinical observations in human malaria: a first malarial episode usually leaves immunity only against the infecting strain, but people with chronic infections in holoendemic areas exhibit resistance against the parasitic species, presumably because multiple past infections raised protection against a large proportion of the antigenic variants of the species.

In addition to the activity of merozoite-inhibiting antibodies, some other defense mechanisms must exist, because monkeys immunized with merozoites in complete Freund's adjuvant develop antibodies only against the inoculated variant, but exhibit protection against other additional variants. Unlike the natural infection, the resistance produced in this way is sterilizing and not of the premunition type. It is possible that this higher efficiency of the immune response is related to the known stimulatory activity of this adjuvant on cell-mediated immunity and phagocytosis. Eventually, the adjuvant can bring up responses to core antigens that normally are too weak to elicit demonstrable immunity. They may be either cell-mediated responses or antibody responses not demonstrable in vitro by conventional tests (opsonization, for example).

This hypothesis is supported by the finding that serum of immune donors transfers less resistance when the recipients are splenectomized than when they are complete. This has fostered the idea that at least part of the inhibition of the merozoites and the neutralization of the sporozoites may be due to an opsonizing activity of the antibodies: removal of the splenic filter would certainly decrease the final effect of any opsonizing stimulation. This notion seems to be logical since antimalarial protection resides fundamentally in the IgG antibodies, which are frequently opsonizing.

In a related experiment, macrophages and infected erythrocytes were enclosed in Millipore chambers and implanted in the peritoneal cavity of resistant or susceptible mice: in resistant animals the macrophages acquired a greater capacity for ingesting and destroying the parasites than the homologous cells in susceptible animals. This finding reveals that humoral substances present in immune mice stimulate the disposal of the plasmodia by the phagocytic cells. It is not completely clear in this case whether the stimulation was mediated by antibodies or by lymphokines, although the short radius of action commonly attributed to the macrophage-activating factor favors the idea of important antibody intervention.

Despite the fact that the proofs are not totally conclusive yet, most specialists assign a preponderant role to macrophagic activity in protection against malaria, and think that the antibodies have an important, although maybe not exclusive, participation in it. Observations in patients of *falciparum* malaria and in rodents infected with *P. berghei* have revealed that their sera also possess antibodies against toxic materials produced during the infection.

Some researchers have tried to establish a cause-effect relationship between the presence of precipitating antibodies and the decline of the parasitemia, because they frequently occur coincidentally. Experimental studies of serum transfer, however, have been unable to demonstrate correspondence between the level of antibodies in a serum and its ability to transmit protection. Actually, it is rather unexpected that the conventional in vitro titration of a serum could predict its activity in vivo when this includes the action of antibodies that neutralize sporozoites and toxins, inhibit merozoites, opsonize parasites, and maybe exert other functions.

Experimental observations in humans, lower primates, and rodents have clearly demonstrated the production of cell-mediated immunity in *Plasmodium* infections. Several instances have shown that thymocytes are essential to mount a protective immune response, but it has been difficult to determine whether these lymphocytes act as effectors of cell-mediated immunity or as helpers in the formation of antibodies. In at least one experiment, it was found that cell transfers transmitted more immunity than serum transfers, and protection was conferred by cells on some occasions in which the corresponding sera were ineffective. These reports do not constitute incontrovertible proof of the participation of cell-mediated immunity in antimalarial acquired resistance, but are suggestive and encourage further research.

Virtually nothing is known about the intervention of the sexual forms of plasmodia in the production of resistance. These forms are antigenically different from the blood schizonts and normally persist in the circulation, although in reduced numbers, after the parasitemia by the asexual forms wanes. It is not known whether the decrease in the number of gamonts is due to a specific immune response against them, or to the action of the antibodies against their precursor merozoites. It is possible that the gamonts are not susceptible to the antimerozoite resistance, in part because of their antigenic differences and in part because of their permanent intracellular habitat.

In one experiment, it was demonstrated that the gametocytes produced immediately after the subsidence of the first parasitemia were not infective for mosquitoes, whereas gametocytes collected later were infective. Based on these results, it has been speculated that gametocytes may have the ability to vary their antigens as the asexual blood forms do.

Artificial Production of Resistance

Malaria still remains as the most lethal parasitic disease of humans and a permanent challenge to our ingenuity. Control of the infection by antivector and chemoprophylactic procedures has been frequently attempted in the past, but with only moderate success. The ecological and economic cost of these rather ineffectual methods and the recent support and encouragement from international organizations have given new impetus to the search for alternative systems of control. Several major lines of investigation are currently attempting to devise adequate immunoprophylactic preparations. Exciting news in this area is bound to occur in the near future. Recent reviews on this subject have been done by Cohen and Mitchell (1978), Miller (1977), and Nussenzweig (1977).

The natural limitations of human experimentation have forced the frequent use of rodents and lower primates as laboratory models. It must be understood, however, that antimalarial immunity is largely dependent on the host-parasite association under consideration: for instance, *P. berghei* is almost invariably fatal in most strains of mice and in young rats, but produces only a mild and transient parasitemia, followed by a strong sterile resistance, in adult rats. *P. yoelii* usually causes a comparatively low parasitemia in mice with complete recovery, but continuous passage in Balb-/c mice or selection of the parasitic strain 17X will kill 100% of the animals. *P. knowlesi* is rapidly lethal in rhesus monkeys, but produces a mild and intermittent parasitemia in kra monkeys. These variations (and factual differences in vaccination, as seen below) warn very strongly against too ready an extrapolation to humans of the findings reported in laboratory models. Also, natural infection of humans with malaria (especially *falciparum* malaria) elicits only partial protection against reinfections, so that the procedures of artificial immunization must aspire to achieve more resistance than the actual infection.

On theoretical grounds malaria vaccination may be attempted with preerythrocytic stages of the parasite or with its blood forms: effective immunity against the preerythrocytic parasites should totally prevent the establishment in the host, whereas inhibition of the stages in the red cells should avoid only the production of parasitemia and of the associated symptomatology. In the long run, the absence of parasitemia will also impede the infection of the vector and, in the absence of nonhuman reservoirs, will eradicate the infection.

Sporozoite vaccination of rodents has been done with living, inactivated, or irradiated parasites; these latter produced up to about 90% protection against challenges after repeated inoculations in mice. Similar immunization of rhesus monkeys against *P. cynomolgi* was rather ineffective, however. Humans were not expected to develop protective immunity to sporozoite inoculations because, unlike the rat, they remain fully susceptible to the natural infection when chemotherapy directed against the blood forms in hyperendemic areas is suspended. Nevertheless, exposure of 5 volunteers to several hundred bites of *P. falciparum*–infected mosquitoes that had been irradiated with 12,000–15,000 rads resulted in resistance to subsequent natural challenges in 3 of them. The resistance was effective against strains other than those used for immunization, but not against other species of plasmodia or against infection with the blood forms of the homologous parasite; it persisted for about 3 months and coincided with the presence of antisporozoite antibodies. A similar experiment with *P. vivax* in 1 volunteer yielded comparable results, but the resistance lasted for 3–6 months. On the other hand, artificial injection in 2 children of 10^6 *P. falciparum* sporozoites irradiated with 20,000 rads, in two doses, did not result in protection against the natural infection. There are too many differences between these two experiments to attempt interpretation of the variation in their results yet.

Immunization against the erythrocytic stages of plasmodia has been attempted by a number of procedures. Resistance has been produced in rodents and monkeys by mild blood infections controlled with antimalaria drugs or with diets poor in PABA. Repeated inoculations of irradiated parasitized erythrocytes have elicited resistance, although somewhat inconsistently, in rodents, and little protection in monkeys. Soluble fractions of parasitized red cells injected with Freund's complete adjuvant protected 9 of 14 monkeys against normally lethal challenges of *P. knowlesi,* but soluble plasma antigen of the same species did not induce protection; circulating *P. berghei* antigen, on the contrary, was protective in mice. Killed parasitized erythrocytes produce little protection generally, but their effect can be boosted by the use of Freund's complete adjuvant or of a sequence of complete plus incomplete adjuvants; since infection with, or inoculation of, *Mycobacterium* has been reported to have an inhibitory effect on the development of plasmodia, it is not clear whether the complete adjuvant assists the antimalarial response in this case or operates by itself. Vaccination of rhesus or of *Aotus* monkeys with erythrocyte-derived merozoites of *P. knowlesi* or *P. falciparum* in Freund's complete adjuvant produced sterilizing resistance to further challenges in a very high proportion of the animals; the protection was species and stage specific but was effective against other variants or strains. By contrast, similar vaccination of mice with *P. yoelii* have been unsuccessful.

Inoculations of interferon inducers or of *Escherichia* endotoxin, Freund's complete adjuvant, Newcastle disease virus (interferon inducer?), *Mycobacterium,* and *Corynebacterium parvum* (*Propionibacterium acnes*) have been often reported to produce nonspecific resistance to subsequent malaria infections, particularly in rodents.

The facts that the asexual blood cycle of at least one human plasmodium (*P. falciparum*) can be reproduced continuously in the laboratory now, that merozoites are readily obtainable from these cultures, that they elicit immunity against more than one strain or variant, and that they can be stored in liquid nitrogen without loss of immunological reactivity make merozoite vaccination a favorite to develop immunization schemes for humans.

Immunodiagnosis

Acute malaria is diagnosed rather easily by the microscopic study of blood samples, but the immunological methods may assist in the implementation of epidemiologic studies, in the identification of blood donors with latent infections, in the retrospective diagnosis of febrile accesses, and in the etiological confirmation of conditions such as hepatosplenomegaly and nephrotic syndromes (Voller, 1976). Because of the availability of comparatively simple methods for the parasitological identification of the infection, the immunological procedures for clinical detection of malaria have not been developed as much as those for other protozoal infections. Only three tests are fairly well established currently: the indirect fluorescent antibody technique (IFAT), the indirect hemagglutination test (IHAT), and precipitation in gel.

Until recently, all clinical tests made ample use of species of plasmodia of lower animals as the antigenic reagent, trusting that the antibodies developed by the patient would react with antigens common to the genus *Plasmodium*. The possibility of maintaining and reproducing the human plasmodia in *Aotus* monkeys has recently permitted the utilization of the homologous species (or a mixture of the human species) to detect antibodies, which has considerably increased the sensitivity of the diagnostic tests. It has also been found that the use of mature blood schizonts gives more sensitivity than the use of other parasitic stages.

The IFAT is specific for the genus *Plasmodium* and does not distinguish consistently among its species. Very occasional cross-reactions with treponema infections, and even rarer ones with trypanosomes and leishmanias, have been observed. Cross-reactivity with *Babesia* is not unusual, however, which is a possibility worth keeping in mind since cases of spontaneous human babesiosis have been found in increasing numbers in the last years. The first antibodies detected correspond to IgM and appear soon after patency; they are replaced by IgG antibodies later. Specific titers usually increase up to 4–6 weeks after infection and then decline, but they are still elevated during latency and augment with each relapse. Since the antibodies normally disappear within a year from the parasitological cure, this test has been utilized to follow the effect of eradication campaigns: lack of antibodies in children born after the campaign should indicate absence of transmission of the infection. Children of mothers with titers of specific antibodies must be excluded until they have eliminated maternal antibodies. Only titers above 20 are commonly regarded as indicative of infection in adult populations. Preliminary studies with an IFAT with soluble antigen have yielded promising results, but further work to obtain reproducible antigens is still needed.

The IHAT becomes positive somewhat later than the IFAT, but both follow approximately the same curve later on. Since the antigens used in the IHAT are frequently contaminated with erythrocytic material of the donor (which may react with antibodies to heterophile antigens), it is recommended that the suspicious serum be adsorbed with noninfected erythrocytes of an individual of the same species as the donor. Conventionally, it is agreed that only titers above 40 indicate infection.

Precipitation in gel has been proposed as a diagnostic tool recently. The low sensitivity of this method makes it inappropriate for individual diagnosis, but this same characteristic may favor its use in epidemiological studies. Besides, there is certain circumstantial evidence that thermolabile antigens demonstrable with this technique may correlate with the presence of acquired resistance. A quicker modification of precipitation in gel, counterimmunoelectrophoresis, is being evaluated currently.

The complement-fixation technique has been abandoned because, despite its ability to differentiate among species of plasmodia, it is relatively insensitive and occasionally yields positive reactions in cases of syphilis and yaws.

Enzyme-linked immunosorbent assay has given promising preliminary results, and radioimmunoassay is being checked at present. Methods such as circumsporozoite precipitation, merozoite inhibition, schizont-infected cell agglutination, opsonization, and others are used fundamentally for purposes of investigation.

Antigens

As with other protozooses of the blood or the tissues, the inherent difficulty of obtaining adequate quantities of intracellular parasites uncontaminated with host materials has been a serious drawback for the study of the antigens of *Plasmodia*. Among the many techniques used (Zuckerman, 1977), the most popular is the disruption of parasitized erythrocytes collected from the peripheral circulation. Because of the low peripheral parasitemia characteristic of *P. falciparum,* this species is usually recovered from placentas of infected women. After years of fruitless work, it was communicated in 1976 that continuous in vitro cultivation of the asexual blood stages of *P. falciparum* had been finally achieved. The prolificacy, synchronicity, and infectivity of the in vitro–grown parasites, however, are reduced in comparison to the naturally grown protozoa. More limited success has also been obtained with cultures of *P. knowlesi.*

Plasmodium presents a formidable array of antigens. Some of them are specific to a species, a strain, a developmental stage, or even a particular relapse in the course of an infection; others are shared by the various strains of the same species, by different species within the genus, and even with other genera of protozoa, notably *Babesia.* The functional antigens appear to have a more restricted specificity than the rest of the antigens because, whereas the serological tests regularly cross-react with the diverse species of plasmodia, the antimalarial protection is specific for the species, the particular stage, and even the variable antigens that elicited its production. Acquired resistance against sporozoites may be more ample because cross-protection has been observed among species of plasmodia in rodents immunized with these stages.

Different studies have identified more than 30 antigens in the human plasmodia; most of them are heterologous substances (proteins, glycoproteins, membrane-associated phospholipids) with molecular weights of 60,000 to 900,000. The examination of *P. falciparum* antigens by gel precipitation has revealed three types of substances: L (labile) antigens, that are destroyed by heating at 56°C for 30 min; R (resistant) antigens, that are destroyed by boiling; and S (stable) antigens, that resist boiling for 5 min. Culture of parasitized erythrocytes with labeled amino acids has shown incorporation of the label only into the L and R antigens, suggesting that the S antigens are not synthesized by the parasite, but may instead be red cell material altered by the parasitism. A number of important correlations have provided circumstantial evidence that links the presence of protection to the existence of IgG antibodies, especially against L antigens (Zuckerman, 1977). Further work in this area is necessary.

Soluble antigens are found in abundance in the circulation of some infected individuals in the case of rodent, avian, and lower primate malaria and in the human infection by *P. falciparum* and *P. malariae*. Inoculation of these antigens in susceptible animals has produced resistance to challenges in birds but not in monkeys. The soluble antigens in *falciparum* malaria correspond to S antigens, have not been shown to be connected with resistance to the reinfection, and appear to be complexed with IgM antibodies in vivo.

A decade ago, it was reported that the schizont-bearing erythrocytes obtained during a recrudescence of *P. knowlesi* were not agglutinated by the serum collected after the first wave of parasitemia. Later studies demonstrated that each recrudescence induced agglutinating antibodies that reacted against the schizont-infected erythrocytes of the same recrudescence, but not with those produced in prior or later relapses. Opsonization experiments yielded similar results. The same phenomenon has been verified subsequently with *P. berghei, P. cynomolgi,* and *P. falciparum* and it is currently thought to be an inherent characteristic of the genus *Plasmodium*. The existence of these "variable antigens," functional and exclusive for each relapse, explains the chronicity of the malarial infection and the presence of circulating parasites in patients whose sera are able to transfer resistance to susceptible individuals. It represents an effective mechanism to elude the host's immune response, which is discussed again in Chapter 7.

Although little is known in regard to the location of the antigens in plasmodia, it is evident that the variable antigens detected by agglutination of schizont-infected red cells must exist on the surface of the erythrocytes. The antigens responsible for the production of sporozoite-neutralizing antibodies must be metabolic products since these are stimulated only by inoculation of living parasites. The species-specific antigens must be soluble substances since the CFT, which utilizes soluble parasitic extracts, is able to distinguish between various species, whereas the IFAT, which uses the complete parasite as antigen, reacts similarly with all the species.

Immunopathology

The anemia characteristic of malaria is frequently much greater than would be expected from the degree of erythrocyte parasitism, and it is often more intense after the parasitemia begins to subside. Several hypotheses have been offered to explain these anomalies, of which an increased erythrophagia and the existence of autoimmune reactions are directly relevant to immunology.

Histopathological observations have documented abundantly the existence of accelerated phagocytosis of parasitized and normal red cells in malarial subjects. This phenomenon may be the result of the nonspecific stimulation of the macrophagic system by the parasite, or the expression of the normal elimination of cells damaged by the protozoan or its products, or the consequence of an immunologically promoted erythrophagocytic hyperactivity. There are indications that all three of these factors may operate in malaria, but the combination of the red cells with agglutinating antibodies, or with antibody classes (IgG) or complement factors (C_3) with receptors for macrophages, would certainly be expected to promote its phagocytic elimination.

Recent studies have demonstrated the presence of IgA agglutinins against trypsinized erythrocytes in the serum of 11 of 38 subjects from Gambia and in none of 179 subjects from the U.S. This finding was interpreted as the result of the exposure in Gambians of erythrocytic antigens that are normally occult, possibly as a result of parasitism by plasmodia. Other authors have found a correlation between the anemia and the presence of IgM antibodies against the patient's own red cells.

The presence of antigen-antibody complexes in the circulation of patients infected with *P. falciparum* is well known and it has been recently verified in *malariae* malaria. There is also evidence that complement is activated by the classic and the alternative pathways in the course of the malaria infection. Most specialists accept that the antigen-antibody complexes may deposit on the erythrocytes with the subsequent activation of the complement system and the binding of its factors on the red cells. Besides the opsonizing and the lytic action of the complement, some of its components induce the formation of antibodies against themselves (*immunoconglutinins*, directed mainly to the bound fraction of C_3); the immunoconglutinins may react with the erythrocytes through the complement factors coating these cells, and may affect them by all the mechanisms open to antibodies. In this respect, it was recently reported that immunoconglutinins were augmented in 6 of 10 malaria patients. Finally, it is possible that the physical deformation of the anti-*Plasmodium* antibodies when they combine with the corresponding antigens in the host may uncover new antigenic determinants that may be recognized as foreign (and therefore antigenic) by the host. In these circumstances, the host may produce antibodies against his own immunoglobulins. This hypothesis is circumstantially supported by the higher abundance of *rheumatoid factor* (antibodies against antigen-bound IgG) in zones of Africa where malaria is endemic.

Not all investigators accept as solidly documented the existence of an autoimmune mechanism in malarial anemia, nor do those that do accept it agree on its major mechanism. It seems, however, that the available evidence is in favor of the immunologically mediated pathophysiology of the anemia. On the other hand, there does not appear to be a good reason to expect that a single mechanism, as opposed to a combination of them, should be the responsible factor.

Another cardinal sign of malaria is the splenomegaly that develops rapidly during the active infection and regresses during latency. The increase in the size of the organ is due fundamentally to the proliferation of lymphomacrophagic elements as a result of stimulation by the plasmodia. This reaction clearly has a protective function since the removal of the spleen aggravates the course of an active infection or produces relapses of latent malarias. In addition to the simple proliferation of the macrophages, the infection must cause an activation of the function of those cells, since the phagocytosis increases up to 10 times more than the increase in the number of phagocytic cells. The "tropical splenomegaly syndrome," observed in highly malarious zones of Africa and New Guinea, coincides with the presence of anti-*Plasmodium* antibodies and subsides with prolonged antimalaria treatment. Several other epidemiological and clinical observations also suggest that this condition is related to the persistence of stimulation by malarial antigens, possibly in subjects with prior immunological defects. Some circumstantial evidence indicates that the host's response in these cases may be predominantly cell mediated.

The cerebral complications of the *falciparum* malaria occur mainly in children between 1 and 5 years of age who have appreciable levels of anti-*Plasmodium* antibodies. Studies in rodents have demonstrated that the cerebral syndrome only happens in animals that have responded immunologically to the parasitism and that it coincides with depressed levels of complement. Its presence in animals is prevented by immunosuppressive manipulations (neonatal thymectomy or administration of antilymphocyte serum). This evidence has fostered the idea that the cerebral manifestations of *falciparum* malaria may be an expression of an immune response of the host against the parasite. Some evidence in human studies supports this notion: a close correlation between reduced levels of the factor 3 of the complement and the presence of cerebral complications in human patients was recently reported; on the other hand, the cerebral forms are rare in children with protein-caloric deficiency who frequently suffer thymic atrophy and depression of immunity. These studies are just beginning, but the evidence gathered so far is highly suggestive.

Another complication seen in malaria are the nephropathies, which include a hemoglobinuria related to massive hemolysis, a "febril proteinuria" associated with the acute infection that regresses on successful antimalaric treatment, or a progressive nephrotic syndrome seen in cases of *malariae* malaria. The hemoglobinuric episodes may be triggered by hemolysis consecutive to the disease or its treatment. Febril proteinuria has been demonstrated in up to 25% of children infected with *P. falciparum,* was found in all of 38 patients with acute *malariae* malaria, and is reported to be slight and transient in *vivax* malaria. A study of 10 patients who showed proteinuria during an acute episode of *falciparum* malaria revealed deposits of immunoglobulins (especially IgM) in the basal membranes of the glomeruli in 9 of them, and malarial antigen in 2. The renal lesions in some cases of *malariae* malaria have a chronic progressive character; in a study of 93 nephrotic kidneys in Nigeria, IgG and IgM were found in 96%, factor 3 of the complement in 66%, and antigens of *P. malariae* in 25%. In contrast, antigens of *P. falciparum* were found in only 1 case. The immunoglobulins eluted from the kidneys of most patients showed anti–*P. malariae* activity.

In the immense majority of the cases of malarial nephropathy, the lesions are attributable to deposits of antigen-antibody complexes in the glomeruli, with the subsequent inflammation triggered by the local activation of complement. This must not be the sole mechanism, however, because IgG antibodies of a subclass unable to activate the complement have been recovered from nephrotic kidneys associated with *malariae* malaria. At present, it is difficult to explain why the initial glomerulonephritis subsides in malarias other than *malariae,* but often follows a chronic course in the latter. The prolonged persistence of the infection, and therefore of the malarial antigens, in this last case does not appear to be a good reason since *P. malariae* materials are infrequently found in the glomeruli of patients with chronic nephropathies. A possibility is that the initial deposit of antigen-antibody complexes in *malariae* malaria causes a lesion of the renal tissues that turns them into autoantigens: the chronic lesion would be sustained then by an immune reaction of the host to his own kidney materials. Even then, an explanation for the absence of the same phenomenon in other malarias is required. A correlation between malaria and the nephrotic syndrome is solidly established, but some authors caution that it is still possible that renal patients may be more susceptible to malarial infections than healthy people, which might explain the correspondence between both conditions equally well.

Apart from autoantibodies against erythrocytes, against antigen-bound immunoglobulins, and possibly against renal tissue, it has been reported that some malaria patients have antibodies directed against nuclear materials and against substances present in extracts of stomach, heart, and thyroid. The antinuclear antibodies are believed to be elicited by the nuclei of the parasites and of the young parasitized red cells released in the body; the antiorgan antibodies are thought to correspond to heterophile antigens present in the parasites. There is no information on their intervention in the pathophysiology of malaria.

There is much evidence currently that infection by a number of plasmodia produces a reduction of the immune responsiveness in humans as well as in animal models (see Chapter 7). This phenomenon may explain why autoimmune diseases are rare, Burkitt's lymphoma is highly prevalent, and children's vaccines are rather ineffective in zones of malarial hyperendemia.

TRYPANOSOMA CRUZI

Trypanosoma cruzi, the etiological agent of Chagas' disease, or American trypanosomiasis, is a flagellate transmitted by contamination of the skin or mucosae of mammals by several hemipteres of the family *Reduvidae* (assassin or kissing bugs; genera *Triatoma*, *Panstrongylus*, *Rhodnius*, etc.). Congenital infections are not rare in nature, and transmission by transfusion of blood of healthy carriers has been reported repeatedly.

Three major forms of the parasite occur: the *epimastigotes*, elongated organisms with a flagellum anterior to the nucleus, which multiply in the gut of the insect intermediate host and in artificial cultures; the *trypomastigotes*, elongated forms with a flagellum that begins at the posterior end and extends the entire length of the parasite and beyond, which are found in the rectum of the vector (from whence they contaminate the skin of the definitive host) and in the blood of the infected mammal; and the *amastigotes* or *micromastigotes*, small, round parasites with no external flagellum that exist and multiply in the nucleated cells of the mammalian host (especially in striated and cardiac muscle, the reticuloendothelial system, and nervous tissue), forming the so-called pseudocysts. New cultivation methods in tissue culture have permitted the growth of micromastigotes and their transformation into trypomastigotes in vitro; a liquid medium with chicken plasma and chick embryo extract has also succeeded in growing micromastigotes outside the definitive host.

In the mammal, the cycles of intracellular reproduction at the micromastigote stage, their release and transformation into trypomastigotes, and the invasion of, and new multiplication in, other cells occur rather rapidly at the beginning of the infection. Destruction of numerous cells and associated phenomena cause clinical symptoms in some cases. With time, the parasite turnover becomes slower and slower, so that the direct damage is less important and the resulting chronic infection is characteristically asymptomatic. No blood forms are demonstrable by conventional procedures during the chronic phase. However, cardiac and digestive lesions, sequelae of the infection, may appear even years after the original invasion by the parasite.

Natural Resistance

In nature, *T. cruzi* affects around a hundred different species of mammals that include primates, carnivores, artiodactyls, rodents, lagomorphs, chiropters, edentates, and marsupials. All the wild hosts and the artiodactyls develop only transient and low parasitemias, without clinical symptoms. Humans and the domestic carnivores, on the contrary, may suffer an acute phase, with serious or fatal symptoms, and a chronic phase with cardiac and digestive sequelae. Some laboratory rodents (rats, mice, hamsters) develop an acute disease similar to that of humans when they are artificially infected, and occasionally they may show some signs characteristic of the human chronic infection.

The sera of numerous mammals (the mouse is an exception) have the capacity to lyse the epimastigotes, but not the trypomastigotes, of *T. cruzi* by a mechanism that apparently involves the activation of complement by the alternative pathway. The significance of this event in the natural history of the infection is not clear since the epimastigotes are not infective for the definitive host in nature. It would be interesting to investigate whether or not the infective trypomastigotes from the rectum of the vector (which are different, at least morphologically, from the trypomastigotes of the mammalian blood) are equally insusceptible to this activity.

Birds are refractory to *T. cruzi* infections. Earlier studies indicated that chicken serum possessed natural antibodies that, in the presence of complement, lysed the culture and the blood stages of the parasite. Recent experiments have demonstrated that *T. cruzi* activates chicken complement in vitro by the alternative pathway, and is lysed in this way even if antibodies are not present. Chicken IgG alone or guinea pig complement do not produce the same effect. Chicken complement seems to be able to act in other hosts also, since the inoculation of chicken serum in mice with acute *T. cruzi* infection reduced the parasitemia considerably. It is possible, however, that this phenomenon is not representative of what happens in all birds and may not be the central reason for their natural resistance to the infection, because pigeon serum does not destroy the blood forms of the parasites as chicken serum does.

Poikilothermous vertebrates are also naturally resistant to *T. cruzi* infection and their sera also destroy the culture and blood stages of the protozoa in vitro.

Studies in mice have verified that different strains show different susceptibility to the infection, but the mechanisms involved are not known. Some recent work showed, for example, that the levels of complement were about equally depressed in infected C3H mice, which died on the 24th day of infection, as in infected C57BL mice, which were still alive on day 45.

Numerous experiments in laboratory models have demonstrated that Chagas' infection and disease are more severe in young animals than in adults. The severe course of the human congenital infection and the frequency of grave lesions in these cases suggest that the same may be true in humans. Abundant observations testify that *T. cruzi* infections are more prevalent and severe in men than in women. This difference has usually been attributed to higher exposure of males, but similar observations in mice and dogs (Goble, 1970) make one wonder whether this explanation is an oversimplification. At least in laboratory animals, the diversity in the course of the infection between the sexes appears not to be due to sexual hormones, because it is present in prepubertal animals and it is not modified by the exogenous administration of the hormones of the opposite sex. Lactation, on the other hand, retards the development of the parasite and favors the survival of the host.

Studies on the effect of cortisone on the infection have yielded contradictory results but, in general, it is considered that corticosteroids aggravate the infection in humans. ACTH administration elevated the resistance of dogs to lethal doses of parasites but did not improve the condition in 2 human patients with Chagas' cardiopathy. Goble (1970) has done a detailed review of the factors that affect the susceptibility to, and the course of, the infection.

The available evidence indicates that the natural resistance depends on circulating substances. The final disposition of the parasites is probably achieved by the macrophage system. Curiously enough, surgical removal of the spleen does not modify the natural insusceptibility: actually, if preexisting humoral factors are able to kill the protozoa, macrophages will act only as scavengers and a reduction of their efficiency will not be expected to influence significantly the course of the infection.

Acquired Immunity

T. cruzi infection of humans or lower animals elicits the production of humoral and cell-mediated responses (World Health Organization, 1974a). Recent studies in mice have shown that seric gammaglobulins begin to increase from the second week of infection to reach a peak (up to 5 times the normal concentration) on the sixth week. Then they decline and approach normality toward the end of the fifth month of infection. IgM reached its peak at the third week and was normal again by the 20th week; IgG began to increase on the third week to reach a maximum on the sixth and seventh weeks: it diminished later on and it was near normal by the fifth month.

Most studies in infected people have revealed normal or slightly elevated concentrations of serum immunoglobulins, although considerable controversy still exists in this regard. Probably the best documented changes correspond to an instance of accidental infection in the laboratory that was treated for 4 months beginning on the 21st day of infection (Hanson, 1976). On exams performed 7, 19, 40, 72, 100, 128, and 159 days after the infection, IgM protein was elevated from the 40th to the 100th day and IgG protein between days 40 and 128; IgM and IgG antibodies were first detected on days 19 and 7, respectively, but both peaked on day 40; IgG antibodies declined after day 72, but were still present at a titer of 256 on day 159; IgM antibodies declined after day 100, but their titer was only 16 on day 159 (Figure 9). It is possible that the arguments about whether Chagas' disease produces hypergammaglobulinemia or not are actually the results of observations made at different times during the course of the infection.

The recent report of immediate-type cutaneous reactions in human patients indicates that skin-sensitizing antibodies are produced during the infection. Increase of IgE protein has been verified in patients, but whether this is actually IgE anti–*T. cruzi* antibodies remains to be investigated.

The existence of cell-mediated immunity has been verified in infected humans and lower animals by cutaneous tests and by tests of inhibition of migration and of lymphocyte transformation with specific *T. cruzi* antigens. The inhibition of macrophage migration is first de-

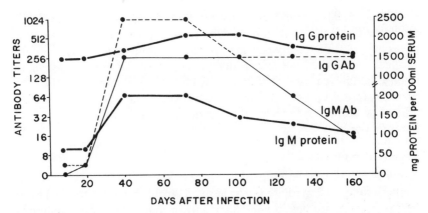

Figure 9.   Activity of specific IgM and IgG antibodies and levels of IgM and IgG protein in the serum of a patient with an accidental infection by *Trypanosoma cruzi*. (Based on data by Hanson, 1976).

tected in mice 60 days after infection with 10,000 or more trypomastigotes, and 90 days after infection with 1000. When the mice were infected with 100,000 epimastigotes from culture, the first positive tests appeared at 150 days of infection. Apparently, cell-mediated immunity is a late manifestation of the infection, which may explain the often negative results of the skin tests in Chagas' disease.

Numerous experimental studies and clinical observations have confirmed that acute nonlethal infections by *T. cruzi* subside spontaneously, leaving a strong resistance to ulterior challenges. Recent work in mice (Hanson, 1976) demonstrated that, after a primary infection, the number of flagellates in the blood and organs increases consistently until the end of the third or fourth week and then it declines rapidly to become undetectable toward the sixth week. Recovered mice are completely resistant to subsequent infections. The clinical experience with human patients indicates that the parasitemia is also transient in this host, and the rarity of repeated episodes of acute Chagas' disease suggests that humans are also refractory to reinfections.

Studies with laboratory animals have shown that, subsequent to a primary infection, mice harbor small numbers of parasites for up to 2 years after the original episode. The use of sensitive diagnostic procedures (xenodiagnosis, culture of samples of blood) has revealed the same phenomenon in people that had an acute syndrome several years earlier and no further chances of reinfection. Restoration of the parasitemia by immunosuppressive treatment in laboratory animals and in a few cases of human infection constitutes an additional proof that the parasites persist for a long period, possibly for life, after a primary infection. As in toxoplasmosis and, to a large extent, in babesiosis, it appears that the immunological protection against *T. cruzi* is of the premunition type.

Mechanisms of Acquired Resistance

Numerous attempts to demonstrate participation of antibodies in resistance by transfer of serum of recovered animals into susceptible ones gave inconsistent or contradictory results in the past. Recent experiments in mice found that the steepest increase in antibody titers coincided with the most rapid decline in the number of parasites in the body of the host and that the peak titers (during the sixth week) corresponded with the disappearance of the flagellates. Transfer of serum of these mice to susceptible mice showed that the serum taken during the sixth week of infection transferred protection consistently, whereas the serum collected during the 12th week was much less effective and the serum obtained during the second week had no effect. These experiments indicate the rather late production of transient protective antibodies and may explain the inconsistency of earlier results.

Transmission of resistance from female mice to their offspring has been occasionally taken as a demonstration of the presence of protective antibodies in the colostrum. Some of these experiments have not considered the transplacental transfer of immunity that occurs in rodents, and, in any case, the recent communication that cell-mediated immunity can be transmitted transmammarily in rodents adds uncertainty to these conclusions.

Experimentally, it has been demonstrated that serum of chronic patients is able to lyse *T. cruzi* in vitro in the presence of complement. Also, the neutralization of the complement activity in vivo increases the parasitemia and the mortality in infected mice. Both experiments indicate that complement plays a role in the mechanisms of acquired resistance to *T. cruzi*. On the other hand, Hanson (1976) found that trypomastigotes enclosed in diffusion chambers in the peritoneal cavity of immune mice were not affected by the immunity, whereas parasites injected intravenously disappeared from the body after 30 min. Since diffusion chambers permit the passage of macromolecules but not of cells, this experiment shows that antibodies or complement alone or in association are not able to destroy the parasite and suggests that their protective action requires the presence of cells to be manifested. It is possible that the primordial function of the antibodies and complement against *T. cruzi* in vivo is to favor the phagocytosis of the parasites.

Several assays to evaluate the participation of the cell-mediated immunity in acquired resistance have been reported. Mice thymectomized or treated with antithymus serum before *T. cruzi* infection developed higher parasitemias and died sooner than control mice. Unfortunately, this experiment does not constitute an irrefutable argument, since the thymus is a source of effector T cells as well as of regulator T cells for the humoral response; its surgical or chemical ablation, therefore, may affect either branch of the immunity. Associated experiments to demonstrate whether production of anti–*T. cruzi* antibodies is conserved or impaired in thymectomized mice have given conflicting results in the hands of different investigators. In any case, it appears that antibodies are synthesized despite the treatment, although their formation may be delayed. Transfer of splenocytes from resistant into susceptible rats also transmitted some resistance, but antibody titers were found in some of the recipient rats, presumably produced by B lymphocytes present in the inoculum.

It has been recently reported that *T. cruzi* grows and multiplies in normal macrophages, but its proliferation is slower and the resulting parasites less infective in macrophages activated by incubation with immune lymphocytes. Several other experiments have emphasized the participation of lymphokine-activated macrophages in protection against Chagas' disease (Nogueira and Cohen, 1977), but some disturbingly contradictory reports are on record: some investigators have found that a previous infection of mice with *Toxoplasma gondii*, known to enhance nonspecific phagocytosis, reduced mortality in a subsequent infection with *T. cruzi* in 60%, whereas others have found that previous toxoplasmosis decreases the parasitemia but does not affect the survival rate of *T. cruzi*–infected mice (Araujo and Nascimento, 1977).

In summary, the evidence currently available indicates that the antibody response has an important protective role in the acute phase of *T. cruzi* infection. The participation of complement is indicated by strong circumstantial evidence and the role of cell-mediated immunity is suggested by some experimental findings. In any case, it appears that the activity of the macrophagic system, stimulated by either branch of immunity, is an essential component of the protective mechanisms.

Artificial Production of Resistance

Several attempts to vaccinate laboratory animals with dead trypanosomes, with their soluble extracts, or with their subcellular particles have given discouraging results: the resistance achieved in most cases has been nil or negligible (Texeira, 1977). The inoculation of avirulent strains of *T. cruzi* in laboratory animals, on the contrary, has often produced considerable protection against subsequent infections with virulent parasites. The impossibility of accurately evaluating the virulence for humans by assays in animal models and the eventuality of reversions of virulence with time make the use of this procedure inadvisable in human populations or in lower animals that can act as reservoirs of the human infection. Despite this, 2 volunteers examined 5 years after vaccination with an avirulent *T. cruzi* strain showed no parasitemia or symptoms attributable to the infection and only rarely had positive serology (Meneses, 1976). Apparently, they eradicated the infection completely and developed no sequelae.

The vaccination with attenuated parasites (by irradiation or by culture with drugs) has given promising results: a recent experiment found that repeated inoculations of mice with 10^7 *T. cruzi* irradiated with 150,000 or 300,000 r did not produce parasitemia, but elicited a significant resistance against subsequent infections with the nonirradiated parasite. This irradiation does not kill the protozoan, but alters its physiology in some way so as to inhibit its ability to invade culture cells and, presumably, the cells of the host.

The fact that dead parasites do not elicit protection indicates that the functional antigens must be secreted or excreted by the living organism. Contribution of somatic materials appears to be necessary, however, because implantation of diffusion chambers with *T. cruzi* in the peritoneal cavity of mice caused the production of antibodies but not of resistance to the infection. It may be that particulate components of the parasitic cells exert an adjuvant effect on the host's responses to excretory/secretory products.

Vaccination with other species of trypanosomes with which *T. cruzi* shares serological reactivity has consistently failed to elicit protection.

Immunodiagnosis

A number of tests have been assayed for the diagnosis of Chagas' disease, particularly during the subacute or the chronic infection, in which blood parasites are virtually absent. The complement-fixation test (CFT, Machado-Guerreiro test) was proposed in 1913 and is still in general use with various modifications. Its efficiency depends heavily on the antigen utilized: currently, the best antigenic preparation seems to be a proteinic fraction obtained from delipidized epimastigotes; a metabolic antigen obtained from cultures gave excellent results, but was too unstable for routine use. The CFT reveals antibodies in humans during the third week of infection or later. In dogs it appears to be more sensitive—10 of 12 dogs were positive between days 6 and 11 of infection. In chronic infections, this test is positive in 90–95% of the patients, and reaches higher proportions in cases with chronic myocarditis. Its specificity is very good when appropriate antigens are used. Some cross-reactivity with sera of *Leishmania* patients has been observed.

The indirect hemagglutination test (IHAT) is comparable to the CFT in sensitivity and specificity, although it reveals antibodies somewhat earlier in the infection. Stabilization of the erythrocytes with glutaraldehyde and conjugation of the antigen with chromium chloride have permitted storage of the lyophilized, sensitized cells at 4°C for up to a year without loss of reactivity. This technique should facilitate the preparation of reagents of adequate sensitivity and specificity in specialized central laboratories and their storage in peripheral diagnosis centers. A "rapid" IHAT has been developed that permits the study of sera in blood banks in only 3 min.

The indirect fluorescent antibody test (IFAT) becomes positive sooner than either of the techniques mentioned above. Its specificity improves markedly when only titers of 40 or above are considered diagnostic. The use of formolized parasites (which keep for 3 or more months at 4°C) and of sera eluted from blood dried on filter paper (which keeps for 30 or more days) has been as effective as the use of fresh reagents and greatly facilitates the study of the infection in populations of areas deprived of adequate medical facilities. An IFAT with soluble antigen conjugated to filter paper has been at least as good as the CFT in preliminary studies. As in the case of toxoplasmosis, the IFAT is particularly appropriate to determine IgM antibodies in infants in cases of congenital infection, in order to distinguish them from the IgG antibodies transmitted transplacentally from the mother. With proper modifications, it is possible to study simultaneously the presence of anti–*T. cruzi* and of anti–*T. gondii* antibodies in the same sample.

A recent trial of enzyme-linked immunosorbent assay showed 98% correlation with the results obtained with the IFAT and yielded negative results with 81 serum samples from Greece (which are presumably free of infection). A delayed skin test has originated reports so conflictive that it has been abandoned in practice. Recent assays with a particulate antigen seem to have yielded better results. An immediate skin test tried in Brazil gave 68% sensitivity and 73% specificity (Carvalho et al., 1977).

Several other tests, such as precipitation, counterimmunoelectrophoresis, direct agglutination, agglutination of sensitized particles on a card, lysis, and a dye test similar to that used in toxoplasmosis, have been assayed on diverse opportunities. Despite the fact that some of them have yielded rather good results, they do not add evident advantages to the more established tests, so their study has not been pursued.

In general, all immunological tests for Chagas' disease give occasional cross-reactions with sera of subjects infected with other trypanosomes or with leishmanias (with which *T. cruzi* shares antigens abundantly). The recent purification of antigen 5, which appears to be exclusive to *T. cruzi*, by Capron's group in France, may improve this situation in the near future (Afchain et al., 1978).

Antigens

For years, the study of the antigens of *T. cruzi* was done with preparations of epimastigotes, which were believed not to appropriately represent the antigens released in the mammalian host by the trypo- and micromastigotes, or with limited quantities of trypomastigotes or micromastigotes obtained from the blood or the tissues of experimental animals. Growth of the parasite in tissue cultures and in special axenic fluid media has recently permitted the production of trypomastigotes and of micromastigotes in vitro, it is hoped in larger amounts. These techniques are expected to provide new and valuable information on relevant antigens in the near future.

It has been taken almost as an axiom that antigens from the forms parasitizing the definitive host must be better diagnostic reagents than preparations from the parasitic forms in the vector or in cultures. Gam and Neva (1977), however, found that the antigens obtained from a mixture of trypo- and micromastigotes are generally more sensitive but not significantly more specific than the extracts of epimastigotes. Moreover, a proteinic fraction of epimastigote extracts was more specific than the trypo-micromastigote preparation, and antigen 5, apparently unique to *T. cruzi*, has been purified from epimastigotes of culture. This phenomenon does not necessarily have to hold true for all cases of infections by protozoa with intermediate hosts, but it is worth considering in the search for adequate diagnostic reagents. Comparison of the extracts of epimastigotes and of the trypo-micromastigotes mixture has demonstrated that they possess exclusive as well as common antigens.

Experiments of cross-adsorption of antibodies have shown the existence of more than three dozen strains of *T. cruzi* that exhibit antigenic differences. The functional antigens, however, must belong to a basic core since numerous assays have indicated cross-protection among strains. Antigenic variation of a single strain in the course of an infection, as occurs with the African trypanosomes (see Chapter 7), has not been demonstrated with *T. cruzi*, and the absence of relapses of the parasitemia in Chagas' disease, comparable to those in sleeping sickness, suggests that antigenic variation does not occur in the parasite of the Americas. Exoantigens of *T. cruzi* have been recovered from culture media but not from the fluids of infected hosts; it is not clear at this moment whether these antigens are metabolic products or simply somatic materials released by dying parasites.

Several analyses of *T. cruzi* antigens have been reported in the literature. In one of them, a lipopolysaccharide that induces resistance but also causes morphophysiological alterations when injected into noninfected animals was purified; it has been called *chagastoxin*. Another study compared two American strains to a Brazilian strain of *T. cruzi*; some indication was observed that the former had fewer antigens than the latter, but it is not known whether this is related to the usual higher pathogenicity of the Brazilian parasites.

Immunopathology

Inoculation of *T. cruzi* in guinea pigs or in monkeys induces the formation of skin-sensitizing antibodies, apparently reaginic, in the respective host, which might suggest the participation of type I allergic reactions in the infection. Immediate skin reactivity and increased seric IgE have been demonstrated in infected persons, and infiltrates with eosinophils are occasionally found in lesions of chronic chagasic myocarditis. The participation of the immediate hypersensitivity in the natural disease has not been evaluated, however.

Inoculation of chagastoxin into people or animals produces fever, leukocytosis, and tissue lesions that resemble those caused by bacterial endotoxins; its role in the symptomatology of the natural infection has not been assessed yet.

The cardiac and digestive lesions found in patients of, or in animals infected with, *T. cruzi* often occur in the absence of parasites in their neighborhood. Among the several theories offered to explain the genesis of these lesions, an autoimmune etiology has been frequently considered. This hypothesis became more tenable with the demonstration that over 80% of the patients of Chagas' disease and of the guinea pigs inoculated with *T. cruzi* had antibodies that reacted with extracts of normal heart. In the guinea pigs, the proportion of animals reacting to cardiac tissue increased as the infection became more chronic. Recent investigations have demonstrated that infected subjects have circulating antibodies that react with the endocardium, vascular structures, and interstitium of striated muscle (EVI antibodies) (Cossio et al., 1977), as well as antibodies directed to neuronal cells, of noninfected hosts. In an accidental infection in the laboratory, the EVI antibodies appeared 8 days later, reached a peak between days 40 and 128, and persisted at low titers at least for 4 years. These findings appeared to definitively support the antibody-mediated pathology of American trypanosomiasis. However, it was later found that although EVI antibodies were present in 11 of 25 chagasic patients, they were also present in 17 of 63 patients of cutaneous leishmaniasis and in 10 of 47 patients with other chronic conditions. The direct relevance of these antibodies to the pathology of Chagas' disease is considerably weakened by this latter report. Texeira et al. (1978) have reported, however, that the migration of leukocytes of chagasic patients is inhibited by extracts of the parasite or by extracts of normal heart, and that lymphocytes of the same patients destroy normal cardiac cells. This research confirms the already known cross-reactivity between *T. cruzi* antigens and normal heart, but also raises the point that the antihost antibodies may only be a side-product of an autoimmune response and that the lesions are actually the result of the activity of cell-mediated immunity. The growing idea among the specialists is that lymphocytes sensitized as a consequence of the infection (by heartlike antigens of the parasite, or by the release of cardiac material normally secluded, or by heart substances altered by the parasitism) continue damaging the heart tissue even after the parasite has left it. Similar mechanisms may explain the production of digestive lesions.

There is no information on the presence of antigen-antibody complexes in the glomeruli of infected animals, as has been reported in other trypanosomiases. The recent finding of Ig, complement, and fibrinogen in the kidneys of rats that had been infected with *T. cruzi* for 6–12 months is suggestive, but it failed to confirm the presence of specific parasitic antigens.

No work appears to have been done in regard to congenital transmission of the parasite. Recent research with mice indicates that the transplacental transfer is facilitated by an elevated pathogenicity of the parasitic strain associated with a reduced phagocytic activity of the host. The relevance of these factors in the human infection remains to be seen.

Some years ago, it was reported that *T. cruzi* infection inhibited the growth of tumor cells in vivo. No further work seems to have been done on this interesting aspect.

Recent papers have communicated that infections with *T. cruzi* depress the host's ability to respond immunologically to nonrelated antigens (Cunningham et al., 1978). This is discussed again in Chapter 7.

SOURCES OF INFORMATION

Aburel, E., Lerbos, G., Titea, V., and Panna, S. 1963. Immunological and therapeutic investigations in vaginal trichomoniasis. Rum. Med. Rev. 7:13–19.

Ackers, J. P., Lumsden, W. H. R., Catterall, R. D., and Coyle, R. 1975. Antitrichomonal antibody in the vaginal secretions of women infected with *Trichomonas vaginalis*. Br. J. Vener. Dis. 51:319–323.

Afchain, D., Fruit, J., Yarzabal, L., and Capron, A. 1978. Purification of a specific antigen of *Trypanosoma cruzi* from culture forms. Am. J. Trop. Med. Hyg. 27:478–482.

Amyx, H, L., Ascher, D. M., Nash, T. E., Gibbs, C. J., Jr., and Gajdusek, D. C. 1978. Hepatic amebiasis in spider monkeys. Am. J. Trop. Med. Hyg. 27:888–891.

Aragon, R. S. 1976. Bovine babesiosis: A review. Vet. Bull. 46:903–917.

Araujo, F. G., and Nascimento, E. 1977. *Trypanosoma cruzi* infection in mice chronically infected with *Toxoplasma gondii*. J. Parasitol. 63:1120–1121.

Augustin, R., and Ridges, A. P. 1963. Immune mechanisms in *Eimeria meleagrimitis*. In: P. C. C. Garnham, A. E. Pierce, and I. Roitt (eds.), Immunity to Protozoa, pp. 296–335. Blackwell Scientific Publications, Oxford.

Balamuth, W., and Siddiqui, W. A. 1970. Amebas and other intestinal protozoa. In: G. J. Jackson, R. Herman, and I. Singer (eds.), Immunity to Parasitic Animals. Vol. II, pp. 439–468. Appleton-Century-Crofts, New York.

Barriga, O. O., and Arnoni, J. V. 1979. *Eimeria stiedae*: Weight, oocyst output, and hepatic function of rabbits with graded infections. Exp. Parasitol. 48:407–414.

Bray, R. S. 1976. Immunodiagnosis of leishmaniasis. In: S. Cohen and E. H. Sadun (eds.), Immunology of Parasitic Infections, pp. 65–76. Blackwell Scientific Publications, Oxford.

Brown, K. N. 1976. Resistance to malaria. In: S. Cohen and E. H. Sadun (eds.), Immunology of Parasitic Infections, pp. 268–295. Blackwell Scientific Publications, Oxford.

Callow, L. L. 1977. Vaccination against bovine babesiosis. Adv. Exp. Biol. Med. 93:121–149.

Cappuccinelli, P. 1975. L'immunodepressione nelle infezioni da protozoi. Gior. Batt. Virol. Immunol. 68:115–129.

Carvalho, E. B., Vaz, M. G. M., Resende, J. M., and Rassi, A. 1977. Intradermoreacão no diagnóstico da doença de Chagas com o antigeno soluvel de reacão imediata de Zeledon e Ponce. Rev. Gaiana Med. 23:41–45.

Chapman, W. E., and Ward, P. A. 1977. *Babesia rodhaini*: Requirement of complement for penetration of human erythrocytes. Science 196:67–70.

Clark, I. A., Cox, F. E. G., and Allison, A. C. 1977. Protection of mice against *Babesia* spp. and *Plasmodium* spp. with killed *Corynebacterium parvum*. Parasitology 74:9–18.

Cohen, S., and Mitchell, G. H. 1978. Prospects for immunization against malaria. Curr. Top. Microbiol. Immunol. 80:97–137.

Cossio, P. M., et al. 1977. Chagasic myocarditis. Am. J. Pathol. 86:533–544.

Cuckler, A. C. 1970. Coccidiosis and histomomiasis in avian hosts. In: G. J. Jackson, R. Herman, and I. Singer (eds.), Immunity to Parasitic Animals. Vol. II, pp. 371–397. Appleton-Century-Crofts, New York.

Cunningham, D. S., Kuhn, R. E., and Rowland, E. C. 1978. Suppression of humoral responses during *Trypanosoma cruzi* infections in mice. Infect. Immun. 22:155–160.

Dubey, J. P. 1977. *Toxoplasma, Hammondia, Besnoitia, Sarcocystis*, and other tissue cyst-forming coccidia of man and animals. In: J. P. Kreier (ed.), Parasitic Protozoa. Vol. III, pp. 101–237. Academic Press, Inc., New York.

Eidelman, S. 1976. Intestinal lesions in immune deficiency. Human Pathol. 7:427–434.

Elsdon-Dew, R. 1976. Serodiagnosis of amebiasis. In: S. Cohen and E. H. Sadun (eds.), Immunology of Parasitic Infections, pp. 58–64. Blackwell Scientific Publications, Oxford.

Farid, L., et al. 1977. Hepatic amebiasis: Diagnostic counterimmunoelectrophoresis and metronidazole (Flagyl) therapy. Am. J. Trop. Med. Hyg. 26:822–823.

Faubert, G. M., Meerovitch, E., and McLaughlin, J. 1978. The presence of liver auto-antibodies induced by *Entamoeba histolytica* in the sera from naturally infected humans and immunized rabbits. Am. J. Trop. Med. Hyg. 27:892–896.

Frenkel, J. K. 1973. Toxoplasmosis: Parasitic life cycle, pathology and immunology. In: D. M. Hammond and P. L. Long (eds.), The Coccidia, pp. 343–410. University Park Press, Baltimore.

Frenkel, J. K. 1977. *Besnoitia wallacei* of cats and rodents: With a reclassification of other cyst-forming isoporoid coccidia. J. Parasitol. 63:611–628.

Gam, A. A., and Neva, F. A. 1977. Comparison of cell culture with epimastigote antigens of *Trypanosoma cruzi*. Am. J. Trop. Med. Hyg. 26:47–57.

Garnham, P. C. C. 1970. Primate malaria. In: G. J. Jackson, R. Herman, and I. Singer (eds.), Immunity to Parasitic Animals. Vol. II, pp. 767–791. Appleton-Century-Crofts, New York.

Goble, F. C. 1970. South American trypanosomes. In: G. J. Jackson, R. Herman, and I. Singer (eds.), Immunity to Parasitic Animals. Vol. II, pp. 597–689. Appleton-Century-Crofts, New York.

Hanson, W. L. 1976. Immunology of American trypanosomiasis. In: S. Cohen and E. H. Sadun (eds.), Immunology of Parasitic Infections, pp. 222–234. Blackwell Scientific Publications, Oxford.

Harris, W. G., and Bray, R. S. 1976. Cellular sensitivity in amoebiasis—Preliminary results of lymphocytic transformation in response to specific antigen and to mitogen in carrier and disease states. Trans. R. Soc. Trop. Med. Hyg. 70:340–343.

Harris, W. G., Friedman, M. J., and Bray, R. S. 1978. Serial measurement of total and parasite-specific IgE in an African population infected with *Entamoeba histolytica*. Trans. R. Soc. Trop. Med. Hyg. 72:427–430.

Hendricks, L. D. 1977. Host range characteristics of the primate coccidian, *Isospora arctopitheci* Rodhain 1933 (Protozoa: *Eimeriidae*). J. Parasitol. 63:32–35.

Honigberg, B. M. 1970. Trichomonads. In: G. J. Jackson, R. Herman, and I. Singer (eds.), Immunity to Parasitic Animals. Vol. II, pp. 469–550. Appleton-Century-Crofts, New York.

Honigberg, B. M. 1978a. Trichomonads of importance in human medicine. In: J. P. Kreier (ed.), Parasitic Protozoa. Vol. II, pp. 275–454. Academic Press, Inc., New York.

Honigberg, B. M. 1978b. Trichomonads of veterinary importance. In: J. P. Kreier (ed.), Parasitic Protozoa. Vol. II, pp. 163–273. Academic Press, Inc., New York.

Horton-Smith, C., Long, P. L., Pierce, A. E., and Rose, M. E. 1963. Immunity to coccidia in domestic animals. In: P. C. C. Garnham, A. E. Pierce, and I. Roitt (eds.), Immunity to Protozoa, pp. 273–295. Blackwell Scientific Publications, Oxford.

Hudson, R. J., Saben, H. S., and Emslie, D. 1974. Physiological and environmental influences on immunity. Vet. Bull. 44:119–128.

Jacobs, L. 1973. New knowledge of *Toxoplasma* and Toxoplasmosis. Adv. Parasitol. 11:631–669.

Jacobs, L. 1976. Serodiagnosis of toxoplasmosis. In: S. Cohen and E. H. Sadun (eds.), Immunology of Parasitic Infections, pp. 94–106. Blackwell Scientific Publications, Oxford.

Kagan, I. G. 1973. The immunology of amebiasis. Arch. Inv. Med. 4(suppl. 1):169–176.

Karim, K. A., and Ludlam, G. B. 1975. The relationship and significance of antibody titers as determined by various serological methods in glandular and ocular toxoplasmosis. J. Clin. Pathol. 28:42–49.

Kozar, Z. 1970. Toxoplasmosis and coccidiosis in mammalian hosts. In: G. J. Jackson, R. Herman, and I. Singer (eds.), Immunity to Parasitic Animals. Vol. II, pp. 871–912. Appleton-Century-Crofts, New York.

Krupp, I. M. 1977. Definition of the antigenic pattern of *Entamoeba histolytica*, and immunoelectrophoretic analysis of the variation of patient response to amebic disease. Am. J. Trop. Med. Hyg. 26:387–392.

Krupp, I. M., and Jung, R. C. 1976. Immunity to amoebic infection. In: S. Cohen and E. H. Sadun (eds.), Immunology of Parasitic Infections, pp. 163–166. Blackwell Scientific Publications, Oxford.

Kua-Eyre, Su Lin, and Honigberg, B. M. 1976. Antigenic analysis of *Trichomonas vaginalis* strains. J. Protozool. 23:18A.

Kulda, J., and Honigberg, B. M. 1969. Behavior and pathogenicity of *Tritrichomonas foetus* in chick liver cell cultures. J. Protozool. 16:479–495.

Kulda, J., and Nohynkova, E. 1978. Flagellates of the human intestine and of intestines of other species. In: J. P. Kreier (ed.), Parasitic Protozoa. Vol. II, pp. 1–138. Academic Press, Inc., New York.

Levine, N. D. 1973. Protozoan Parasites of Domestic Animals and of Man. 2nd ed. Burgess Publ. Co., Minneapolis.

Levine, N. D. 1977. Nomenclature of *Sarcocystis* in the ox and sheep and of fecal coccidia of the dog and cat. J. Parasitol. 63:36–51.

McLeod, R., and Remington, J. S. 1977. Influence of infection with *Toxoplasma* on macrophage function, and role of macrophages in resistance to *Toxoplasma*. Am. J. Trop. Med. Hyg. 26:170–186.

Maekelt, G. A. 1972. Immune response to intracellular parasites. I. *Leishmania*. In: E. J. L. Soulsby (ed.), Immunity to Animal Parasites, pp. 343–363. Academic Press, Inc., New York.

Magliulo, E., et al. 1976. *Toxoplasma gondii*: A new diagnostic approach based on the specific binding of erythrocytes to lymphoid cells. Exp. Parasitol. 39:143–149.

Mahoney, D. F. 1972. Immune response to hemoprotozoa. II. *Babesia* spp. In: E. J. L. Soulsby (ed.), Immunity to Animal Parasites, pp. 302–341. Academic Press, Inc., New York.

Mahoney, D. F. 1977. Babesia of domestic animals. In: J. P. Kreier (ed.), Parasitic Protozoa. Vol. IV, pp. 1–52. Academic Press, Inc., New York.

Mahoney, D. F., Wright, I. G., and Ketterer, P. J. 1973. *Babesia argentina*: The infectivity and immunogenicity of irradiated blood parasites for splenectomized calves. Int. J. Parasitol. 3:209–217.

Manson-Bahr, P. E. C. 1971. Leishmaniasis. Int. Rev. Trop. Med. 4:123–140.

Markus, M. B. 1978. *Sarcocystis* and sarcocystosis in domestic animals and man. Adv. Vet. Sci. Comp. Med. 22:159–193.

Marquart, W. C. 1976. Some problems of host and parasite interactions in the coccidia. J. Protozool. 23:287–290.

Mattern, C. F. T., and Keister, D. B. 1977. Experimental amebiasis. Am. J. Trop. Med. Hyg. 26:393–411.

Meneses, H. 1976. A vacinacão de seres humanos com vacina viva avirulenta de *Trypanosoma cruzi*. Rev. Assoc. Med. Bras. 22:252–255.

Miller, L. H. 1977. A critique of merozoite and sporozoite vaccines in malaria. Adv. Exp. Med. Biol. 93:113–120.

Miller, L. H., and Carter, R. 1976. Innate resistance in malaria. Exp. Parasitol. 40:132–146.

Morgan, B. B. 1946. Bovine Trichomoniasis. Rev. ed. Burgess Publ. Co., Minneapolis.

Nogueira, N., and Cohn, Z. 1977. *Trypanosoma cruzi*: Uptake and intracellular fate in normal and activated cells. Am. J. Trop. Med. Hyg. 26(suppl):194–203.

Nussenzweig, R. H. 1977. Immunoprophylaxis of malaria: Sporozoite-induced immunity. Adv. Exp. Med. Biol. 93:75–87.

O'Connor, G. R. 1970. The influence of hypersensitivity in the pathogenesis of ocular toxoplasmosis. Trans. Am. Ophthalmol. Soc. 68:501–547.

Ogilvie, M. B., and Rose, M. E. 1978. The response of the host to some parasites of the small intestine: Coccidia and nematodes. In: Colloque INSERM-INRA, Immunity in Parasitic Diseases, pp. 237–248. INSERM, Paris.

Preston, P. M., and Dumonde, D. C. 1976. Immunology of clinical and experimental leishmaniasis. In: S. Cohen and E. H. Sadun (eds.), Immunology of Parasitic Infections, pp. 167–202. Blackwell Scientific Publications, Oxford.

Remington, J. S., and Krahenbuhl, J. L. 1976. Immunology of *Toxoplasma* infection. In: S. Cohen and E. H. Sadun (eds.), Immunology of Parasitic Infections, pp. 235–267. Blackwell Scientific Publications, Oxford.

Rezai, H. R., Ardehali, S. M., Amirhakimi, G., and Kharazmi, A. 1978. Immunological features of Kala-Azar. Am. J. Trop. Med. Hyg. 27:1079–1083.

Riek, R. F. 1968. Babesiosis. In: D. Weinmann and M. Ristic (eds.), Infectious Blood Diseases of Man and Animals. Vol. II, pp. 220–268. Academic Press, Inc., New York.

Ristic, M. 1970. Babesiosis and theileriosis. In: G. J. Jackson, R. Herman, and I. Singer (eds.), Immunity to Parasitic Animals. Vol. II, pp. 831–870. Appleton-Century-Crofts, New York.

Rose, M. E. 1972. Coccidia. In: E. J. L. Soulsby (ed.), Immunity to Animal Parasites, pp. 366–388. Academic Press, Inc., New York.

Rose, M. E. 1976. Coccidiosis: Immunity and the prospects for prophylactic immunization. Vet. Rec. 481–484.

Seah, S. K. K., and Hucal, G. 1975. The use of irradiated vaccine in immunization against experimental murine toxoplasmosis. Can. J. Microbiol. 21:1379–1385.

Shaw, J. J., and Lainson, R. 1977. A simple prepared amastigote Leishmanial antigen for use in the indirect fluorescent antibody test for Leishmaniasis. J. Parasitol. 63:384–385.

Stahl, W., Matsubayashi, H., and Akao, S. 1965. Effect of 6-mercaptopurine on cyst development in experimental toxoplasmosis. Keio J. Med. 14:1–12.

Texeira, A. R. L. 1977. Immunoprophylaxis against Chagas' disease. Adv. Exp. Med. Biol. 93:243–280.

Texeira, A. R. L., Texeira, G., Macedo, V., and Prata, A. 1978. *Trypanosoma cruzi*-sensitized T-lymphocyte mediated ^{51}Cr release from human heart cells in Chagas' disease. Am. J. Trop. Med. Hyg. 27:1097–1107.

Tsang, C. L., and Lee, Y. C. 1975. Studies on the preventive and therapeutic effects of antiserum against *Isospora felis* infection in puppies. J. Chin. Soc. Vet. Sci. 1:37–41.

Voller, A. 1976. Serodiagnosis of malaria. In: S. Cohen and E. H. Sadun (eds.), Immunology of Parasitic Infections, pp. 107–119. Blackwell Scientific Publications, Oxford.

World Health Organization. 1974a. Immunology of Chagas' disease. Memoranda. Bull. WHO 50:459–472.

World Health Organization. 1974b. Serological testing in malaria. Memoranda. Bull. WHO 50:527–538.

World Health Organization. 1974c. Symposium on malaria research. Bull. WHO 50:143–372.

Zuckerman, A. 1975. Current status of the immunology of blood and tissue protozoa. I. *Leishmania*. Exp. Parasitol. 38:370–400.

Zuckerman, A. 1977. Current status of the immunology of blood and tissue protozoa. II. *Plasmodium*. A review. Exp. Parasitol. 42:374–446.

Zuckerman, A., and Lainson, R. 1977. Leishmania. In: J. P. Kreier (ed.), Parasitic Protozoa. Vol. I, pp. 57–133. Academic Press, Inc., New York.

Chapter 4

Immune
Reactions to Parasitic Nematodes

THE NEMATODES OF THE DIGESTIVE TRACT

The nematodes found in the digestive tract constitute a biologically complex group: some of them are restricted exclusively to the digestive tube (e.g., *Trichostrongylus*), others undergo limited migrations into the intestinal wall (e.g., *Oesophagostomum*), others migrate extensively and for long periods before arriving eventually at the alimentary canal (e.g., *Toxocara*), and others only spend a comparatively brief time in the digestive system (e.g., *Trichinella*). It is often difficult to determine which responses are elicited by the enteral and which by the extraenteral forms of the migratory nematodes. Furthermore, recent findings have emphasized that the larvae of many parasites traditionally thought to be exclusive to the intestinal lumen (e.g., *Trichostrongylus*) may invade the intestinal wall in the case of massive infections. This section is concerned mainly with those nematodes that do not undergo extensive somatic migrations and with the digestive phase of those that do so.

NATURAL RESISTANCE

No attempts appear to have been made to systematize the components of natural resistance to gastrointestinal nematodes, so the comments here are on some factors that obviously play a role in this phenomenon. In every case, the infective form of a nematode penetrates its host in a resting stage that must experience several moults and periods of growth before reaching its reproductive phase. In some cases, like the ascarids, the oxyiurids, and the trichurids, the moults must be preceded by the hatching of the egg in the host.

The biochemical events responsible for hatching or moulting are strictly dependent on the parasite, but they must be triggered by *stimuli from the host.* Hatching of *Ascaris* eggs requires proper concentration of CO_2 in a reducing environment and adequate conditions of temperature, pH, and ionic concentration. Release of the infective larvae of *Haemonchus* from their external cuticles necessitates similar factors but in different proportions. Evidently, the precise harmony of these elements exists only in a few animal species, which will be the only ones able to sustain the initial invasion by the parasite; any other species will be naturally resistant to infection by that particular parasite. Infective larvae ingested by these "abnormal hosts" will be eliminated unchanged or will die in the digestive tract of the host.

The further development of the nematode to adulthood also requires successive biochemical signals from the host to trigger the necessary physiological events in the parasite. Occasionally, some abnormal hosts are able to provide the initial stimuli for invasion but lack the further signals for growth; in these cases, the parasite may persist in the host as a larval stage, constituting a *larva migrans* (see "Nematodes of the Tissues," below). From the biological point of view, these larvae represent a failure in the evolution of the parasite, since they cannot reach the reproductive stage necessary for dissemination of the species.

Even if a parasite encounters the appropriate signals for invasion and development, some *physiological characteristics of the host* may prevent its definitive establishment. These physiological factors have not been well characterized yet, but a few examples are known: it has been demonstrated that incubation of *Trichinella* larvae with extracts of stomach or small intestine of chicken killed more than 50% of the larvae; sequential incubation with both extracts killed over 95% of the parasites (Barriga, 1980). Work with *Ascaridia* has demonstrated that old chickens develop a heat-stable substance in their intestine that is toxic for the nematode. The lethal effect of the composition of some biles on *Echinococcus* is reviewed with the cestodes (Chapter 5). Other factors, such as pH of the various digestive compartments, intestinal emptying time, composition of the diet, and hormonal influences, are also believed to play a role in the susceptibility of the different host species to colonization by the parasite.

Practically all parasites establish some sort of contact with the host that is harmful to the host's cells and triggers diverse degrees of *nonspecific inflammation*. It is logical to expect that the vascular, metabolic, and physicochemical alterations characteristic of the inflammatory process will modify the tissues of the host in such a manner that the microecological conditions will change and the permanence of the parasite will be endangered. The elimination of some intestinal parasites has been presumptively connected with an increased secretion of mucus in some cases.

In a few occasions, hosts normally insusceptible to a given parasite have been able to sustain complete infection with it after treatment of the host with immunosuppressive drugs. It is not clear at the moment whether this phenomenon should be regarded as an indication that some form of *nonspecific immunity* participates in the natural resistance, or whether the treatment simply reduces the cellular component of the inflammatory response.

Other elements of natural resistance are less well defined, but numerous observations have verified their roles in the susceptibility to the infection. The *age of the host* is one of them. With few exceptions (*Dirofilaria immitis, Babesia*), young individuals show a much greater susceptibility to parasitic infections than their adult counterparts. However, it has often been difficult to determine how much of the resistance of the adult is due to true age resistance and how much has been induced by previous experiences with the parasite. At present, there are reasonably convincing proofs that age resistance exists in *Ancylostoma caninum* and *Toxocara canis* of the dog, *Strongylus vulgaris* of the horse, *Oesophagostomum radiatum* of cattle, *Haemonchus contortus* of sheep, *Trichinella spiralis* of rats, and oxyurids of mice. On the contrary, studies with *Trichuris vulpis* in dogs and with *Ostertagia ostertagi* in cattle have demonstrated that the adults are as susceptible as the young hosts, if they are maintained in an environment free of the specific infection. Clinical and epidemiological observations in man also suggest the presence of age resistance, but in this case is even more difficult to distinguish between true age resistance and acquired immunity or behavioral components that make infection less probable in the adult.

The real causes of age resistance are unknown, but two hypotheses have been frequently proposed. Older animals have more goblet cells and globule leukocytes (degranulated mast cells?) in their intestinal mucosa, which may interfere with the development and establishment of gastrointestinal nematodes. Age is also related to a higher polymerization of the acellular elements of the skin, which may better resist the activity of the penetration enzymes of the skin-invading parasites.

The *sex of the host* also plays some role in natural resistance. Studies with oxyurids of mice have shown that males are infected twice as frequently as females in nature. The same observation has been made in dogs infected with *Toxocara canis*. In both cases, the differences between sexes become evident by the time of puberty which suggests intervention of the sexual hormones. Similar phenomena in humans have been often attributed to different behavior, which influences the risks of infection for each sex.

Another possible factor of natural resistance is *immunological cross-reactivity*. Gnotobiotic guinea pigs (raised in the absence of common animal and vegetal parasitic contaminants) are susceptible to infection with nematodes of rats and mice that do not affect conventional guinea pigs. It is probable that humoral or cellular immune responses, directed against antigenic determinants shared by murine nematodes and common contaminants, may be effective enough to set a limit to the infection by the parasite.

Microecological factors may also play a role in natural susceptibility. *D. immitis* and other canine filariids infect lower primates rather easily under experimental conditions, despite the fact that they are of exceptional occurrence in humans. It is possible that the hairless human skin does not provide the appropriate conditions for the infective stage to persist long enough to achieve penetration.

Finally, some experimental observations have indicated that *intraspecific genetic characteristics of the host* may influence natural resistance. Studies of *H. contortus* infections in two strains of sheep given identical infective doses demonstrated considerable variation in the parasitic load and in the intensity of the symptoms developed in each group; some researchers have found direct correlation between natural resistance and the presence of hemoglobin A. Experiments with *Trichuris muris* showed that, whereas some mice expelled the parasites within conventional periods, other mice retained the infection much longer; cross-breeding between both groups demonstrated that the ability to reject the infection depended on a few dominant genes. There is no proof that genetic make-up plays a role in the susceptibility of humans to nematode infections, but the much greater resistance of the Negroid race to ancylostomiasis suggests that such a phenomenon may also exist in humans.

ACQUIRED IMMUNITY

Most parasitic infections begin with the activation of the infective stage of the parasite, which proceeds to eclosion, excystment, exsheathment, or moulting to initiate the invasion. All these physiological events entail the production of enzymes by the parasite and often involve deposition of other parasitic elements (e.g., cuticles) in the tissues of the host. An esterase, a chitinase, and possibly a protease have been identified in hatching eggs of *Ascaris*; species-specific leucine aminopeptidases have been found in exsheathing larvae of *Haemonchus* and *Trichostrongylus*.

The rapid period of growth that follows invasion and precedes adulthood in the new host comprises new changes of cuticles and an extremely active metabolism, with the consequent production of enzymes and catabolites. Finally, the accelerated production of eggs characteristic of most nematodes also suggests a very rapid metabolism that must release a great amount of enzymes and catabolites in the body of the host. The finding of antibodies against parasites of exclusively intestinal location indicates that even antigenic substances deposited in the digestive lumen can find their way to the immune system.

Thus it is evident that, since the beginning of the infection, the host is constantly subjected to antigenic stimulation by parasitic substances. The variety of antigens produced, their changes with the diverse developmental stages, the different location of their release in the host body and the particular physiological status of the host at the time of stimulation constitute splendid opportunities to put in play the multiple capacities of the host immune system.

Modern immunological thinking does not support the old notion that the immune response to animal parasites must be inherently different from the immune responses to other pathogens; on the contrary, all evidence indicates that the lymphoid system has limited possibilities of response that are differentially triggered whenever the appropriate stimulus operates. The diversity of the immune responses to different species or stages of parasites is nowadays attributed to variations in the quality and quantity of the antigens produced and to differences in the location and opportunity of their presentation to the immune system.

For didactic purposes, the diverse manifestations of immunity on parasites and on parasitic infections are discussed as individual events. It is convenient to keep in mind, however, that these phenomena are frequently simultaneous and additive, so that the final effect may be far more dramatic than is suggested by their presentation as single occurrences.

Recent reviews on the immunity of infections by digestive nematodes have been written by Sinclair (1970), Thorson (1970), Ogilvie and Jones (1973), and Wakelin (1978).

Host Immune Responses

Production of antibodies in nematode infections has been known for a long time, and cell-mediated immunity is beginning to be found with increasing frequency. Eosinophilia was considered until recently to be a secondary manifestation of type I allergy, but recent findings have shown that its control and influence in parasitic infections are considerably more complicated.

Production of Antibodies   Antibody production has been found in every nematode infection in which it has been investigated. It has been determined on several occasions that the precocity and the intensity of the humoral response are directly dependent on the magnitude of the infective dose administered.

In some cases, the antibodies are directed preponderantly against the adult worms (e.g., *Hyostrongylus* and *Oesophagostomum* in swine), (Castellino, 1970), whereas in other cases they react mainly with larval stages (e.g., *Ostertagia* in sheep). On occasion, reinfections trigger typical secondary responses (e.g., *Hyostrongylus*), but other parasites fail to elicit a prolonged immunological memory (e.g., *Haemonchus*).

The identification of the classes of antibodies produced in nematodiases has been successfully attempted in only a few instances. In the infection of rats with *Nippostrongylus brasiliensis* it has been possible to demonstrate IgM, IgG, and IgA antibodies in the serum, in the digestive secretions, and on the parasites. In trichinous rats, IgM, IgG, and IgA antibodies have been found in the serum and the digestive mucus. In sheep infected with *H. contortus*, IgM and IgG antibodies were identified in the serum and IgA and IgG antibodies in the gastric mucus. IgE antibodies are very common in nematode infections, but they seem to be stimulated particularly by parasites that reach an intimate contact with the tissues of the host. Specific antibodies, the class of which has not yet been determined, are found in infections by nematodes restricted exclusively to the intestinal tract, such as *Hyostrongylus*, *Oesophagostomum*, and *Trichuris*.

Evidently, antibody synthesis is a general phenomenon in nematodiases, even in cases in which the contact between the parasite and the host seems to be too restricted to guarantee the proper antigenic stimulation. The exact biological significance of the presence of specific antibodies has been more difficult to assess, however. It has been impossible on numerous occasions to relate the presence of antibodies to the spontaneous reduction of the parasitic load or to the protection against reinfections.

In infections of rats with *Nippostrongylus* it has been found that IgG, but not IgM or IgA, antibodies afford protection against a challenge. In trichinellosis, it was reported on one occasion that incubation of muscle larvae with complete immune serum or with its IgM fraction, but not with the IgG fraction, reduced the infectivity of the parasites. Other investigators have been unable to confirm these findings or to detect metabolic changes in larvae incubated with immune serum. In sheep vaccinated with irradiated *Haemonchus* larvae, the resistance to the challenges has been related to IgG antibodies in the serum and to IgG and IgA antibodies in the abomasal mucosa. The regular presence of IgA in most exocrine secretions and its protective activity in many viral and bacterial infections of the mucous membranes prompted several attempts to investigate its role in gastrointestinal nematodiases. Most of them have failed to demonstrate a protective effect (Dobson, 1972).

Immunoglobulin E   Production of IgE and IgE antibodies (or their equivalents in lower animals) is so characteristic of helminth (and arthropod) infections and has been the subject of so many speculations that it deserves particular mention. Recent studies by Ishizaka et al. (1976) with the rat nematode *N. brasiliensis* have synthesized and organized most of our knowledge in this regard.

The elevation of nonspecific IgE in the serum begins about 10 days after infection, reaches its peak toward the end of the second week in conjunction with the spontaneous expulsion of the parasites, and declines rapidly thereafter. A reinfection causes an increase of similar magnitude, but one that reaches its peak in only a week and persists longer. Production of IgE antibodies is not a constant feature of the primary infection, but, when it happens, it does not show strict correspondence with the levels of nonspecific IgE. In many cases IgE antibodies reach their peak in the third week of infection, but they are undetectable in the first or fifth week. A reinfection, however, elicits a typically secondary response, even if IgE antibodies were not detected after the primary infection. IgE anamnestic responses are uncharacteristic of the rat stimulated with injectable antigens, so this suggests that the parasite possesses a particular stimulatory factor for this response.

Examinations of lymphoid organs of infected animals showed the presence of IgE-producing cells in mesenteric lymph nodes, in the spleen, and in the bone marrow; the precocity and intensity of the respective responses diminished in the same order. This variation suggests that the degree of stimulation of the lymphoid tissue depends on the quantity of parasite products available to it. On the other hand, no differences were found between infected and control animals in the number of cells producing other classes of immunoglobulins, which indicates that the parasite stimulates specifically the production of IgE.

Other experiments showed that the actual production of IgE as well as the establishment of an immunological memory required the presence of T lymphocytes. The former event only occurred in infections with the living worm, whereas the latter could also be induced by inoculation of parasitic extracts. Additional evidence indicated that the regulation of IgE production is more dependent on the activity of helper and suppressor cells than is the regulation of the production of IgG.

Evidently, the production of nonspecific IgE and of IgE antibodies responds to different parasitic stimuli, since they do not coincide in time. A possible explanation is that the parasite actually produces two sets of substances: one with a nonspecific adjuvant effect on IgE production, and another that behaves as the actual parasite antigen. Synthesis of specific IgE antibodies would occur only when both sets reach the host immune system at the same time; the effect of the adjuvant alone, in association with unrelated antigens, would be detected as production of nonspecific IgE. The hypothesis of an IgE adjuvant is supported by the fact that *Nippostrongylus* (and other nematode) infections given shortly after the injection of an unrelated antigen stimulate the preferential synthesis of IgE antibodies against the antigen.

Lack of stimulation of IgE antibody production by inoculation of extracts of the parasite may be due to a minimum time period necessary for the adjuvant to act. This minimum time, known to be required by routine adjuvants, may not be provided by an extract likely to be rapidly degraded by the host. In actual infections, the adjuvant may be permanently replenished by the metabolizing parasite. Since priming of lymphocytes for immunological memory is most likely an event that precedes the actual antibody formation, the fleeting effect of the extract may be effective in this regard. Only *Ascaris* extracts have been demonstrated to be as effective as the living parasite in stimulating IgE formation; it would be interesting to know whether the composition of this extract permits a longer permanence in the body of the host than extracts from other nematodes.

As implied above, it is possible that the increment in nonspecific IgE so frequently reported in helminth infections may be nothing more than the stimulation by the parasite of the IgE response to components constantly present in the host, such as antigens from foodstuff and from commensal germs. Without knowing which antigens to use, these antibodies would certainly be detected as nonspecific IgE. Despite the almost universal production of IgE in helminthiases, there is little evidence at present that this immunoglobulin or the corresponding antibodies play a protective role in these infections. Only in the classic form of the "self-cure" phenomenon (*Haemonchus*-type) and in schistosomiasis are there indications that IgE antibodies may have some participation in the elimination of the parasites. On the other hand, no IgE antibodies are produced in some nematodiases, such as trichuriasis, despite which the adult parasites are regularly expelled from the intestine of the hosts. There are convincing proofs, however, that IgE antibodies may add to the symptomatology of the infection: the urticaria of fascioliasis, the anaphylactic shock of hydatid cyst ruptures, the edema of trichinellosis, and the epigastric pain in ascariasis are a few examples.

Some authors have associated asthma and other allergic disturbances with the presence of intestinal parasites. This association is rather negated by the rarity of atopic diseases in Africa, which is otherwise notorious for its high prevalence of intestinal helminthiases. It is possible on a theoretical basis to defend a regulatory role of the parasites in the production of allergic conditions. Helminths may stimulate heterologous IgE responses by the mechanisms explained above when their number is moderate or their presence is temporary; massive or permanent infections, on the contrary, may induce levels of homologous IgE antibodies so elevated that the immunoglobulin may saturate the corresponding receptors on mast cells and basophils, actually blocking their combination with IgE antibodies against other antigens. In this way, helminths might enhance or suppress other allergic responses, depending on the intensity and persistence of the parasitic infection.

Production of Cell-mediated Responses Cellular immunity has been demonstrated in infections with *Ascaris*, *Ancylostoma*, *Trichinella*, *Capillaria hepatica*, and *Nippostrongylus* (Larsh and Weatherly, 1975). Among the strictly intestinal nematodiases, it has been verified directly in infections with *Trichostrongylus colubriformis* in the guinea pig and indirectly in infections with *Trichuris muris* in mice. Also, there is evidence of its presence in infections of lambs with *Chabertia*, *Haemonchus*, and *Oesophagostomum* and in infections of calves with *Ostertagia*.

Cell-mediated immunity may be as prevalent as antibody formation in nematodiases, and its comparative scarcity may be only a representation of the greater difficulties of verifying its existence and of the interests of the investigators in the past. Recent work has indicated that cell-mediated responses have a critical role in the spontaneous elimination of some intestinal nematodes (see "Expulsion of Adult Parasites," below). At least one report has claimed that transfer of resistance to *Trichostrongylus axei* has been achieved by injections of lysates of leukocytes of infected sheep into susceptible lambs. The full biological significance of cellular immunity in parasitology is far from being completely perceived at present and constitutes an almost virgin field of research.

Eosinophilia As is IgE production, eosinophilia is a common feature of most helminth and arthropod infections, especially of those in which the parasite achieves a rather intimate association with the tissues of the host. Recent investigations have greatly expanded our knowledge and understanding of the participation of eosinophils in parasitic infections. A comprehensive review has been published recently by Butterworth (1977) on this subject.

Most commonly, peripheral eosinophilia is detected during the second week of a helminth infection and reaches its peak a few days later. A challenge infection caused a quicker and more intense rise of the eosinophils, which soon suggested that an immunological background was involved in the reaction. It is known now that the eosinophilic response can be stimulated by Type I allergy, by cell-mediated reactions, and by the parasites themselves.

Degranulation of mast cells and basophils in the course of Type I allergies constitutes a major but not exclusive mechanism of eosinophilia. In addition to the classic mediators of Type I allergy (e.g., histamine, slow-reacting substance of anaphylaxis, or SRS-A, heparin, chymase, platelet activating factor, or PAF), these cells release two low molecular weight peptides and two intermediate molecular weight peptides, collectively known as eosinophil chemotactic factors of anaphylaxis (ECF-A), with chemotactic activity for eosinophils. The former peptides also act as inactivators of these cells so that, more than stimulators, they must be regarded as regulators of the eosinophilic activity. Also, the tissue alterations produced by the classic mediators secondarily generate lipids with chemotactic action on polymorphonuclear leukocytes, preferentially on eosinophils (ECF-T).

Studies with *Trichinella* and *Schistosoma* infections have recently demonstrated the existence of a further eosinophil chemotactic factor released by stimulated T lymphocytes (ECF-L) during reactions of cell-mediated immunity. This lymphokine, originally known as "eosinophil stimulation promoter," stimulates the eosinophil precursor cells in the bone marrow and attracts eosinophils and monocytes.

Finally, the parasites themselves may induce eosinophilia by at least two different mechanisms. Substances with direct chemotactic activity for eosinophils have been found in extracts of *Ascaris* and of *Taenia taeniformis*; also some parasites (e.g., *Schistosoma, Babesia, Trypanosoma cruzi*) are able to directly activate the complement system with production of peptides C3a and C5a, which possess chemotactic activity for polymorphonuclear leukocytes.

A very recent report has indicated that the eosinophils of people with helminth infections have considerably more receptors for IgG and for complement than the eosinophils of noninfected people. This finding may indicate that parasitic infections play a further role in the maturation of these cells.

Two major functions have been so far assigned to the eosinophilic cells in regard to parasitic infections: a regulatory action on the manifestations of Type I allergy and a lethal effect on some parasites. In regard to the first function, eosinophils restrict the expression of allergy by producing histaminase and phospholipase D and by phagocytizing SRS-A, mast cell and basophil granules, and antigen–IgE antibody complexes. The histaminase and the phospholipase D degrade the histamine and the PAF, respectively. These substances, together with the SRS-A, which is eliminated by phagocytosis, are the major inducers of the tissue alterations that occur in Type I allergy. The basophilic granules contain heparin and chymase, which also contribute to tissue damage. Antigen–IgE antibody complexes are the initiators of the degranulation of mast cells and basophils. Thus, these activities of the eosinophils contribute to prevent or to neutralize the actions of tissue-damaging components in Type I allergy.

In connection with their effect on parasites, some recent in vivo studies with specific antieosinophil serum revealed a strong correlation between the absence of eosinophilic cells and the lack of protection against infections with *Trichinella* or *Schistosoma* in laboratory animals. Further experiments demonstrated that eosinophilic cells were able to damage schistosomulas and *Schistosoma* eggs when incubated in vitro in the presence of specific antibodies (see Chapter 5).

These findings are extremely important, but it is still too early to draw conclusions on the general biological significance of eosinophils in parasitology, especially because some of the reports are still conflicting. Adults or larvae of *Nippostrongylus* incubated with eosinophils and specific antibodies become covered with eosinophilic cells, but do not show any evidence of losing infectivity for rats; inoculation of guinea pigs with antieosinophil serum, however, increases their susceptibility to *Nippostrongylus* as compared to normal controls, although it does not affect the development of resistance to the infection.

A further intriguing occurrence that requires explanation is the occasional association between absence of eosinophilia and strong resistance to reinfections with *Ascaris* or with *Trichinella*.

A summary of the regulation and functions of the eosinophils as we known them today appears in Figure 10.

Effects of the Host Immune Response on the Parasites

The difficulties of reproducing in vitro the complex physicochemical interactions characteristic of the mammalian tissues have considerably impaired the precise determination of the intimate mechanisms that affect parasites in the immune host. However, numerous observations

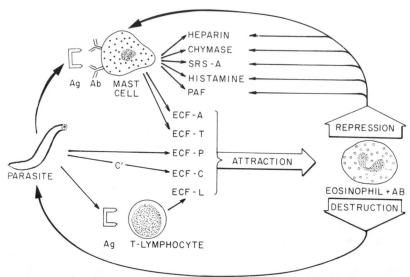

Figure 10.   Regulation and function of the eosinophils in parasitic infections. Eosinophils are attracted to particular locations by the activity of eosinophil chemotactic factors (**ECF**) that are produced directly by the mast cells in the course of anaphylactic reactions (**A**) or by alteration of tissue lipids (**T**) during the same reactions. It has also been demonstrated experimentally that extracts of parasites (**P**), substances released during complement activation (**C**), and secretions of T lymphocytes stimulated by specific antigens (**L**) contain similar factors. Once on location, the eosinophils produce substances that neutralize the activity of some mediators of hypersensitivity (histamine, PAF) or phagocytize elements that contribute to the allergic reaction (antigen–IgE antibody complexes, heparin, chymase, SRS-A). In addition, with the assistance of specific antibodies the eosinophilic cells are able to destroy some parasites. See text for other abbreviations. (Modified from Goetzl and Austen, 1977.)

in vivo and their occasional confirmation in vitro have identified a few general consequences of the host immune response on some parasites and, in a few instances, have suggested their possible mechanism of production.

Inhibition of Growth ("Stunting")   Studies with many nematodes (*Dictyocaulus, Trichinella, Ancylostoma, Ascaris, Ostertagia, Trichostrongylus*) have shown that parasites grow more slowly in immune hosts. Often their final size as adults is considerably reduced as compared to the same parasites growing in nonimmune hosts. When parasites developing at a reduced rate are transferred to a nonimmune host, they resume their normal rate of growth and attain the size typical for the species. This phenomenon indicates that the stunting is attributable to the immune response of the host. Further studies demonstrated that the inhibition of growth takes effect even when the parasites are placed in the immune host within Millipore chambers that permit the passage of macromolecules but not of cells. This is satisfactory proof that the stunting is mediated by antibody activity rather than by a cell-mediated reaction. A curious manifestation of this phenomenon is the fact that females of *Haemonchus* and *Ostertagia* occasionally do not develop full vulvar flaps in immune hosts.

Inhibition of growth is a different event than inhibition of development: in the first case the parasites are smaller but reach adulthood; in the second, the ontogenic evolution of the parasite is stopped for prolonged periods at some stage prior to adulthood.

Inhibition of Fecundity   Observations in a large number of parasitic infections have shown that the ratio of eggs per female drops in direct relationship with the age of the infection or the parasitic load. Laboratory experimentation with *Nippostrongylus*, *Trichinella*, and *Ostertagia* have demonstrated that the parasites recover their normal fecundity when transferred to nonimmune hosts. Undoubtedly, reduction of egg production is also influenced by other factors, such as age of the parasites and intraspecific competition, but few specialists question that the host immune response plays a major role in this phenomenon.

Modulation of Enzymatic Activity   In vitro studies have revealed that the sera of dogs that have undergone multiple *Ancylostoma* infections have the property of inhibiting the proteolytic and lipolytic activity of the esophageal extracts of the same parasite. Probably the same phenomenon occurs in vivo and affects the viability of the nematode, since inoculation of extracts of *Ancylostoma* esophagus in dogs confers some protection against homologous challenges.

Immunization of guinea pigs with *Ascaris* malic dehydrogenase reduces the pulmonary lesions caused by a subsequent *Ascaris* infection. This has also been taken as an indication that antienzyme antibodies are able to reduce the viability of the migrating larvae.

Infection of rats with *Nippostrongylus* induces the formation of antiacetylcholinesterase antibodies. The same has been verified in infections with *Trichostrongylus*, *Ostertagia*, and *Oesophagostomum*, but the corresponding antibodies have not been found in infections with *Toxocara*, *Haemonchus*, *Cooperia*, and *Trichuris*. In every instance, the antibodies were specific for the genus rather than the species of parasite utilized in the respective experiment.

In *Nippostrongylus* infections it has been demonstrated that the antienzyme antibodies stimulate the hyperproduction of acetylcholinesterase by the worms, which may be related in a yet unknown way to the phenomenon of adaptation to immunity (see below). Alterations of the morphological structures of the parasites in immune hosts have been described in this same infection. Some workers believe that these are mediated by the neutralization of enzymes in the nematode, which causes a condition similar to malnutrition.

Inhibition of Metabolism   In vitro experiments have shown that third-stage larvae of *Nippostrongylus*, *Oesophagostomum*, and *Chabertia* incubated in serum or in digestive mucus of immune animals utilize considerably less oxygen than the same larvae incubated in serum or mucus of normal animals. On the other hand, the incorporation of labeled phosphate by adults of *Nippostrongylus* in the intestine of infected rats begins to decline after the seventh day postinfection, reaches one-tenth of its prior values on the 10th day of infection, and stays at this level until the spontaneous expulsion of the parasites, about the 15th day postinfection. Although strict evidence that this is an immunological phenomenon is still lacking, its dynamics strongly suggest so.

Direct Damage to the Parasites   More than a decade ago it was reported that lymphocytes adhered to *Ascaris* larvae in laboratory animals resistant to the infection. Although it is widely assumed that these lymphocytes must affect the parasite in some still undetermined way, there is no current proof that this is the case.

Recent work has demonstrated that eosinophils are deleterious to *Schistosoma* and to *Trichinella*. Damage to schistosomulas and to *Schistosoma* eggs has been demonstrated in vitro (see corresponding section). Also, a relationship has been found between antienzyme antibodies and malnutritionlike damage of the worms in infections of rats by *Nippostrongylus*. The same relationship between production of antibodies to the parasite and evidence of structural damage of the cells of the nematode has been verified in several cases of spontaneous expulsion of the adult parasites (see below).

Adaptation to Immunity   Recent reports have indicated that *Nippostrongylus* grown in rats immunized against the homologous parasite survive longer when transferred to nonimmune rats than worms grown in normal rats do. It has been proposed that this phenomenon corresponds to an adaptation to the host immunity in which the parasites diminish their antigenicity. The "adapted" worms are smaller and produce more and different acetylcholinesterases than normal worms; some authors feel that the elevated synthesis of diverse isozymes must be triggered by specific antibodies and be related to the adaptation in some way. This hypothesis may seem far-fetched, but regulation of cell functions by interactions of molecules (antibodies in this case) with receptors on the external cellular membranes is a rather common phenomenon in biology. This is the way hormones, neurohumoral transmitters, antigens, and many drugs operate in the organism.

Effects of the Host Immune Response on the Course of the Infection

Three major events that affect the course of many nematode infections have been attributed to the immune response of the host: larval arrest, lactational rise, and expulsion of adult parasites. Recent work has demonstrated that the participation of the host immunity in these phenomena is more limited than originally believed, but it is still substantial in many instances.

The immunological mechanisms responsible for these events are not inherently different from those that affect the individual parasites; in fact, it is possible that the major difference is our way of detecting the consequences of the immunity, on individual parasites in one case and on the infection in the other. At any rate, the importance of these manifestations to the survival of the parasitic species and to its transmission to new hosts is so great that they deserve separate discussion.

Larval Arrest   It has been observed in many species of nematodes that a proportion of the infective larvae administered to a susceptible host proceeds to develop to adulthood immediately while another proportion may remain arrested, as larvae or juveniles, for prolonged periods of time. This phenomenon, compared to the diapause of insects by some authors, has been variously called larval arrest, arrested development, inhibition of larval development, or hypobiosis. An extensive review on the subject has been written by Michel (1974), and Schad (1977) commented on its role in the regulation of nematode populations.

Larval arrest has already been verified in more than 20 species of nematodes that affect man and domestic animals. It occurs at the third larval stage in the ascarids, the ancylostomids, *Nippostrongylus*, and *Trichostrongylus*; at the fourth in the trichonematids and most trichostrongylids; and affects the juveniles (fifth stage of some researchers) in the genus *Dictyocaulus*. The proportion of larvae that becomes arrested may vary from none to 100% of the infective dose administered, according to the particular circumstances. The length of the arrest has not been defined exactly; some workers think that it would be variable, and others believe that it must be predetermined. The available reports mention periods of several weeks to a few months as the most common.

Early experiments of infections of sheep with *Haemonchus* or with *Nematodirus* compared the proportion of larvae that arrested when a dose of infective larvae was given all at once to that when the dose was given in successive small amounts. Many more larvae became arrested in the second case, which was attributed to the production of a stronger immunity by the repeated antigenic stimulation represented by the multiple doses. Along the same lines, immunodepression of sheep with corticosteroids or with irradiation prior to infection with *Ostertagia* greatly reduced the proportion of larvae that arrested as compared to that in normal animals. These findings and others have been taken as evidence that arrested development is directly connected with the ability of the host to produce an energetic immune response against the invading parasite.

The observation that antihelmintic treatment to remove the adult nematodes (since arrested larvae are insusceptible to conventional antihelmintics) stimulated a part of the hypobiotic larval population to resume development fostered the notion that the adult parasites provided the antigenic stimuli that maintained the inhibiting immunity.

The importance of the host immunity in the induction and maintenance of larval arrest was weakened by the finding that a proportion of larvae entered in hypobiosis even during primary infections, before the host had time to mount an immune response. Also, immunosuppressive treatment of cattle did not change the proportion of *O. ostertagi* larvae that became arrested.

Noting that more larvae appeared to arrest in infections acquired during late autumn, Armour and Bruce (1974) investigated the effect of the environmental temperature on induction of larval arrest: incubation of infective *O. ostertagi* larvae at 5°C for periods of 3 to 19 weeks resulted in a higher proportion of arrested larvae on infection than when the larvae were fresh or had been maintained at constant 18°C. These findings have been considered as an indication that the cold weather of autumn acts on the infective larvae still in the pastures to trigger some arresting mechanism that will be expressed in the host after the infection. Similar findings have been reported in relationship to *H. contortus*.

Epidemiological observations in different parts of the world support the role of environmental factors in the induction of larval arrest, at least in the case of *Haemonchus* and *Ostertagia* infections. The precise triggering stimuli seem to vary, however, according to the various climatic conditions; cold, drought, photoperiodicity, or their combinations may be of importance in different areas. Larval arrest appears not to occur in zones with weather conditions that favor the survival of the free-living stages of the parasites during the whole year.

Other studies have demonstrated that the offspring of parasites that spent winter in the host (presumably in hypobiosis) are more prone to arrest than the offspring of parasites that spent winter in the pastures. This finding suggests that the ability of a population to arrest depends on its genetic make-up, and some investigators have defended the thesis that larval arrest is a property codified by the genetic material of the parasite. It seems equally probable, however, that the genetic characteristics of a subpopulation are responsible for the capacity of the parasites to respond or not respond to the stimuli triggering the hypobiosis rather than inducing arrest by themselves on every occasion. In fact, this alternative would provide more of the plasticity necessary for successful evolution.

Larval arrest in humans has been little studied, but Loos reported in the 1900s that *Ancylostoma* larvae remained arrested in the tissues of old people, and Brumpt in the 1950s found prepatent periods abnormally prolonged in some cases of human ancylostomiasis. A recent experimental infection with *Ancylostoma duodenale* in a human volunteer showed a prepatent period of 22 weeks instead of the 6 weeks considered normal for this species. Limited studies in India suggest that larval arrest in human ancylostomiasis is triggered by environmental factors (Banwell and Schad, 1978).

The recent report that *Strongyloides stercoralis* is transmitted transmammarily in humans may indicate that this species also arrests in humans. For this mechanism of transmission to have any biological significance, it is necessary for the parasite to wait in the host until lactation occurs, which clearly implies a period of arrest. In two other instances of parasites transmitted transmammarily (*Toxocara canis* and *A. caninum* of dogs) hypobiosis has been demonstrated satisfactorily. In the case of intestinal infections of humans with *S. stercoralis*, it has been found that treatment with immunosuppressive drugs reactivates asymptomatic infections. However, it is not clear whether this is due to resumption of larval development or to restoration of female fertility.

There is considerable controversy at present among researchers in regard to which is the primordial mechanism responsible for induction of larval arrest. Each of the three major hypotheses—immunological inhibition, environmental stimuli, and genetic make-up—has ardent defenders. These discussions are bound to continue for some time in the future, since all three theories have been confirmed in different host-parasite systems and occasionally, more than one mechanism has been demonstrated in a single system.

It does not appear necessary, however, to accept one mechanism with exclusion of the others. As Michel (1974) put it, "punctuality— defined as being at the proper place at the proper time—is the essence of successful parasitism." Prolonged survival of a parasite in the external environment or its quick transmission to a fresh host are more easily achieved at certain times of the year (spring or autumn) or in particular physiological conditions (postpartum or lactation, when a new crop of susceptible hosts is available). It is entirely conceivable that evolutionary pressures must have favored any mechanism that synchronized the production of infective elements by the parasite with these periods or conditions.

Since the best conditions for survival or transmission occur only once or twice a year in nature, and many gastrointestinal nematodes (the trichostrongylids in particular) maintain the peak of their fertility for only a few weeks, the synchronization must include a period of waiting in the comparatively favorable environment inside the host. This waiting period also must take place at a larval stage in order to avoid the senescence of the parasite and the consequent reduction in the number of infective elements produced.

There is a further evolutionary reason that must have favored any mechanism inducing larval arrest: persistence in time of the parasitic species and the host species depends on each partner being able to tolerate its associate without destroying it. Because of their reduced metabolic activity, hypobiotic larvae are less pathogenic than actively metabolizing parasites and are less susceptible to the host's defense processes. Again, any mechanism that induces the formation of parasites that are relatively inert from the biological standpoint should have evolutionary advantages and be selected for preferentially.

It is completely plausible, then, that multiple and different biological mechanisms directed toward the synchronization of the production of infective elements and/or the reduction of the host-parasite conflict have evolved convergently toward the same purpose. Arrested larval development may represent only the common end of those various mechanisms. It is probable that future research will show that all three factors—immunological, environmental, and genetic—and maybe others operate simultaneously, in diverse proportions, under different circumstances.

Lactational Rise   Many observations, begun in the 1930s, have shown that the number of eggs per gram of feces passed by sheep infected with gastrointestinal nematodes increases greatly from 4 to 8 weeks after parturition and falls precipitately after a couple of weeks. Since most of the original observations related this phenomenon to spring and, in southwestern Australia, to summer, it was called spring or seasonal rise. Later studies showed that it followed parturition consistently, so the name postparturient or periparturient rise gained favor. Recently, its relationship with lactation has been demonstrated unequivocally, so that the name lactational rise seems to be more exact (Connan, 1976).

At present, lactational rise has been reported in infections with *Haemonchus, Ostertagia, Trichostrongylus, Cooperia, Nematodirus, Bunostomum, Hyostrongylus, Oesophagostomum,* and other nematodes in sheep, cattle, swine, rabbits, and 20 species of ungulates of a zoo in Zurich. Some epidemiological studies in India indicate that it also occurs in human ancylostomiasis in relation to seasonal climatic changes. Although spring or lactational rise appears to be a general phenomenon that includes many species of nematodes in numerous species of hosts, most of the investigations have been done in sheep. As is the case with arrested larval development, research on lactational rise is full of controversial reports: most experimental findings have been confirmed by some authors and denied by others.

The immediate cause of lactational rise was originally determined by Morgan et al. (1951) by performing serial necropsies in infected sheep and finding a direct correlation of the number of adult parasites in the digestive tract with the quantity of eggs in the feces. Several later trials have confirmed this correspondence. Nevertheless, a recent critical study found that the increase in egg production was disproportionately high when compared to the increase in the number of adult parasites: this report indicates that the adult parasites not only expand in number during lactational rise but that their fecundity is also enhanced.

Determination of the remote causes of the enlargement of the adult parasite population has proved to be more elusive. A favorite approach to explain lactational rise has been to regard this event as a complement to larval arrest: it was proposed that the stress of parturition reduced host immunity, which, in turn, allowed the hypobiotic larvae to resume development to the adult stage. Observations that rise in egg production occurred to some extent in nonpregnant ewes and in males during the spring forced a search for alternative explanations. It was hypothesized, then, that the lack of ingestion of infective larvae during winter (absence of antigenic stimulation) or the poor quality of winter feeding (physiological deterioration) caused immunity to wane so that the hypobiotic larvae were released. These explanations were considerably weakened, however, with the findings that sheep that lambed in autumn also showed the postpartum rise in egg production. Besides, the phenomenon could not be induced by the administration of immunosuppressants or prevented by feeding concentrates to sheep that lambed in early spring.

Since some studies had failed to find arrested larvae during winter in sheep belonging to flocks that consistently experienced lactational rise in spring, it seemed necessary to evaluate the actual participation of hypobiotic larvae in the rise of egg production. A group of sheep was repeatedly treated with antihelmintics in an attempt to eliminate adult as well as hypobiotic parasites and then were left on pastures together with a group of untreated sheep. Examinations at the proper time revealed that the lactational rise was completely comparable in treated and untreated ewes. The authors concluded that arrested parasites did not play a critical role in the rise, and that most of the adult parasites responsible for it must have originated from newly acquired infections. Conversely, other experiments have shown that sheep kept in environments that preclude new infections from before parturition experience a lactational rise equal to that in sheep maintained in a pasture.

An important contribution was the finding that weaning immediately after parturition prevented the occurrence of lactational rise. All authorities now accept that the rise in egg production seen after parturition is directly related to lactation, but the controversy about its intimate mechanism still continues. Injection of prolactin or induction of its production by administration of stilbestrol or acetylpromazine in virgin ewes reproduces the rise only weakly, which suggests that the hormone does not act directly on arrested larvae.

Recently there has been a revival of the hypothesis that waned host immunity plays a major role in lactational rise, this time centered on cell-mediated immunity. It has been known for some time that cellular immunity is depressed in rats during lactation, and lately it has been reported that cell-mediated reactions are reduced in sheep undergoing lactational rise. Proof that both events are causally related is still lacking, however.

It seems fair to say that the only thing known for sure in respect to lactational rise is that it appears to be a complex phenomenon for which the mechanism of production has not been determined yet. Recently Michel (1974) has taken a wholly commendable approach by proposing that lactational rise may have multifactorial causes. Acquisition of new infections, resumption of the development of hypobiotic larvae, reduction of the normal rate of expulsion of adults from the digestive tract, and enhancement of fecundity have all been satisfactorily demonstrated to occur in connection with this phenomenon, and may contribute to it in diverse proportions, under different circumstances. Also, this notion expands the older idea that lactational rise is a complement to arrested larval development and presents it as an adaptative trait in itself. It seems more plausible that an adaptation as important for the survival of the parasitic species as lactational rise must draw from several sources rather than restricting itself to the simple regulation of another adaptation such as hypobiosis.

All the proposed factors of lactational rise may be influenced to a greater or lesser extent by the host immunity. Some investigators, however, believe that induction of larval arrest is mediated by environmental signals and that its termination occurs after a fixed period of time, which is codified by the genetic material of the parasite and probably was acquired during evolution.

The occurrence of a small rise in egg production in nonpregnant ewes and in males only in spring makes one wonder whether lactational rise and spring rise may be different adaptive phenomena, the former related mainly to the production of a new generation of fully susceptible hosts and the latter to the presence of the best season for survival of infective larvae in the external environment. Whereas the signal for lactational rise is an event normally connected with the new generation of hosts (lactation), the signal for spring rise may be an event normally related to spring, such as prolonged periods of daylight, which are known to influence the hormonal balance of the homoiotherms. As in the case of larval arrest, the importance of a rise in egg production at the appropriate opportunity is so great for the persistence of the parasite in time that convergent evolution of more than one mechanism toward this end seems probable.

At any rate, elevation of egg production by carriers of the infection, either in spring or shortly after parturition, facilitates transmission of the parasites to new hosts and, consequently, survival of the parasitic species. In the first case, the eggs find mild weather that favors the formation of a high proportion of infective larvae and its persistence outside the host, fresh grass is growing and the animals eat it avidly, and the winter months have allowed specific immunity to wane considerably in the animals that had experienced previous infections. In the second case, the infective larvae find a new generation of hosts that, by virtue of their age and their inexperience with the parasite, are fully susceptible to the infection. In most natural instances, both situations coincide in time in such a way that they assure that an adequate proportion of infective elements finds its way to establish the new generation of parasites.

For a veterinarian in charge of the health of a flock, spring or lactational rise poses special problems. Although the rise lasts only for about 2 weeks in any individual sheep, its duration in the entire flock may be close to 2 months, during which period the contamination of the pastures by the eggs passed by the ewes reaches dangerous levels. Lambs do not eat much grass until weaned, so that they normally acquire rather light infections at this time, but the eggs that they pass as a consequence of these infections contribute to increase even more the already considerable contamination of the fields. A cyclic situation develops thereafter: the lambs, which are still fully susceptible, acquire the infection at an increasing rate according to their appetite and to the degree of pasture contamination, and they keep adding more and more to this contamination from their own infections. Usually the heat and/

or the drought of summer restrain the development of eggs in the pastures of temperate areas, but the arrival of autumn, with rains and moderate temperatures, favors an explosive formation of infective larvae with the consequent massive infections and production of clinical disease in the flock.

Most veterinary epidemiologists recommend treatment of ewes at about parturition time to eliminate the adult parasites responsible for the lactational rise and the spring contamination of the pastures. It is probable that the timing of this treatment will be less strict in the future because of the development of new antihelmintics that can act on the arrested stages. In the absence of massive infections, treatment of lambs is commonly postponed until weaning, when the new animals are moved to clean pastures: treatment at this time will eliminate most of the adult parasites in the lambs and will delay and reduce considerably the contamination of the new fields.

Not enough is known about spring rise in human nematodiases yet, but it is probable that this phenomenon will be of importance in the measures of control of helminthiases in human populations once more information becomes available.

Expulsion of Adult Parasites   Most researchers are accepting now that, under normal circumstances, the gastrointestinal nematodes of adult hosts subjected to regular reinfections are in a constant process of turnover; some adult parasites are eliminated because of senescence or the host immune response, and new adult worms arise from hypobiotic larvae or from newly acquired infections. The total population of parasites at the different developmental stages remains fairly constant, however. Lactational rise, as seen above, is a phenomenon that is superimposed on the normal pattern of evolution of the nematode infection.

The "Self-cure" Phenomenon   The first evidence of immune expulsion of gastrointestinal nematodes was observed by Stoll in 1929. He noticed that sheep infected with *H. contortus* that were allowed to graze in contaminated pastures showed a violent suppression of egg production within a few days. This suppression was often accompanied by elimination of adult worms and by a strong resistance to reinfections. He called this phenomenon "self-cure." Extensive research done mainly in the 1950s produced the basis for our present understanding of the mechanisms of self-cure.

Soon it was found that self-cure could only be induced with living infective larvae: administration of death parasites or its extracts or adult worms did not cause expulsion of the preexisting population. Since the expulsion begins about the third day of the new infection, in coincidence with the time of moulting from the third to the fourth larval stage, it was postulated that the phenomenon was triggered by substances released during this moult. This thesis was supported by the demonstration that self-cure could be induced by introduction of exsheathing fluid directly into the abomasum of infected sheep.

The observation that self-cure occurred only in sheep that had experienced repeated infections suggested an immunological background for the event. This notion grew stronger when it was reported that self-cure was consistently followed by the appearance of specific, circulating antibodies against exsheathing fluid or extracts of infective larvae. Also, an inverse correlation was demonstrated between the quantity of eggs in the feces and the titers of complement-fixing antibodies in the serum.

The presence of edematous lesions in the location of the parasites being expulsed and of elevated levels of histamine in the blood of animals undergoing self-cure suggested that Type I allergy might play a role in the phenomenon. The presence of allergy was confirmed by the finding that these animals showed cutaneous hypersensitivity of the immediate type against exsheathment fluid, and its importance was assessed by the demonstration that antihistaminic treatment often prevented the expulsion of the parasites.

The current ideas about the mechanism of self-cure in *Haemonchus* infections of sheep is that repeated infections sensitize the animals against antigens contained in the moulting fluid in such a way that they form reaginic antibodies. Since the allergens are produced by larval stages during moulting, adult parasites may persist in the host without inducing any reaction. When a large new infection takes place, however, the production of moulting fluid during the early infection initiates a Type I allergic reaction in the abomasum that results in the typical tissue alterations of the local anaphylaxis. The immediate cause of expulsion of the worms is believed to be the changes in the physicochemical constants of the tissues, possibly assisted by motor activity, that renders the habitat inadequate for the continuous presence of the parasite. Based on the reported correlation between the suppression of egg production and the titers of complement-fixing antibodies, some authors think that manifestations of the Arthus phenomenon (Type III allergy) may also be involved.

This interpretation implies that, although the initiation of the self-cure reaction is an event that requires stimulation by a specific antigen, the effector mechanism is nonspecific and may affect more than one species of parasites. In fact, administration of infective larvae of *Haemonchus* to sheep harboring *Haemonchus*, *Ostertagia*, *T. axei*, and *T. colubriformis* resulted in the expulsion of all four species. Administration of larvae of *T. colubriformis* to sheep infected with *Haemonchus* and with *T. colubriformis*, however, caused only the homologous species to be expelled. The postulated explanation for this conflict is that, *Haemonchus* being a gastric parasite, the moulting fluid of the respective larvae moves down with the intestinal content and triggers anaphylactic reactions in the small intestine; since the moulting fluid of *T. colubriformis* is released in the small intestine, it does not have the chance to elicit the gastric anaphylaxis that would eliminate *Haemonchus*.

The most apparent characteristic of the *Haemonchus*-type expulsion of parasites, or self-cure phenomenon, is the requirement of a new larval invasion to provide the provocative stimulus. So defined, self-cure has been reported in *Cooperia* and *Dictyocaulus* infections of cattle, in *Haemonchus*, *Ostertagia*, and *Trichostrongylus* infections of sheep, in *Trichonema* infections of horses, in *Ascaris* and *Hyostrongylus* infections of swine, in *Toxocara* infections of dogs, in *Brugia pahangi* infections of cats, and in *Ascaris* infections of humans. The wide variations of egg production in human ancylostomiasis suggests that it may also occur in this infection. It is probable that the same phenomenon takes place in many other nematode infections for which reports have not been filed. It is far from clear, however, whether anaphylaxis is the major mechanism of expulsion in all cases.

It has been recently reported that self-cure of *Haemonchus* infections can be elicited by ingestion of freshly growing grass, and this was interpreted as demonstration of the presence of anthelmintic substances in the grass. The antigenic cross-reactivity of *Haemonchus* with soil nematodes was not considered, however.

Expulsion of Nippostrongylus brasiliensis   For some time it was thought that the findings in respect to *Haemonchus* expulsion were of general validity for all nematode infections. Recent research with *N. brasiliensis* showed that this is not the case. *N. brasiliensis* migrates from the skin of the rat through the lungs to the small intestine where ovoposition starts on the sixth day of infection. The adult parasites are spontaneously eliminated 12 to 15 days after infection. Ogilvie et al. (1977) have reviewed some recent work on the immunology of this parasite and listed previous publications.

Taliaferro and Sarles (1939) had already concluded that effective immunity against *Nippostrongylus* necessitated the activity of antibodies and the assistance of a cellular component. Modern techniques permit the isolation of humoral or cell-mediated responses in rats by transferring B cells or T cells from immune rats into immunosuppressed, isogeneic animals. Using these type of methods it has been possible to demonstrate that the expulsion of *Nippostrongylus* indeed requires a sequence of humoral and cellular responses.

The antibody activity becomes manifested toward the ninth day of infection and causes changes in the metabolism and in the secretion of enzymes of the parasites and degenerative alterations of their intestinal cells. The major immunoglobulin involved seems to be a subclass of IgG, and complement appears not to participate. Despite the evident damage, the parasites remain in the intestine until a cell-mediated response appears. At that time, they are rapidly eliminated. On the other hand, the exclusive presence of a cell-mediated response without previous antibody-mediated damage of the worms causes a very sluggish expulsion. In agreement with these experimental findings, the elimination of the parasites is very slow in young and in lactating rats, which are known to possess a deficient cell-mediated immunity.

The actual mechanism of expulsion is not known, however. An early hypothesis assumed that T lymphocytes damaged the parasites directly, but this notion had to be ruled out, since expelled worms are able to continue living when transferred to nonimmune rats. Another hypothesis maintained that cellular immunity allowed antibodies to leak into the intestinal lumen and reach the parasites; this is also improbable since antibody-mediated damage to the worms is previous to the manifestation of cell-mediated immunity and is insufficient to provoke expulsion. Based on the observation that antihistaminics such as promethazine prevented the elimination of the parasites, it was proposed that lymphokines would attract basophilic cells whose vasoactive amines would produce local alterations of the tissues, making them incompatible with the presence of the worms; this notion lost some value when it was demonstrated that promethazine also depresses cell-mediated immunity and that the maximum local concentration of basophilic cells did not coincide with the period of expulsion. Experimentally, it has been possible to induce the elimination of *Nippostrongylus* by administration of prostaglandins, and to prevent it by treatment with prostaglandin antagonists; this suggests that prostaglandins may play an important role in the elimination of the worms. However, the same results could not be confirmed by others or reproduced in *T. colubriformis* infection of guinea pigs or in *Trichinella* infections of rats.

Although we do not have at present a clear perception of the probable mechanism of expulsion of parasites in the *Nippostrongylus* system, the combination of antibody and cellular immunity creates a wealth of possibilities. Antibodies can affect the parasites directly or by activation of the complement system; alternatively, the combination of determined antibodies with specific antigen on the surface of mast cells or the activity of complement components could result in tissue modifications conducive to elimination of the worms. Apart from their role as regulators of the immune response, T lymphocytes may affect the parasites directly by means of cytotoxic cells similar to those described in tumors systems or by means of lymphokines. These latter also have the characteristic ability of attracting a large variety of cells that may exert a direct action on the worms (e.g., eosinophils, neutrophils, macrophages, K cells), or may produce antimicrobial enzymes and vasoactive amines with activity on the parasites or on the local tissues. The ubiquitous prostaglandins may have also some influence through their ability to cause tissue alterations or to regulate cellular function, particularly of mast cells and lymphocytes.

Figure 11, suggested by an original idea of Befus and Podesta (1976), summarizes these possibilities within an integrated view. It is interesting to keep in mind that none of these mechanisms has to operate independently and that, in all probability, the final result observed in the host is a combination, possibly synergistic, of several factors.

The most evident characteristic of the *Nippostrongylus*-type expulsion is its spontaneous production; the required antigenic stimulation is provided by the parasites already present in the host. From this point of view, the same phenomenon has been demonstrated to occur in *Trichinella* infections of rodents, *T. muris* infections of mice, and *Trichostrongylus colubriformis* infections of guinea pigs. In all three cases the sequence of a humoral and then a cell-mediated immune response has been satisfactorily confirmed. Still limited observations suggest that the same kind of expulsion happens in *Trichuris ovis* infections of sheep, *Trichuris vulpis* infections of dogs, *Strongiloides ratti* infections in rats, and *Oesophagostomum radiatum* infections in cattle.

As in the classic self-cure reaction, the effector phase of the *Nippostrongylus*-type expulsion appears to be nonspecific: infections with *Trichinella* accelerate the elimination of a previous infection with *Trichuris muris* in mice and reduce the establishment of an ulterior infection with *Nippostrongylus* in rats. An attractive verification of the specificity of the triggering stimulation and of the nonspecificity of the effector activity has been produced with trichostrongylids of sheep. Infection with *Trichostrongylus colubriformis* elicits strong immunity against the homologous parasite, some resistance against challenges with *Trichostrongylus vitrinus*, and no protection to infections with *Nematodirus*; if reinfection is attempted with the three species simultaneously, however, the same strong resistance to *Trichostrongylus colubriformis* is expressed against the other two species.

A disturbing note is the report that resistance against *Nippostrongylus* and *Trichinella* in rats and in mice is transmitted from mothers to offspring by milk. Since suckling rats and mice are deficient in cell-

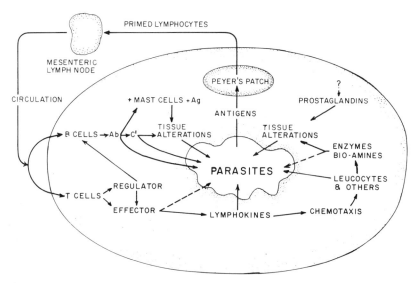

Figure 11.   Possible immunological effector mechanisms in the expulsion of digestive nematodes. Production of parasite antigens results in the presence of specifically committed B and T cells in the digestive mucosa. Antibodies produced by B cells may affect parasites directly or through complement activation; this latter and the activity of anaphylactic antibodies combined with antigen on mast cells may also cause alteration of the physicochemical constants of the tissues, which may become inadequate to sustain the presence of the parasite. T cells act as regulators of the immune response and as effectors of cell-mediated immunity; these latter may affect the parasites through direct cytotoxic activity or by production of lymphokines. Lymphokines may act on the parasites themselves or attract cells that, in turn, may attack the worms (e.g., eosinophils) or produce compounds with deleterious activity on the tissues and probably on the worms. Prostaglandins are known to cause tissue modifications and to be able to regulate cell functions, particularly on mast cells and lymphocytes.

mediated responses and both humoral and cellular immunity are required to expel the parasites, we must assume that, besides antibodies, milk transfers T lymphocytes or some element that replaces them.

Expulsion—A Self-regulatory Mechanism?   The biological significance of the immune expulsion of nematodes appears to be critical for the survival of the parasitic species. A fundamental condition for successful parasitism is the mutual tolerance between parasite and host. Certainly a population of parasites that accumulates in its host without restraint will soon reach a density that will endanger the life of the host and, consequently, the survival of the parasitic species. The phenomenon of expulsion of parasites, either spontaneously or induced by reinfection, suggests that nematodes have developed mechanisms of self-regulation during evolution that maintain the size of the population within limits compatible with the life of the host. The fact that they utilize the host immunity to this end is only an example of how remarkable biological integration is.

Forced by the rigorous demands of scientific research, most contemporary investigations of nematode immune elimination have made use of heavy or massive infections in fully immunocompetent hosts. The general result has been the production of strong and effective immune reactions that frequently eliminate the parasites completely. The situation seems to be somewhat different in nature: most animals that normally are subjected to constant reinfections possess a moderate number of gastrointestinal parasites but rarely are totally free of them. Even in those cases of human nematodiases in which the contribution of new infections to the preexisting parasitic load is negligible or can be ruled out, a small number of parasites persists in the digestive tract for prolonged periods. It seems evident that the mechanisms of elimination of gastrointestinal nematodes in nature are less efficient than in their artificial reproduction in laboratories.

There are at least two ready explanations to account for this difference. It is well known in immunology that the administration of large amounts of antigen to a subject whose lymphoid system is still immature often results in tolerance to the particular antigen. Periodic administration of the antigen can maintain tolerance almost indefinitely. In most cases of animal parasitism the host is subjected to infections from very early in life, prenatally in some instances. Also, the inability of sheep to produce immunity to *Haemonchus* and *Trichostrongylus* before 4–6 months of age and the slow development of effective immunity to *Ostertagia* in cattle, which requires 5 months of constant infection, suggest that the maturation of the immunocompetence to gastrointestinal nematodes is a sluggish process. It is not unreasonable to speculate that precocious infections in only partially immunocompetent hosts may produce some degree of tolerance, perhaps directed toward certain antigens important for protection, in such a way that later manifestations of immunity will have a reduced efficacy.

On the other hand, Dineen (1963) has interpreted some instances of antigen sharing among host and parasite as an indication that parasites tend to develop "fitness antigens" during evolution (see Chapter 7). These are antigens rather similar to the host's components that would elicit an immune response only after reaching a threshold concentration. This postulate would explain why a small population of parasites does not promote expulsion, whereas a large population may induce immune reactions that will be maintained only until the number of parasites is again below the reactive level. This mechanism should favor the presence of light infections such as the ones we see in nature.

Practical Aspects of Immunity to Gastrointestinal Nematodes

Everyday observation indicates that mammals do develop resistance
to reinfection with digestive nematodes. This resistance is manifested
in various ways: fewer parasites succeed in establishing definitively
in the host; many of those that do become physiologically impaired in
a way that prevents the full expression of their pathogenic and reproductive capacities; the rate of expulsion of adult parasites is increased;
and, eventually, the actively metabolizing stages are eliminated. It is
now known with certainty which of the individual manifestations of
immunity is responsible for each of these effects. Furthermore, it is
possible that the immunological mechanisms of defense are not stable,
but instead vary according to the diverse circumstances of the infection. Recent work with *Nippostrongylus* in rats revealed that the parasites are eliminated from the intestine after 12 to 15 days in a first
infection, they are expelled as soon as they get to the intestine (about
the sixth day) in a second infection, and most of them are destroyed
before reaching the intestine in a third infection. Since the adults expelled after 12 to 15 days are still viable enough to survive when transferred to nonimmune rats, we must assume that the destruction during
migration in a third infection is due to a mechanism inherently different
from the mechanism of intestinal elimination.

The success obtained by vaccination with enzymes of parasites
in the cases in which it has been attempted suggests that neutralization
of the activity of essential enzymes may be an effective manner of
attacking the parasite. Certainly interference with enzymes involved
in the feeding processes of hematophagous nematodes will explain satisfactorily the reduction of the pathogenicity of the infection.

Although immunity to gastrointestinal nematodes may develop
rapidly in some instances (e.g., *Nematodirus* in sheep), in most cases it
is a rather slow process. Five months of constant exposure to infection
are required for cattle to produce effective immunity against *Ostertagia*; although most dogs are born already infected with *Toxocara*,
evidence of protective immunity is commonly seen only after 3 months
of age; sheep are unable to mount an immune response against *Haemonchus* or *Trichostrongylus* during their first 4–6 months of life. Resistance to these infections approaches premunition, since it wanes
very rapidly in the absence of parasites. In ovine haemonchosis there
is no evidence that a primary infection establishes immunologic memory; in bovine ostertagiasis, reinfections produce a typical anamnestic
response, but the arrested larvae remaining in the host during winter
do not seem able to maintain the immunity. In bitches harboring ar-

rested larvae of *Toxocara*, immunity against the intestinal infection is very effective, but in this case, besides conventional immunity, there are factors related to age and sex that have not been evaluated completely yet.

Immunity against these organisms appears to be the result of a delicate physiological balance, since episodes of physiological stress, such as periods of malnutrition, parturition, lactation, accelerated production, and concurrent diseases, tend to break the herd immunity and again result in clinical manifestations of parasitism. Massive infections by the specific parasite against which the host is resistant occasionally may also overcome the established immunity.

Veterinarians in charge of animal populations must be aware of these peculiarities of the immunity to gastrointestinal nematodes: it takes a long time to build up, its maintenance is precarious in the presence of stress conditions, and it wanes rapidly in the absence of regular reinfections. Nevertheless, proper management and the use of antihelmintics when emergency situations arise will keep the immune mechanisms operative, which eventually may be the ultimate goal in the control of parasitoses.

There is no reason to think that control of human digestive parasitism on a populational basis will not respond to a similar immunoepidemiologic approach, but there is still a long way to go in gathering information before any rational plan of attack based on immunology could be formulated. The picture of immunity against digestive nematodes that slowly is emerging seems to indicate that, contrary to traditional views, every parasite stimulates the production of protective responses at a characteristic rate and with a specific efficacy. It seems important, therefore, to regard the infection by each parasitic species as an individual event for purposes of immunological control.

ARTIFICIAL PRODUCTION OF RESISTANCE

The great economic and health importance of gastrointestinal nematodiasis of man and domestic animals prompted investigators to try methods of vaccination soon after production of resistance against these infections was recognized. Clegg and Smith (1978) have done a detailed review of the use of nonliving immunizing preparations.

After a successful attempt to induce immunity in rats by injections of killed larvae of *Strongyloides ratti* in the 1930s, many other assays with killed parasites or with extracts were either ineffective or produced only negligible protection from challenges. These results suggested that only living parasites produced the antigens important for protection in quantity or quality sufficient to induce effective immunity.

Recent studies of vaccination of guinea pigs with extracts of *Trichostrongylus colubriformis* have shown that preparations of whole worms may produce an important degree of resistance, but the source of antigens must be carefully selected: homogenates of adults or of fourth-stage larvae gave strong immunity, but extracts of third-stage larvae were ineffective.

The early observation that serum of immune animals produces immunoprecipitates at the natural orifices of nematodes when incubated together indicated that antibodies were directed against secretions or excretions of the parasite and suggested that they might interfere with vital functions of the worm. Attempts to vaccinate with parasite secretions obtained by in vitro incubation soon followed. This method has been successful in a number of nematode infections, such as *Nippostrongylus* in rats, *Trichinella* and *Trichuris* in mice, *Trichostrongylus colubriformis*, *Ascaris suum*, and *Dictyocaulus viviparus* in guinea pigs, *Strongyloides papillosus* in rabbits, and *H. contortus* in sheep. The origin of the protective antigens may differ markedly among the different nematodes, however; the important antigens of *Haemonchus* are released during moulting, whereas those of *Trichostrongylus colubriformis* are secreted by the excretory and esophageal glands present in the fourth-stage larvae and in the adults. In *Trichinella* and *Trichuris*, protection has been produced by inoculation of the granules present in the periesophageal glands (the stichocytes), the normal function of which is not known. In the case of *Ascaris*, some protection has been obtained by vaccination with malic dehydrogenase, with aminopeptidase, and with aldolase of parasitic origin.

A comparative look at the results, however, seems to indicate that protection is more effectively induced against those parasites that migrate extraenterally than against the parasites of exclusively digestive location. As an example, six daily injections of secretions of adult *Trichinella* caused a marked reduction in the number of muscle larvae on challenge, but a longer course of injections was required to show a protective effect against the establishment of adults in the intestine.

It seems logical to assume that the migratory parasites are more subjected to the action of the immune response, in virtue of their more intimate contact with the tissues of the host, than the parasites restricted only to the digestive lumen. However, the site of stimulation with the antigens, which possibly may direct the locations where the immune response is expressed preferentially, may be critical in this regard according to recent findings: an experiment showed that induction of protection against *Nematospiroides dubius* (*Heligmosomoides polygyrus*) infections necessitated 10 times more larvae when the dose was administered subcutaneously than when it was given orally. This report supports the notion emerging from recent findings that effective immunity may occasionally be local rather than systemic. There is a growing consensus among the specialists that the immune responses that occur in the gastrointestinal tract are often produced locally. Despite these encouraging results, the present methodology for in vitro culture of parasites is not sufficiently advanced yet to provide secreted antigens in the quantities necessary for practical use.

The reports of cross-protection among infections with nonrelated parasites in laboratory animals encouraged a novel approach to immunization against parasitosis—the attempt to produce immunity against pathogenic nematodes by infection with or inoculation of free-living worms. Vaccination of puppies with the free-living nematode *Eucephalobus* reportedly elicited some resistance against challenges with *A. caninum*, and infection of sheep with the free-living nematode *Rhabditis axei* produced 75% to 90% reduction of the development of *Dictyocaulus filaria* larvae used for challenge. Curiously the latter experiment gave inconclusive results when repeated with guinea pigs. This exciting line of research does not seem to have been pursued lately. The selection of the proper nematode to elicit effective immunity requires actual immunization and challenge, since serological cross-reactivity is not a valid indication of cross-protection: extracts of *T. canis*, *H. contortus*, and the free-living nematode *Caenorhabditis briggsae* react with anti-*A. suum* serum, but none of them confers resistance in guinea pigs against challenge with *Ascaris*.

The use of limited infections in order to produce immunity without causing pathology is not a satisfactory method for most gastrointestinal nematodiases, since the prolonged period required for an effective immune response will allow contamination of the pastures in the interim. Utilization of infective larvae irradiated in such a way that they will develop to sterile adults will solve this problem, however. This approach has been very successfully applied to the control of infections by *A. caninum*, *Dictyocaulus*, and *Syngamus tracheae*, as is explained in the section on tissue nematodes.

Limited studies of vaccination of sheep with irradiated larvae of *H. contortus* and *T. colubriformis* achieved a high degree of protection against the homologous challenge. Unfortunately, lambs do not become immunocompetent to these nematodes until they are about 4–6 months old, which largely invalidates this procedure in practice. On the other hand, vaccination procedures of sheep that do not include control of *Haemonchus* and *Trichostrongylus* would be of little use in most circumstances.

IMMUNODIAGNOSIS

Since the immune responses to gastrointestinal nematodes develop rather slowly and commonly appear after the prepatent period is over, there has been no incentive to devise methods of immunodiagnosis of practical application. For research purposes, virtually any sensitive test will be adequate to demonstrate immune responses.

IMMUNOPATHOLOGY

The participation of IgE antibodies in the production of symptoms and the possible influence of helminth infections in the regulation of allergic conditions have already been mentioned in relation to production of antibodies. The role of immunity in the genesis of tissue damage is still under study, as explained in connection with the mechanisms of expulsion of the parasites from the digestive tract. Based on histological evidence, the damage caused by *Oesophagostomum* is believed to be due to a cell-mediated reaction.

Recent studies have shown that several helminthiases are able to reduce the capacity of the host to mount an immune response to antigens not related to the parasite (see Chapter 7). All of these are tissue-dwelling organisms; the only strictly digestive helminth that has produced circumstantial evidence of being able to affect the host immunoresponsiveness is *H. contortus*, and confirmation of this finding is sorely needed.

NEMATODES OF THE TISSUES

THE MIGRATORY LARVAE

A number of nematode larvae are able to invade and survive in abnormal hosts, but do not develop beyond the infective stage. Apparently, these hosts provide the physicochemical stimuli required to trigger the invasion by the larva and the materials for the subsistance of the parasite, but lack the proper signals to direct its further growth. Traditionally, these parasites and the conditions produced by them have been called *larvae migrans*.

In most cases, larva migrans are a source of discomfort or mild disease, but they infrequently produce dramatic systemic consequences. For these reasons, the corresponding infections have remained mostly within the realms of human medicine, although they are frequent in domestic animals. The major conditions of humans in the Americas are *cutaneous larva migrans*, a dermatologic affliction caused by the migration of the larvae of *Ancylostoma braziliensis* of carnivores and, less frequently, by other ancylostomids; *visceral larva migrans*, a systemic infection produced by *T. canis* of dogs and, occasionally, by related ascarids; *anisakiasis*, a gastrointestinal infection by larvae of *Anisakis* and related nematodes that normally parasitize marine mammals and whose larvae develop in sea fish and invertebrates; and *abdominal angiostrongylosis*, the invasion of the arteries of the intestinal wall by the larvae of the rat nematode *Angiostrongylus* (*Morerastrongylus*) *costaricensis*. Other larva migrans, notably *Angiostrongylus cantonensis* and *Gnathostoma* spp., either do not occur in the Americas or are sporadic findings.

A few intestinal nematodes of humans, such as *Ascaris*, *Ancylostoma*, and *Strongyloides*, have a phase of migratory larva as a part of their usual life cycle in the normal host. This does not constitute larva migrans in the usual context, but, since the immunology and clinical picture of this phase are quite different from the corresponding events of the intestinal stages, these larvae are incidentally included in this discussion.

Natural Resistance

Little is known about the natural resistance to migratory larvae. The factors influencing the susceptibility to infection with intestinal nematodes that possess a migratory phase were discussed in the preceding section. In the case of larva migrans, these conditions are sporadic by definition, and it is often difficult to establish their definitive diagnosis; as a consequence, we do not know what proportion of the population at risk actually acquires the infection. In these conditions, any attempt to evaluate natural resistance would be widely speculative.

There is no doubt that a large part of the susceptibility to larva migrans must be connected to *behavioral characteristics of the host:* cutaneous larva migrans will occur only when naked skin comes in contact with contaminated soil; visceral larva migrans requires oral ingestion of eggs in the dirt; anisakiasis will result from the consumption of raw seafood, and so on. Individuals who regularly avoid these activities will not have the ecological opportunity to acquire the respective infection.

There is some evidence that *age resistance* may develop against migratory larvae. Older rats are more resistant to the skin-penetrating nematode *Strongyloides ratti* than are young animals, which has been demonstrated to be related to the higher polymerization of the acellular elements of the skin of the former. The susceptibility of dogs to a primary infection with *A. caninum* or with *T. canis* diminishes with age: in older animals the infective larvae become disseminated in the somatic tissues instead of migrating to the intestine, and do not proceed in their development. Sexual hormones must participate in this phenomenon, because the resistance occurs earlier and is stronger in females, and wanes during late pregnancy and lactation. Acquired immunity has a similar inhibitory effect on the migrating larvae, which brings up the question of whether the natural resistance is actually an increased ability of older animals to mount a rapid and effective immune response or is simply a manifestation of nonimmune mechanisms. Studies with *S. vulgaris* of horses have demonstrated that 3-year-old animals exhibit a resistance to primary infections that prevents the completion of the life cycle of most of the infected larvae administered.

Actually, we do not know whether age resistance also occurs in abnormal hosts. Young mice are more susceptible to *T. canis* infections than older mice and visceral larva migrans of humans affects almost exclusively children below 4 years of age. This latter phenomenon, however, is usually attributed to the greater chances of infection in children. Serological tests and cases of ocular toxocariasis have shown that the infection also exists in adults, but it is not known whether these are persistent parasites acquired as infants or what proportion of the population at risk these individuals represent. Cutaneous larva migrans occurs in subjects of any age, but the ratio between exposure and actual infection is not known, so that no reasonable conclusion can be drawn. It has been often stated that the inability of *A. braziliensis* to penetrate beyond the stratum germinativum of the human skin is a manifestation of natural resistance. This may well be so, but transcutaneous infections with this species are not too efficient in dogs either; besides, there is evidence that this parasite may actually reach the lungs of human patients occasionally (see below).

At least some of the natural resistance appears to be due to *parasite factors*. Studies with infective larvae of *A. braziliensis, A. ceylanicum,* and *A. caninum* demonstrated that the first species penetrated human skin more readily than the other two. When these experiments were repeated with puppies, the natural host, the same phenomenon was verified. This finding suggests that the ease with which the parasite penetrates the skin is an inherent characteristic of the larva, little affected by the host species. This may explain why the cutaneous infection with *A. braziliensis* is much more frequent in humans than the equivalent infection by *A. ceylanicum* or *A. caninum:* a brief exposure that will permit the invasion by the former species may be insufficient for the latter ones to initiate a successful infection.

Larva migrans appears to be an adaptative trait of some parasitic species in which humans are unwilling participants. The presence of dormant infective larvae in prey animals (e.g., rodents, fish) rescues the parasite from the vicissitudes of the climate and directs it irremediably to the final predator host (e.g., carnivores, cetaceans). As an efficacious mechanism of species survival, this characteristic must have been selected for during evolution. Humans became accidental intruders in this cycle who certainly have not added significantly to the odds of the parasite completing its life history in the last few thousand years. Evolutionary selection may explain why these parasites developed such a broad specificity that allows them to recognize the physicochemical conditions of the digestive tract of many species as adequate signals for invasion. It would be of interest to study whether the infective larvae of larva migrans–producing nematodes are activated by basic physiological events likely to exist in ample groups of prey animals.

Acquired Immunity

The immunity of cutaneous larva migrans, a benign and transient, although uncomfortable condition, has not been studied with any thoroughness. Some authors believe that symptoms serious enough to cause the sufferer to seek medical attention would not occur on primary infection, but only in previously sensitized individuals. Patients affected by *A. braziliensis* commonly present a pruritic creeping eruption that advances a few millimeters per day; the lesions show an abundant infiltration of eosinophils and the patient usually exhibits peripheral eosinophilia; subsequent infections are more severe. These characteristics have been interpreted, and probably rightly so, as manifestations of acquired immunity. The lesions normally subside spontaneously in less than 3 months, which has been traditionally attributed to development of immunity lethal to the larvae. It is equally probable, however, that the larvae die because of metabolic exhaustion.

Recent studies cast some doubts on the efficacy of a possible immune reaction to kill the larvae in the epidermis: in a series of 26 patients, 9 (35%) showed a pulmonary syndrome with eosinophils or Charcot-Leyden crystals in the sputum that followed the spontaneous cure of the skin condition. This and other findings suggest that some larvae may continue their migration to the lungs after a period in the skin. No parasites have been recovered from the intestine in these cases, so it must be assumed that the larvae that reach the lungs are retained, killed, or expelled at this level. The characteristics of the cutaneous and pulmonary manifestations suggest that Type I allergy is an important component of the immune response to this parasite.

Cutaneous larva migrans caused by *A. caninum* is more transient (about 2 weeks) and the lesions are often nonlinear papules and bullae with indentated edges, rather than the linear eruptions typical of *A braziliensis*. The notion has been offered that *A. caninum* may penetrate deeper in the skin than *A. braziliensis*, thus provoking qualitatively different structural alterations. This interpretation does not agree with the dynamics of the skin migration of both species as determined experimentally. On the other hand, *A. caninum* appears to elicit stronger reactions in the skin of its normal host than *A. braziliensis*, since transcutaneous infections in dogs with the former species result in a very small proportion of adult parasites developing in the intestine. It is possible that the shorter course and the different pathology caused by *A. caninum* are due to the fact that this parasite stimulates the production of predominantly different immune responses (Type III allergy?) in the host.

A type of cutaneous larva migrans, commonly called *larva currens*, in which the parasite progresses very rapidly (up to 10 cm/hr) has been found in consistent association with intestinal infections by *S. stercoralis* (Stone, Newell, and Mullins, 1972). Although the larvae have not been definitively identified yet, this association and the fact that the lesions typically begin in the perianal region have convinced most authors that these are instances of cutaneous strongyloidiasis. On a number of occasions, it has been reported that larva currens appears subsequent to or during an episode of intestinal strongyloidiasis, and may recur periodically for up to 40 years. Several theories have been proposed to explain this cyclic invasion of the skin. Recent findings of recurrence of intestinal strongyloidiasis, sometimes fatal, in patients undergoing immunosuppressive treatment have demonstrated indirectly that *S. stercoralis* may persist for long periods as an asymptomatic infection in humans, apparently kept inhibited by the host's immune response. This phenomenon may correspond to the somatic larval arrest of *T. canis* and *A. caninum* (see below). The episodes of cutaneous strongyloidiasis may be caused by larvae that accidentally arrive in the skin and find a reduced efficiency of the inhibiting immunity in the avascular layers. They may be the cutaneous (and active) counterpart of the somatic arrested larvae that are kept dormant by an effective immune response.

Immune responses to skin-invading larvae also develop in the normal host, although more slowly than in abnormal ones: a primary infection of a man with infective larvae of *Strongyloides ransomi* of pigs initially resulted in the appearance of vesicles in the skin; after repeated infections the subject developed intense inflammation to the challenge, with formation of pustules. The repeated skin infection of a human volunteer with the human hookworm *Necator americanus* did not produce cutaneous symptoms initially, however, but a creeping eruption developed after later administrations. Primary infections of the normal hosts with *S. ransomi* of pigs, *S. papillosus* of sheep, and *S. ratti* of murides caused no external evidence of dermatitis, although some neutrophilic infiltrate occurred near the larvae after 24 hr. A fourth exposure in lambs produced slight cutaneous erythema, and a pustular dermatitis appeared after the fifth and subsequent exposures. After the 11th exposure, every larva was surrounded by an intense area of inflammation containing eosinophils, neutrophils, lymphocytes, and giant cells: active phagocytosis and destruction of the larvae in the inflammatory foci were observed. Clinical observations of hookworm disease in humans and dogs also indicate that the skin reactions are more severe with the reinfections, most probably because of secondary immune responses. Although there is no solid evidence that the skin reactions are lethal to the larvae in human and canine infections, the studies mentioned above and the clinical experience suggest that this may be so.

In regard to somatically migratory larvae, possibly our most detailed knowledge is that of the immune response to *A. suum* in laboratory animals, particularly through the work of Soulsby's group in Philadelphia (Khoury and Soulsby, 1977). Primary infection of guinea pigs demonstrated the production of humoral and cell-mediated responses in the lymphoid structures draining the organ invaded by the parasite at that time, with a minimal systemic distribution of the antigens, during the first 12 days of infection. The antibodies produced were predominantly IgM and IgE in the regional lymph nodes and IgM in the spleen. Subsequent infections produced more rapid and intense, but still predominantly local, responses: IgA was the major antibody in the mesenteric lymph nodes, IgG in the hepatic node, and IgE in the mediastinal node.

A. suum does not correspond to visceral larva migrans in the guinea pig, since the parasite moults at least once in this case. In any case, these studies are restricted only to the early infection (12 days), and differences have been found with similar infections in mice. Extrapolation of these findings to the human infection, therefore, must be only tentative.

Human infection by *T. canis* produces low fever with hepatomegaly, eosinophilia, hypergammaglobulinemia, and symptoms of localized tissue damage. The latter is particularly frequent in the liver, but can also affect other organs (e.g., lungs, heart, kidney, central nervous system). Ocular lesions commonly are sequelae that do not coincide with the acute syndrome. Besides production of nonspecific immunoglobulins, the infection results in formation of a variety of antibodies of the IgM, IgG, and IgE classes and in manifestations of cell-mediated immunity. Histological studies of the lesions have demonstrated the formation of granulomas with an abundant eosinophilic infiltration around the larvae.

The infection of pigs with *T. canis* may be an appropriate model for the human infection: larvae and an eosinophilic-lymphocytic infiltration were seen in the ileum of this host 2 days after a massive infection (250,000 embryonated eggs); the inflammation subsided and disappeared by the 32nd day. Destruction, hemorrhage, and leukocytic inflammation were seen in the liver beginning on the second day; by the fourth day eosinophilic granulomas started to form around the larvae to become established by the end of the first week; after a couple of months, the granulomas had resolved and there was accumulation of fibrous tissue with some infiltration by eosinophils. Larvae in the lungs were found on the fourth day; by the eighth day foci of necrosis and of eosinophilic-lymphocytic infiltration and the beginning of granuloma formation were observed, although these latter did not constitute completely until the 32nd day; healing was apparent by the 49th day and encapsulation of the remaining larvae was completed by the 64th day. Myocardial lesions followed approximately the course of the infection in the lung, but were less eosinophilic. Infections by *T. canis* in sheep and dogs have produced essentially similar results (Poynter, 1966).

It appears from these findings that the intensity of the lesions is proportional to the local concentration of antigen and, by their characteristics, that they might be the result of Type I plus Type IV allergic responses. In these animal models, the primary infection subsides spontaneously in about 2 months, whereas the human disease commonly lasts 3 to 18 months. This may be an indication that the clinical condition in humans actually occurs only in the case of reinfections, the immune response of the host being exacerbated by the previous sensitization.

Despite the occasional observation of remains of parasites in the host's lesions, there is evidence that the immune response is not lethal to most of the larvae: living, infective parasites are recovered from the tissues of rodents several months after the original infection; infected bitches may transmit the parasites of an infection congenitally to at least 3 successive litters; and larvae are known to have survived 7 years in the livers of two infected monkeys. Serological surveys in adult human populations have found about 10% positive results and many of the cases of ocular toxocariasis occur in adults; presumably these are instances of persistent infections acquired during infancy.

Protection against reinfections is arduous to study because of the difficulty of differentiating between the parasites from the original infection and those from the challenge. Several attempts at immunization by infections with eggs have indicated that little or no protection against challenges is produced. An experiment of immunization of mice with large amounts of parasite extract in Freund's complete adjuvant, however, resulted in a significant resistance to a challenge infection. This may indicate that the parasite normally releases functional antigens in a quantity insufficient to stimulate an effective protective response. Since production of functional antigens has been connected with the moulting process in several nematode infections, the failure of *T. canis* to develop beyond the infective stage in the cases of somatic migration may account for its reduced immunogenicity.

On the basis mainly of work with *A. suum* in guinea pigs and with *A. caninum* in dogs, it is often implied in the literature that the host's immunity arrests the development of *T. canis* in reinfections. This phenomenon effectively occurs in the natural host, but there is no proof of its immunological nature. The wide array of mammalian species that will carry arrested *T. canis* larvae after a primary infection (including adult dogs) instead suggests that hypobiosis is a characteristic of the *T. canis* infection that depends on nonimmune mechanisms, since immunity rarely reaches such an effective level in such a short time.

The immunity to canine hookworms has been recently reviewed by Miller (1971) and that to human hookworm by Banwell and Schad (1978).

The best controlled experiments in immunity of dogs to *A. caninum* have been performed by using irradiated infective larvae to elicit immunity, and normal infective larvae to determine the protection produced. Two infective routes have been tried: subcutaneous inoculation, in which all the larvae migrate extensively in the body before reaching the intestine; and oral administration, in which most of the larvae develop in the gut without any extended contact with the somatic tissues. Combination of immunization and challenge by both routes has demonstrated that subcutaneous vaccination is a more effective inducer of immunity than oral vaccination, and that subcutaneous challenges are more affected by the preexisting immunity than oral challenges. These findings suggest that the migratory parasites rather than the intestinal forms are the stages that are more efficient at producing immune protection and the most susceptible to it. Since the main difference between a primary infection with normal and one with irradiated larvae appears to be that most of the parasites will reach the intestine in the first case and only about a fourth of the dose will do so in the second case, the findings with irradiated larvae are probably valid also for the natural infection.

The protection afforded by the vaccine is not affected by removal of the few parasites that developed in the intestine. In agreement with the suggestion above, this resistance therefore must be maintained by parasite stages arrested in the somatic tissues.

Clinical and experimental observations have demonstrated that an infected bitch will pass decreasing numbers of worms to her successive litters (mainly through the milk of the first 2–3 weeks), even if she is kept in a contaminated environment. This has been interpreted as an indication that the larvae of the reinfections do not accumulate in the infected bitch and implies that they are destroyed by the immunity elicited by the primary infection. Immune protection does not seem to be effective, however, against the arrested larvae of the first or the few first infections, since they survive in the female and are reactivated during late pregnancy and lactation to cause perinatal infections of the puppies. The reason for this difference is not known, but it might be related to the reduced metabolism (and consequent diminished production of functional antigens) of the dormant larvae.

Besides causing arrested development and possibly destruction of the migratory larvae (for both of which there is at least circumstantial evidence), the protection from *A. caninum* reinfections is evidenced by a strongly reduced parasite burden in the intestine, by a diminished ability of the few adult worms developed to suck blood, and by the increased ability of the host to compensate for blood losses (bone marrow stimulation by the immunity-producing infection?). Some reports indicate that the fecundity of the few females that achieve establishment and the fertility of their eggs are also affected.

Human infection by hookworms (especially *Necator americanus*) is known to induce antibody formation (including IgE) and cell-mediated immunity (detected by lymphocyte blastogenesis). There is no solid evidence, however, that a primary infection affords protection from subsequent infections. In endemic areas, worm burdens increase with the age of the patient in children, remain rather stable from the second to the fifth decade of life, and then increase again in older people. This and the low level of infection characteristic of treated and reinfected adults have been interpreted by some authors as evidence of immune protection. In one experiment, however, no proof of resistance was observed in a volunteer challenged periodically with *N. americanus* over a 4-year period.

Fluctuations of egg output in infected human populations related to climate, abnormally prolonged prepatent periods in naive and previously infected volunteers, and reports of infections in infants that are best explained by assuming transmammary transmission have suggested to some researchers that arrested development occurs in human ancylostomids. According to the scant information available, this phenomenon would be triggered by climatic conditions rather than by the host's immunity (see "The Nematodes of the Digestive Tract"). *Ascaris lumbricoides* infections in humans also induce antibody production (especially IgE) and cell-mediated immunity, but there is no evidence of protection to reinfections in humans, although resistance occurs in pigs infected with *A. suum*. Unlike *T. canis*, the granulomas formed around *Ascaris* migratory larvae are lethal to the parasite.

The recent findings that human strongyloidosis is transmissible transmammarily and is reactivated by immunosuppressive treatment point very strongly to the existence of immunologically mediated hypobiosis in this infection. The significance of larval arrest and of the larval reactivation in the perinatal period has already been commented on in connection with the nematodes of the digestive tube.

The immunology of *Anisakis* infections in humans is just being studied. Smith and Wootten (1978) have published a review that includes this subject. *Anisakis* infection stimulates the formation of a variety of antibodies, including IgE. There is some preliminary experimental evidence that cell-mediated immunity may also occur. Nothing is known yet about resistance to reinfection, although, at least in some cases, challenges appear to produce more pathology than the primary infection.

Mechanisms of Acquired Resistance

Protection against migrating *T. canis* larvae in the dog appears to be directed against the infective larval stage, since intraintestinal infection of previously infected and treated puppies with infective larvae did not result in patency, whereas fourth-stage larvae administered in the same way completed their development to adults. Besides, the serum from immune dogs formed a precipitate in the oral orifice of infective larvae, but not in the fourth-stage parasites. (This experiment considered the infective form to be a second-stage larvae, which is now known to correspond to a third-stage parasite).

Repeated immunization of mice with small doses of eggs of *T. canis* followed by a massive challenge reduced the number of parasites recovered from the immunized mice to 30–43% of the parasites found in nonimmunized controls. The experimental mice, however, had 100–350 times more larvae in their livers than the control animals. This was interpreted as evidence that most of the immunity against *T. canis*, at least in the abnormal host, acts in the liver, arresting the migration and further development of the parasite. This finding coincides with clinical observations in children infected with this parasite: the most common organ location in them also appears to be the liver.

Protection against *A. suum* infections has been demonstrated in pigs by transfer of a gammaglobulin fraction of the colostrum and in pigs and guinea pigs by transfer of serum. It has also been reported recently that *A. suum* causes heavier infections and more pathology in guinea pigs naturally or artificially deficient in complement than in conventional animals. This work indicates that protection against *A. suum* larvae is mediated by antibodies, probably by a mechanism that involves complement.

Later work has shown that transfer of either serum or lymph node cells from infected guinea pigs into inbred, susceptible guinea pigs afforded protection, with the cells transmitting more protection than the serum. This result does not necessarily mean that the resistance is produced by a cell-mediated reaction, since antibody-producing cells from the donors can live in the recipients, providing a more sustained humoral response than the simple injection of serum.

Injections of somatic extracts of various stages of *A. suum* parasitic larvae did not elicit immune protection in guinea pigs, but supernatant of cultures where third-stage larvae had moulted did. This result indicates that only metabolic antigens, most likely associated with the moulting process, are protective.

Artificial Production of Resistance

Most attempts to produce resistance against migratory larvae by immunization with extracts of the parasites have given poor results. Vaccination of guinea pigs with supernatant fluid of *A. suum* cultures has given rise to controversial reports, but it seems that most investigators have obtained more protection with this than with somatic extracts. Some resistance has also been produced by inoculation of parasite enzymes. Immunization work is being currently done with *A. suum* in guinea pigs using metabolic antigens of moulting third-stage larvae. Some protection against homologous infection has been obtained in mice by inoculation of large amounts of *T. canis* egg extract in Freund's complete adjuvant.

Over two decades ago, it was reported that the repeated inoculation of esophageal extracts of *Ancylostoma caninum* in puppies reduced the worm burden of a challenge in 25%. This interesting work has not been continued, but apparently the resistance observed then was directed against the hematophagous adults rather than against the larvae.

The most effective vaccination procedure against this group of nematodes that is available at present is the infection of dogs with irradiated *A. caninum* larvae. Only about 25% of the larvae irradiated with about 40,000 r of x rays reach adulthood when administered subcutaneously to *Ancylostoma*-naive dogs. The adults resulting from this infection are sterile and soon are eliminated. It appears that a large proportion of the irradiated larvae dies in the tissues of vaccinated animals, since only a few irradiated parasites are transmitted congenitally to the litter of vaccinated bitches. Nevertheless, the parasites remaining in the dog afford a satisfactory resistance against reinfection: a single vaccination with 1000 irradiated larvae reduces the parasitic burden that results from a subsequent challenge in 37%, and two doses separated by about 4 weeks reduces the development of a challenge to adults, and the congenital transmission of a challenge to the litter, in about 90%.

The vaccine is recommended for puppies between 6 and 10 weeks old, in good health, and free of hookworms or other parasites. Effective results have been obtained also with puppies 3–4 weeks old and even younger, and with adult dogs that have not had a chance to become resistant before moving to an enzootic area. The immunity produced by the vaccine lasts at least for 7 months. It is more than likely that natural infections will periodically restimulate resistance during this period in any area where the vaccine is necessary. Irradiated canine hookworm vaccine was commercially produced in the U.S. until a few years ago but, despite its efficacy, marketing considerations forced its cancellation.

Immunodiagnosis

Immediate-type skin tests and a variety of serological procedures are generally positive in patients of ascariasis and hookworm disease, but, since comparatively simple procedures are available for the parasitological diagnosis of these infections once patency occurs, no interest in immunological methods of diagnosis has existed.

The parasitological diagnosis of strongyloidiasis is more involved and less reliable, which has stimulated some work on the indirect identification of the infection. A recent trial of the immediate-type skin test in 20 patients with positive coprology revealed 16 infections with a larval delipidized antigen, 6 with a polysaccharide fraction, and 20 with a protein fraction.

The condition that has attracted most of the work by far is visceral larva migrans by *T. canis*. The immunodiagnostic research of this infection in humans is particularly difficult, since the condition is rarely identified parasitologically and the diagnosis is commonly only a clinical presumption. Also, the ascarids share a large number of antigens among themselves and with other helminths (Torres and Barriga, 1975), which makes nonspecific reactions almost a rule. Although the cross-reacting antigens do not induce protection, some work with rabbits suggests that the immune system reacts preferentially to them. In addition, all ascarid infections of humans are acquired by ingestion of contaminated soil, so that infections other than *T. canis* in serologically positive patients cannot be excluded on epidemiological grounds.

The situation is such that the specificity and sensitivity of a given test in humans has to be measured by the presumptive clinical diagnosis, which in turn relies on immunological results to support the clinical conclusion. Immediate-type skin tests with extracts of adult *T. canis* have been used rather extensively in epidemiological investigations as well as for clinical diagnosis. Delayed skin responses have been reported in patients with the ocular form of the disease.

A number of serological tests, mostly with extracts of adult worms (e.g., precipitation in fluid, in agar, or on larvae; complement fixation; passive agglutination of antigen-sensitized erythrocytes or inert particles; fluorescent antibody techniques), have been utilized with moderate success due to the presence of cross-reactions with other ascarids, with ancylostomids, and with other helminths. On occasion, the cross-reactive antibodies have been removed by adsorption of the suspected serum with *Ascaris* extracts prior to the performance of the test.

Recently, it has been found that metabolic products of *T. canis* larvae are much more specific antigens than the extracts of adult parasites. Glickman et al. (1978) recently tried this preparation in four different tests with sera of 110 presumed cases of visceral larva migrans. They found that the sensitivities of indirect hemagglutination (IHA), bentonite agglutination, double diffusion in agar, and enzyme-linked immunosorbent assay (ELISA) were 18%, 26%, 65%, and 78%, respectively. The specificity of all of them was at least 92%. Positivity of any test, except IHA, correlated very well with the presence of the clinical disease, and negativity of the ELISA correlated very closely with its absence.

The immunological diagnosis of anisakiasis has also been attempted by a variety of tests (e.g., intradermal, complement fixation, hemagglutination, immunofluorescence), with the same problems of cross-reactivity characteristic of the ascarids. At present, it appears that the most sensitive and specific procedure is the fluorescent antibody technique with hemoglobin from the perienteric cavity of *Anisakis* larvae as antigen.

Human abdominal angiostrongyliasis has been diagnosed by precipitation in gel using serum of infected rats as the antigenic preparation (Sauerbrey, 1977).

Immunopathology

From the histopathology and the exacerbation of the lesions on reinfections, it appears that the host's immune response is a large component in the damage caused by migratory parasites. When respiratory symptoms are detected in human hookworm disease, they usually coincide with the development of the parasite in the intestine, which also suggests that they are the product of a hypersensitivity reaction. The more severe pulmonary pathology produced by the migrant stages of *Ascaris* and *Strongyloides* in humans, as compared to ancylostomids, has been attributed to the fact that the former moult in the lungs and the latter do not: moulting represents the production of potent or persistent antigens (moulting fluid and larval cuticles) that must elicit a stronger immune response than the simple passage of the larvae. Experiments with rabbits have also demonstrated that reinfections with *Anisakis* are more likely to produce clinical symptoms, probably because the primary infection sensitized the host to parasite antigens in such a way that subsequent invasions are less well tolerated.

Several migratory larvae, particularly *T. canis*, appear to act as immunological adjuvants and stimulate the production of gamma-globulins of unknown specificity. It is possible that this immunopotentiation triggers allergic diseases, since 22% of 50 asthmatic children had positive serology for *T. canis* in Montreal, whereas only 4.5% of 940 adults gave similar results. *T. canis* larvae may also be responsible for lesions of the central nervous system, since the same survey found 16% positive tests among 43 epileptic children and 20% among 63 epileptic adults (Viens, 1977). Comparable results have been reported from Great Britain.

Ascaris is particularly effective as a stimulator of IgE production, which may play an important role in the symptomatology of the infection: in a study of 4 human patients those 2 that showed elevated IgE and IgM levels had the most severe symptoms. Also, cutaneous and respiratory manifestations are not uncommon in this infection. Bronchial constriction has been reported in *Ascaris*-infected pigs subsequent to the inhalation of *Ascaris* aerosols.

A recent report from Australia claimed that lymphocytes of *A. caninum*-infected dogs have a reduced reactivity to the T cell mitogen, phytohemagglutinin (Kelly, Kenny, and Whitlock, 1977) (see Chapter 7).

THE PULMONARY NEMATODES

Several nematode species have become adapted to the vertebrate lungs as a definitive habitat. Many of them are found rather sporadically or have a restricted geographic distribution, so that little research has been done on their immunology. The species *Dictyocaulus filaria* of sheep and *Dictyocaulus viviparus* of cattle, on the contrary, are widespread and economically important nematodes that have commanded the attention of investigators. Although *Muellerius capillaris* and *Protostrongylus rufescens* of sheep and goats and *Metastrongylus elongatus* and *Metastrongylus pudendotectus* of swine are less significant as agents of disease, their high prevalence and cosmopolitan distribution have also won them some attention.

D. filaria live in the bronchi of sheep and goats, where they lay eggs that are swallowed and hatch in the small intestine of the host. The first-stage larvae are evacuated with the feces, mature to third-stage infective larvae in a few days, and are ingested by a new host with the herbage. The ingested larvae exsheath in the small intestine, penetrate its wall and travel by the lymphatic circulation to the mesenteric lymph nodes, where they accumulate from the second to the sixth day of infection and moult to the fourth stage. These larvae continue their journey by the bloodstream through the heart to the lungs, where they break into the alveoli starting around the seventh day, and move up to the bronchi. Adults are present by the 18th day and patency begins around the 26th day of infection.

The life cycle of *D. viviparus* is essentially similar to that of *D. filaria*, although there are some questions about whether the cattle parasite actually dwells in the mesenteric lymph nodes, as the sheep lungworm does.

The life histories of *Mullerius* and *Protostrongylus* are similar to that of *Dictyocaulus*, but they require land mollusks as intermediate hosts to develop the infective larvae.

The development of *Metastrongylus* species follows a comparable pattern, but eggs rather than larvae are passed by the definitive host and earthworms are necessary as intermediate hosts.

Natural Resistance

A number of investigators have found that it is rather difficult to produce patent infections with *D. filaria* in sheep. Better rates of success are obtained when the animals are younger than $5\frac{1}{2}$ months and when the total infective dose is divided into several periodic administrations. Despite the fact that these findings suggest the existence of age resistance and the importance of a massive antigenic challenge in the production of protection, none of them is observed in natural field infections. The significance of the early events of the infection in the resistance to the parasitism has been demonstrated by the inoculation of infective larvae orally and intravenously in susceptible sheep; whereas only about 33% of the administered dose developed to adults in the first case, around 80% of the larvae attained maturity in the second instance. The circumvention of the migration from the intestinal wall to the mesenteric lymph nodes greatly improves the efficiency of the infection then. Since the parasites are present in these locations for only the first 6 days of infection, which appears to be insufficient for the development of an effective immunity, any antiparasitic action exerted at this level during a primary infection may be regarded as an expression of natural resistance.

Sheep undergoing a primary infection by *D. filaria* normally begin to expel the parasites from their lungs after the 12th day of infection. Since the rate of loss of worms was not affected by administration of cortisone, it was proposed that this elimination was the result of a nonimmune mechanism, possibly related to natural resistance. Subsequent findings of the variable effect of cortisone on the immune response, which in any case will not interfere with the preexisting levels of antibodies, weaken this argument considerably.

Studies with Soay and Blackface lambs in Scotland have demonstrated that the former sustained a heavier and more severe infection than the latter. Although these results suggest the participation of the genetic makeup of the host in the resistance to the current infection, it is not clear whether the difference must be attributed to preexisting mechanisms of natural insusceptibility or to a particular ability to mount a rapid and effective immunity. The experimental infection of goats with *D. filaria* takes more readily, has a shorter prepatency, produces more larvae, and causes a more severe disease than the homologous infection in sheep. Possibly the goat rather than the sheep is the primary host for *D. filaria*.

Attempts to infect cattle with *D. filaria* have achieved patency only rarely, but the disease has been often recorded and precipitating antibodies have been detected after these incomplete infections. The presence of antibodies in these cases suggests that the reduced susceptibility of cattle is the consequence of an immune response rather than of natural resistance.

Guinea pigs are often used as laboratory models for *Dictyocaulus* infections. However, some authors have reported the finding of hepatic lesions in *D. filaria*–infected guinea pigs, which indicates migration through the portal system and, possibly, the presence of third-stage larvae in the lungs. It is not known whether this divergency may have some influence in the final outcome of the immunity, but it certainly suggests that the guinea pig infection is different from the one in the sheep, so that restraint should be exerted in extrapolating results from one host to the other. At any rate, guinea pigs are considerably less susceptible than sheep: whereas about 25% and 90% of the administered dose (orally and intravenously, respectively) are present in the lungs of sheep on the 12th day of infection, only 12% and 40% of the larvae are recovered from the lungs of guinea pigs on the same date. Besides, guinea pigs have eliminated all parasites by day 16, whereas sheep are near the peak of the parasitic burden at this time (Rose, 1973).

    D. viviparus affects cattle and closely related species (deer and camel). Sheep sustain the infection only to the fourth larval stage, but the mechanisms of protection must be immunological, because the incomplete infection produces resistance to a later challenge with *D. filaria*. *D viviparus* will also grow in guinea pigs, but its development is faster than in cattle and some work has suggested that the larvae may not dwell in the lymph nodes of this species, as occurs in bovines. In fact, the controversy in regard to the early migration of *D. viviparus* sprang from findings in guinea pigs.

    Metastrongylus species of swine are somewhat less specific. Besides pigs, natural infections are occasionally found in domestic and wild ruminants and a few cases have been recorded in humans. Rats, guinea pigs, and monkeys can be infected experimentally. Pigs appear to have age resistance to *Metastrongylus*, since passage of eggs terminates in 8-month-old animals regardless of the age of the original infection.

    Nothing is known about the mechanisms of natural resistance to pulmonary nematodes, but many of the general principles discussed under "Nematodes of the Digestive Tract" must be applicable to this case.

Acquired Immunity

Reviews that include immunological aspects of the pulmonary nematodiasis have been written by Poynter (1968) and Rose (1973). Some important information also appears in the workshop edited by Urquhart and Armour (1973).

    Infection by lungworms results in several manifestations of immunity. Antibodies have been verified by complement-fixation tests,

by precipitation techniques, by hemagglutination, by immunofluorescence, and by immediate-type skin tests. Although the class of the antibodies involved can be inferred from the technique utilized for their
demonstration in some cases, it appears that only IgM and IgG classes
have been identified on a couple of occasions. IgE antibodies have
been demonstrated indirectly by specific tests.

Cutaneous manifestations of cell-mediated immunity have been
demonstrated in infected guinea pigs by skin tests, but the same procedure in infected sheep and cattle has produced only immediate-type
reactions. The kinetics of the reaction in cattle raises suspicions of the
occurrence of an associated Arthus phenomenon. It is not known at
the moment whether the ruminants fail to produce a cell-mediated response to lungworms or earlier local reactions in these cases washed
out the antigen before the cellular reaction had a chance to appear. An
elevated peripheral eosinophilia and a high phagocytic index have been
determined in infected sheep and cattle. In all the cases investigated,
reinfections gave origin to anamnestic reactions.

The facts that dictyocaulosis is an infection of young animals and
that it disappears spontaneously after a while in most of them suggested
long ago that the condition stimulated a strong protective immunity.
Actually, the parasites of a primary infection in sheep begin to be eliminated from the lungs by the 12th day of infection, and most of them
have been already rejected by the 20th day. The infection normally
disappears by itself after a couple of months, and only a few carrier
animals will maintain an infection for about 6 or 7 months. The evolution of the infection in cattle is similar.

Mullerius and *Protostrongylus* elicit a much weaker protective immunity against the current infection than *Dictyocaulus*; the life span
of the worms in the host is counted in years and reinfections appear
to accumulate, with the result that older animals are more infected than
young ones.

Swine *Metastrongylus* species are also eliminated from the lungs
of infected pigs during the first weeks of a primary infection, but less
efficiently than *Dictyocaulus* in ruminants. Peak egg production is
found 5–9 weeks after infection, and patency continues at least until
the seventh month.

Experimentally, it has been determined that the resistance against
challenges with *Dictyocaulus* develops slowly after a primary infection
and that it may take 2 or 3 months to reach its maximum efficiency.
It declines gradually from then on, and is absent after 12 months in
cattle and after 26 months in sheep. The manifestations of immune
protection to *Dictyocaulus* challenges are multiple: some reports indicate that many larvae of the challenge are trapped and destroyed in

the mesenteric lymph nodes of sheep, but other authors have not confirmed this; the growth of the parasites is greatly retarded during the
first 10 days of infection; the number of worms recovered from the
lungs between days 11 and 29 of infection is much lower than in the
controls; many of the worms that become established are inhibited in
their development and remain as juveniles (fifth-stage larvae of some
authors); and those few parasites that reach adulthood are frequently
stunted. The effect of the acquired immunity on the fecundity of the
adults is controversial; some investigators have found a considerable
reduction of the number of larvae per worm in reinfections, whereas
others report that this effect is minimal.

Experimental infections of sheep with *Mullerius* have failed to
produce complete protection against reinfections, although it appears
that fewer worms become established in these animals as compared
to infection-naive sheep.

Studies with *Metastrongylus* species have shown that pigs infected
with 500 eggs in one dose did not develop protection to challenges, but
a single dose of 2500 eggs caused retarded development of the parasites
of reinfections; 540 eggs divided into 7 doses and administered over
40 days (which must approach the pattern of natural infection) reduced
the number of worms that developed on challenge, however (Rose,
1973). Other experiments have shown that resistance is manifested by
expulsion of the worms from the lungs rather than by inhibition of
development. This contradiction may be due to the different efficiency
of the various immune mechanisms brought into action with diverse
experimental methodologies.

Mechanisms of Acquired Resistance

The stages of the parasite that elicit immunity have been investigated
by infections with different developmental stages and by termination
of the infection at diverse levels with anthelmintics. The results have
indicated that each of the various stages of *Dictyocaulus* produces some
immune resistance.

Similar studies have been conducted to determine which stages
of *Dictyocaulus* are affected by the protective response. Histopathological examinations of challenged, resistant sheep showed that the
parasites of the challenge reach the mesenteric lymph nodes, but very
few of them were found in the lungs. Some work has demonstrated
that the larvae in these animals are retained in the nodes for up to 15
days and that their growth is severely retarded. Bypassing of the mesenteric nodes by administration of the infective larvae intravenously
also resulted in reduced recovery of worms in resistant lambs as com-

pared to controls. These and other works indicate that infective larvae are retained, inhibited, and possibly destroyed in the mesenteric lymph nodes during the early phase of the infection and that the parasites that succeed in getting to the lungs are destroyed in, or expelled from this organ in immune animals.

Some researchers have reported transfer of resistance to susceptible sheep by inoculation of serum from immune animals, and Movsesijan in Yugoslavia maintains that he found an inverse relationship between antibody titers and resistance in infected sheep (Urquhart and Armour, 1973). These findings suggest that the protection is antibody mediated; however, other investigators have failed to find correlation between antibody levels and resistance, and the appearance of antibodies does not coincide with the beginning of the manifestations of protection. These latter arguments are rather weak, since the antibodies detected in vitro are not necessarily the same ones affording protection in vivo. At any rate, if the antibodies are protective, probably this activity is associated with IgG, since it has been demonstrated that a primary infection triggers the production of IgM followed later by IgG antibodies, whereas a reinfection produces mainly IgG antibodies.

A recent report from Italy (Casarola and Orlandi, 1977) communicated that normal guinea pig macrophages attach to *D. filaria* infective larvae when these have been sensitized with serum of immune sheep. This phenomenon does not involve complement and depends on an opsonizing Ig that attaches to the cuticle of the worms. In view of the recent advances in regard to the immune protective mechanisms in schistosomiasis and the known stimulation of the eosinophilic and phagocytic cells in dictyocaulosis, it would be interesting to investigate the protective effect of antibody-cells associations in this infection.

The cell-mediated response does not appear to have been examined in the natural hosts.

Artificial Production of Resistance

Attempts to vaccinate sheep or cattle with dead parasites or with their extracts have produced some degree of resistance, but this has been considerably less effective than that elicited by the actual infection. Injection of *D. viviparus* excretory/secretory products in guinea pigs produced variable protection to challenges; the fact that much better results were obtained with repeated inoculations or by the use of adjuvants suggests that the results might have been better had more antigen been available.

Since the actual infection elicits a very strong protection against reinfections, Jarret and collaborators tried and developed a method of immunization of cattle with irradiated infective larvae (Urquhart and Armour, 1973). Soon a similar vaccine became available commercially in Great Britain, and has been extensively used in Europe. A similar vaccine against *D. filaria* of sheep has been produced by Yugoslavian workers.

It has been found that the most effective schedule of immunization is to irradiate the larvae with 40,000 r and to give two doses of 1000 larvae each, separated by 4 weeks, to calves 2 months of age or older. Later experimental work has demonstrated that reduction of the doses to 250 or 500 larvae still produces considerable, although slightly decreased, protection. Shortening the interval between doses to 2 weeks did not affect the production of resistance. It must be remembered, however, that these results were obtained under experimental conditions, and that they may not hold in the field as well as the older and well-proven method.

Studies with *D. filaria* in sheep have verified that practically all the irradiated larvae die during the first 5 days of infection and none reaches adulthood. The same appears to be true for cattle. The resistance produced by the vaccination of calves is manifested during the first 10 days of a challenge infection (see above), but the parasites that survive this phase are less affected in their further development than when the resistance was evoked by a natural infection. This supports the idea that immune resistance to dictyocaulosis involves two steps, one expressed in the mesenteric lymph nodes and the other in the lungs. The reduced life span of the irradiated larvae appears to stimulate fundamentally the lymph node level, largely avoiding the pulmonary phase that is also the cause of the pathology.

Recent work in Czechoslovakia has demonstrated that a first dose of 1000 irradiated *D. filaria* larvae produces only IgM antibodies in sheep, a second dose 5 weeks later induced the formation of mainly IgG antibodies, and a challenge with nonirradiated larvae 3 weeks later resulted in the production of solely IgG antibodies. A primary infection with 1000 nonirradiated larvae, however, induced the formation of IgM antibodies around the second week, followed by production of IgG antibodies a week later. Evidently, the stimulation of the immune response by irradiated larvae is not identical to that produced by a natural infection, although it is quite efficient to elicit protection.

The finding of anti–*D. viviparus* antibodies in calves 10–30 days old suggests colostral transmission from the mothers. These antibodies appear to interfere little with the effect of the vaccination, however, since suckling calves vaccinated when they were 3 weeks old had 87% resistance to a later challenge. Pail milk–fed controls developed 97% immunity under the same conditions. Vaccination of 8- and 12-week-old calves produced 99% resistance to the challenge (Benitez-Usher, Armour, and Urquhart, 1976). Although it appears convenient to wait until the calves are 2 or 3 months old before vaccination, earlier vaccination will be effective if necessitated. Vaccination of 6-, 8-, 16-, and 18-week-old lambs produced only 22%, 63%, 91%, and 95% resistance, respectively.

The post-vaccinal protection develops rather slowly: sheep vaccinated with two doses of irradiated *D. filaria* larvae eliminated 38% of a challenge given 3 days after the second dose and 86% of a challenge given at 15 days, and did not accept reinfection 2 months after the vaccination. In practice, vaccinated animals are kept away from contaminated environments at least for 15 days after the second dose.

A large number of reports testify to the efficacy of *D. viviparus* vaccination (and a lesser number to that of *D. filaria*). The immunity produced is not absolute, however; a few of the larvae picked up in the field will make it to the adult stage and will maintain the contamination of the pastures. This situation is beneficial for vaccinated animals, since it will provide boosters to the rather transient effect of the vaccine, but it prevents the grazing of nonvaccinated ruminants in the same fields. The veterinarian must decide, given the local conditions, whether to use the vaccine and accept a contaminated environment or to attempt the eradication of the infection.

Massive infections can break the immunity produced by the vaccine and cause disease. Nevertheless, the proportion of animals affected and the severity of the symptoms is much less among vaccinates than among nonvaccinates: in a study of heavy pasture contamination, 80% of the control calves died, whereas the disease was fatal to only 18% of the vaccinates. Gastrointestinal parasitism, fascioliasis, and infectious pneumonia interfere with the protection afforded by the vaccine. Unfortunately, the shelf life of the vaccine appears to be only 2 to 4 weeks at 4°C.

Heterologous vaccination has been attempted with several parasites. Infection of cattle with *D. filaria* and of sheep with *D. viviparus* causes some protection against challenges with the heterologous species, without reaching patency most of the time. The heterologous parasite causes disease, however, so that no real advantages over vaccination with the homologous worm exist. Of the 22 to 23 antigens demonstrated in *D. viviparus* and *D. filaria* extracts, only two or three are species specific; all the others are shared by both parasites.

In Russia, immunization of sheep with *Aphelenchus avenae* (a parasite of plants) or with *Neoaplectana glaseri* (a parasite of insects) has been reported to produce 79–89% resistance to a *D. filaria* challenge a month later, but the protection diminished to 10–18% in the next 3–4 weeks. *D. filaria* shares two antigens with each of these worms. Vaccination with these organisms could be effective only in currently contaminated fields and with such a timing that the animals would be exposed to the immunizing effects of the natural infection during the third or fourth week following the vaccination. An advantage of vaccination with these species is the comparative ease with which they can be raised and kept in the laboratory.

The free-living nematode *Rhabditis axei* produced 75–90% protection against *D. filaria* in sheep in one attempt, but similar experiments with guinea pigs were inconclusive. Vaccination of sheep with *Nippostrongylus brasiliensis* or with *Ascaris suum* made them more susceptible to a later challenge with *D. filaria*.

Immunodiagnosis

There are a number of works on the serology of lungworm infections, but this does not appear to have excited great interest in the practical use of these methods for diagnosis. Immunological techniques may be useful to identify the disease during the prepatent phase (which may be lethal for heavily infected animals), to diagnose it during patency (Denev found that serology of infected pigs was more sensitive than coprology), and to evaluate the state of resistance (high-titered serum in nonpatent animals presumably indicates past infection and, therefore, probable resistance).

Immediate-type skin tests become positive in infected cattle, sheep, and swine during the second week of infection. They show no correspondence with the production of specific precipitating or hemagglutinating antibodies and no cross-reactions with *Haemonchus contortus* or *Oesophagostomum columbianum* infections. However, they are positive in sheep vaccinated with *Aphelenchus avenae* or with *Neoaplectana glaseri* (see above). Delayed-type skin tests have produced positive reactions in guinea pigs infected with *D. filaria*, but have not been reported in domestic animals.

Antibodies that precipitated in the natural orifices of infective larvae have been reported in the serum of cattle, sheep, and swine infected with their specific lungworms. This reaction did not affect the viability or the subsequent ability of the larvae to cause infection, however.

Complement-fixing antibodies are detectable in the three host species beginning during the third week of infection or earlier, and persist at least for 150 days in sheep.

The indirect hemagglutination test has produced positive reactions in infected cattle and swine. In the latter, the first reactions appeared during the second or third week of infection and persisted for 4–5 months.

The indirect fluorescent antibody technique has been used to detect infection in all three species of domestic hosts. The respective antibodies begin to be demonstrated at the start of the second week of infection and become positive in all animals during the fifth week postinfection. They persist for more than 3 months in sheep and swine. Most of the fluorescence appears in the walls of the reproductive and digestive organs of the adults and in the cuticle of the larvae, which indicates that these are the locations of the antigens concerned with diagnosis.

It was recently reported that extracts of *Aphelenchus avenae* and of *Neoaplectana glaseri* are as specific for detecting *Dictyocaulus* infections as the homologous antigens; they do not react with anti-*Haemonchus*, anti-*Trichinella*, and anti-*Echinococcus* sera, but show some cross-reactivity with anti-*Fasciola* antibodies.

Immunopathology

There are few studies directed to evaluating the role of immunology in the pathology of lungworm disease, but a few known facts suggest that this may be important. It has often been reported that there is no relationship between larval counts in the feces and the severity of the disease. In addition, severe respiratory symptoms may occur in the virtual absence of parasites in the lungs during early prepatency or during postpatency. All these findings point to a pathogenic mechanism other than the direct activity of the worms.

Most of the pathology of the early disease and a very considerable part of the later symptoms are produced by eosinophils obstructing the bronchioles and small bronchi, or infiltrating the bronchiolar wall, respectively. That this reaction is immunological is suggested by the hyperplasia and eosinophilic infiltration of the mediastinal lymph nodes, and confirmed by the boosting effect of a reinfection on the eosinophilia and on the lymph node changes.

The necrotic foci of the alveolar walls seen from the fifth day of infection onward are generally attributed to the mechanical penetration of the larvae into the alveoli. The possible occurrence of a local Arthus phenomenon does not appear to have been examined, although the lesions show intense infiltration with neutrophils and infected cattle present skin reactions that are reminiscent of the cutaneous manifestations of the Type III allergy.

A serious complication of lungworm disease is pulmonary edema, the etiology of which, immunological or otherwise, does not seem to have been studied yet.

The so-called lymphoreticular broncho-occlusive lesions are characteristic of reinfections. They consist of infiltrates of giant cells, eosinophils, macrophages, plasma cells, and lymphocytes, surrounding a dead parasite. With time, these lesions develop germinal centers and take on the aspect of lymph nodes. The immunological nature of these lesions was verified by reproducing them, during a primary infection, in calves that had received serum from immune animals but not in calves abstained from the immune serum.

A clinical syndrome of respiratory distress that occurs in cattle when moved from dry pastures to irrigated fields has been called "acute bovine pulmonary emphysema" in North America and "fog fever" in Great Britain. Its etiology is not known, but some authors have advanced the idea that it could be a hypersensitivity reaction to reinfections with *D. viviparus*. A team of researchers from Glasgow recently infected 7 cows that had shown the condition 9 to 57 days previously and were positive to *D. viviparus* by skin tests with 30,000 homologous infective larvae. Since they could not reproduce the syndrome and the necropsy 30 days later showed no evidence of lesions typical of the illness, they concluded that there was no support for this hypothesis. It must be said, however, that use of 30,000 larvae in a single dose seems rather excessive to mimic natural conditions. Also, if the disease was actually a hypersensitivity reaction to *Dictyocaulus* infection, its natural occurrence shortly before experimental infection might have produced a temporal desensitization of the animals. Acute bovine pulmonary emphysema may be a complex syndrome involving multiple etiologies, but I believe that an allergy to *Dictyocaulus* has not been excluded yet. The presence of abundant tissue and peripheral eosinophilia, the existence of skin-sensitizing antibodies, and the effective expulsion of the parasites from the lungs of infected animals certainly suggest that Type I allergy does occur in dictyocaulosis; its role in the pathology of the disease should not be dismissed without exhaustive research.

TRICHINELLA SPIRALIS

The trichina is a small nematode that has been a compelling enigma for many years for experimental biologists as well as health professionals. It combines a fascinating natural history with an unusually high prevalence in some parts of the world (more than one-sixth of the U.S. human population was infected in the 1930s). A ubiquitous parasite, it may infect almost any mammal, although humans, swine, and rats are particularly important as hosts in medicine. At different times in its life cycle, *Trichinella spiralis* behaves as an intestinal parasite, as a migrating larva, or as a permanent resident of the intracellular environment.

First-stage larvae ingested with the muscles of an infected host are released by the digestive fluids of the new host, mature to adult worms in the small intestine, mate, and produce a new generation of larvae between days 4 and 13 of infection. During the moultings to reach the adult stage and for the duration of the larviposition, the parasites penetrate the intestinal mucosa, assuring a rather intimate contact with the tissues of the host.

The newborn larvae migrate mainly through the lymph-blood circulation and penetrate the skeletal muscle fibers very rapidly. This invasion causes profound metabolic and morphologic changes in the muscle cell, which appear to modify its functions mainly toward sustaining the life of the parasite. Once inside the muscle fibers, the larvae form an intracellular cyst composed mainly of collagen that is usually complete by the fourth week of infection. Phenomena of calcification may occur with time, but the parasite survives for years in old, calcified cysts.

The immunology of trichinellosis has been the subject of inumerable papers and I do not attempt to summarize them here. I limit myself to discussion of the areas of most relevance to public health and refer the interested reader to the reviews by Larsh (1970), Catty (1976), and Warren (1978). These papers constitute excellent introductions to the subject and to the pertinent literature. The immunology of the intestinal stages of *T. spiralis* (Love, Ogilvie, and McLaren, 1976) has already been considered among the digestive nematodes, and is mentioned here only when it helps in understanding the immune phenomena that affect the parenteral phases of the infection.

Natural Resistance

Trichinella spiralis is a parasite with broad specificity that may affect a large number of mammalian species and, occasionally, may survive for limited periods in the muscles of birds. Tanner recently studied the infection in nine strains of mice and found no significant differences in susceptibility. Other studies, however, have shown that the size of the muscle cysts varies from one host species to another, which may indicate diverse degrees of resistance to the development of the parasite.

A curious phenomenon is the elevated susceptibility of the Chinese hamster to the intestinal infection associated with a considerable resistance to the muscle infection, as compared to the golden hamster or other rodents. The resistance to the tissue phase can be reduced by depressors of the cellular response, but not by inhibitors of humoral immunity. Although it seems evident that the destruction of the muscle-invading larvae in the Chinese hamster is mediated by an infiltrate of mononuclear and eosinophilic cells, it is not clear whether this phenomenon should be attributed to innate or to acquired resistance.

Several investigators have found that parasites recovered from diverse geographic locations exhibit different infective and pathogenic capacity for various species of hosts, and diverse sensitivity to cold. These characteristics, used by some writers to create new species or varieties of *Trichinella*, may correspond to a process of selection of subpopulations of the parasite to peculiar ecological conditions.

It must be mentioned that strictly herbivorous animals (such as rabbits) do not have the ecological opportunity to become infected in nature, but they are fully susceptible when infected artificially. Birds, however, digest the infective larvae, and intramuscularly injected parasites survive uncoiled for only short periods in them. Some limited research had suggested that the resistance of birds to the tissue phase of trichinellosis was related to their high body temperature; recent research does not support this notion.

Acquired Immunity

There is a formidable amount of information currently available on acquired immunity to trichinellosis. Most of the older findings have been confirmed, corrected, or expanded in recent papers by Despommier et al. (1977), and Grove et al. (1977), among others.

In brief, infection with *T. spiralis* elicits humoral and cell-mediated immune responses. Antibodies of the classes IgM, IgG, IgA, and IgE have been detected in the circulation. Cellular immunity has been demonstrated by a variety of techniques, including skin tests. Recent reports and my own work indicate that the acute infection also produces a nonspecific activation of the macrophages; chronic infections cause inhibition.

The most evident expressions of acquired resistance to *T. spiralis* are accelerated expulsion of adult worms, reduced fecundity of the females, and diminution of the number of larvae that invade the muscles.

Evidence has accumulated that protection against the intestinal stages and against the migratory and muscle worms in laboratory rodents are different and independent events; a primary infection artificially restricted solely to the enteric phase reduced in 87% a second infection given orally, but only in 15% when the second infection was given intravenously. These results indicate that most of the immunity produced by the intestinal parasites remains confined to the intestine, since circumventing this organ by means of intravenous administration of the parasites reduces greatly the efficacy of the protection. The infection with muscle stages, on the contrary, protected completely against a subsequent intravenous injection of newborn larvae.

In another experiment, abolishment of the eosinophilic response by inoculation of antieosinophilic serum did not change the dynamics of the expulsion of the adults, but increased significantly the number of larvae in the muscle. This work not only demonstrated differences in the protective immunity to enteral and parenteral worms, but also suggested that eosinophils have an important role in the control of the muscle infection.

Considerable work has been done to identify the intimate mechanisms responsible for the protective immunity to *T. spiralis*, but many of the results obtained are conflicting. In regard to the extraenteral parasites, some researchers have found that whole serum of infected animals, or its IgM or IgG fractions, destroyed or reduced the infectivity of the muscle larvae. Other investigators have been unable to confirm these effects. On the other hand, the administration of niridazole, which is a suppressant of cell-mediated but not of humoral immunity, did not reduce the number of muscle or intestinal parasites that developed on infection.

At present, most of the evidence indicates that neither the humoral nor the cell-mediated responses alone are able to explain completely the immunological resistance to trichinellosis, although some effect of each one of them has been demonstrated on repeated occasions. Recent work by Kazura and Grove (1978) may contribute to explain older results; they found that incubation of newborn larvae in peritoneal exudate cells and serum of mice infected for 5 weeks resulted in the larvae being covered by cells and destroyed. Neither serum nor cells alone, or cell suspensions containing less than 2% eosinophils, reproduced this effect. Decomplementation of the serum by heat did not affect the final result of the incubation. The conclusion was that the newborn larvae are killed by the combined activity of cells and antibody, the latter possibly facilitating the attachment of the cells on the parasite. Complement does not intervene in the reaction, but eosinophils are essential in a yet unidentified manner. Additionally, this work demonstrated that newborn larvae are sensitive to immune mechanisms (adults and muscle larvae were not affected by the same incubation procedure), which is in opposition to previous reports that claimed that newborn larvae did not elicit or were not affected by the immune response.

The encysted muscle parasites must be immunogenic, because the specific immune responses of the host persist for years after the infection, but they do not appear to be affected by the immunity since they survive for several years in immune hosts. However, antigens obtained from the encysted larvae, presumably derived from the granules of the periesophageal glands (*stichocytes*), will induce immune responses that expel the adult worms or reduce their fecundity. It is possible that the cysts constitute effective barriers to the passage of antigens to the exterior and to the introduction of immune elements (antibodies and cells) to the interior. In this latter respect, the muscle fiber by itself may be enough of a deterrent to penetration.

From the practical point of view, it would appear that the immunity to the intestinal parasites is the most relevant protective mechanism, since this phase is an obligatory stage in the natural infection but it is not infective for other hosts. Resistance to the development of adult parasites in a second infection in mice persists at least for 40 weeks, although the adult worms of the primary infection are expelled during the beginning of the third week. Since the continuance of an effective immunity for 37 weeks after the parasitological cure is unusual in intestinal parasites, and considering that immunization with products of muscle larvae induces protection from colonization by adult worms, it is probable that the persistent immune resistance expressed at the intestinal level is maintained by antigens produced by the parasites encysted in the muscles.

Artificial Production of Resistance

The literature describes many attempts to produce immune protection to trichinellosis by the use of various procedures. Immunization with dead larvae or with their whole extracts has produced weak resistance to the infection only on some occasions. Many investigators have been unable to find any evidence of benefit with such treatments.

Inoculation of excretions and secretions of parasites maintained in vitro, either muscle larvae or adult worms, has reduced the parasitic load in the muscles and frequently has produced stunting and rapid expulsion of the adults of a subsequent infection. These phenomena are not as intense or consistent as the resistance produced by the actual infection, but demonstrate that effective vaccination with nonliving material may be achieved.

Strong emphasis has been placed in the last decade on the protective properties of the material obtained from the secretion granules of the periesophageal cells of adults and muscle parasites. Vaccination with relatively large doses of this material (100 μg per mouse) reduced the number of adults that developed in a subsequent infection, whereas smaller doses (10μg per mouse) inhibited the fecundity but not the number of adult worms. Both doses resulted in diminished numbers of parasites in the muscles, but at this time it is not known whether the immune response is effective against migrating and muscle larvae or only reduces their production by the adult females. Originally, it was reported that parenteral inoculation of newborn larvae (which have not formed such secretion granules yet) did not elicit protection to a challenge infection. It was then assumed that the protective antigens were related to these granules. The recent report that incubation with peritoneal cells and serum of infected animals kills newborn larvae indicates that antigens other than those in the granules are also relevant to protection.

At present, no experimental procedure has been able to reproduce the immune protection to trichinellosis with the strength and consistency of actual infections. However, the participation of the parenteral parasites in infection-induced immunity has not been properly assessed yet.

Implantation for 7 days of freed muscle larvae, enclosed in chambers that permit only the passage of macromolecules, in the peritoneal cavity of rats induced a long-lasting protection to subsequent infections. This experiment demonstrated that muscle parasites are able to produce functional antigens, but the question of whether the larvae surrounded by the muscle cyst are equally effective producers of antigens still remains.

Newborn larvae were not immunogenic in one experiment, but their sensitivity to the host immune response, as reported recently, indicates that they possess antigens that react in the protective immune reactions. Most likely, therefore, they would be able to produce resistance to challenges in natural infections. At any rate, the use of persisting parenteral parasites to induce resistance to trichinellosis will defeat the ultimate purpose of immunization, which is to avoid the infection in swine in order to prevent its transmission to humans.

Infections restricted to the intestinal phase by using irradiated parasites that do not attain sexual maturity or by terminating the infection with anthelmintics before larviposition, have produced considerable protection to challenges when the original infection lasted long enough for adult worms to develop (2 or 3 days).

It should be mentioned that resistance to trichinellosis has often been directly related to the degree of inflammation elicited by the parasite. Some authors even maintain that inflammation is the immediate mechanism of protection. Doubts are now cast on this notion, since administration of niridazole reduces the cellular infiltrate around the muscle larvae but does not affect the number of adult or larvae that develop in mice.

At present, it appears that abbreviated intestinal infections and selected parasite antigens produce enough protection to be useful in areas of high prevalence of swine trichinellosis. There seem to be no studies comparing the efficiency and cost of vaccination of swine with those of the current inspection of pork in abattoirs. However, it is likely that the time in which this confrontation will be appropriate is rapidly approaching.

Immunodiagnosis

Because of the protean nature of clinical trichinellosis, identification of the parasite by direct or indirect means is essential to the diagnosis. Since Strobel (1911) proposed the use of the complement-fixation test, a large number of procedures for the immunological diagnosis of *T. spiralis* infections have been tried. A comprehensive review was published by Kagan and Norman (1970) and a briefer update by Kagan (1976).

The complement-fixation test (CFT) yields positive reactions in infected patients a little earlier than the bentonite agglutination test (BAT) (see below), but wide variations in the results, possibly attributable to differences in techniques and in antigenic preparations, have reduced its popularity for the diagnosis of trichinellosis. The Center for Disease Control in Atlanta, Georgia, has recently proposed a standardized procedure for the CFT; its use in trichinellosis may revive this method's status.

Precipitation tests have been assayed using living larvae as antigen or using extracts of the parasite in fluid media or in gel. In general, their sensitivity to detect the infection has been too low to compete successfully with other tests available. Counterimmunoelectrophoresis was proposed recently, but, in a study in parallel with BAT, it diagnosed only half as many infections as the latter. An ingenious uroprecipitation test has been extensively researched by Polish workers: with this method, antigens of the parasite, rather than antibodies to it, are investigated in concentrated urine of the patient by reacting it with high-titered precipitating anti–*T. spiralis* serum produced in rabbits. The test is more reactive early in the infection, which is desirable on clinical grounds, but it was too insensitive and inconsistent for practical use.

A number of agglutination tests (frequently called floculation tests) with the antigen adsorbed on inert particles (e.g., bentonite, charcoal, latex, cholesterol) have been employed since the late 1940s. The possibility of using these tests on a slide or even with the sensitized particles adsorbed on a card makes them particularly adequate for small laboratories. The BAT has become the preferred method for many specialists: it was used in the diagnosis of 84% of the cases of human trichinellosis reported in 1971–1975 in the U.S. Properly used, it is very specific and its sensitivity is better than 95% when titers of 5 are considered diagnostic. Its main drawback is that usually it appears positive after 3 or more weeks of symptomatic infection. At this time, the diagnosis often has only an academic value.

The indirect hemagglutination test (IHAT) is not used extensively in trichinellosis despite its exquisite sensitivity. An inconvenience of this procedure is that, if it is made sensitive enough to detect early infections, it will also detect antibodies remaining from old infections that may not be related to the current disease. However, there are indications that the use of the IHAT with properly selected antigens may reveal the infection before the onset of symptoms, without interference from residual antibodies (Barriga, 1977).

The indirect fluorescent antibody test (IFAT) has been occasionally used in the diagnosis of human and swine trichinellosis and was demonstrated to be highly sensitive to detect the infection. Some problems remain, however, in the selection and preservation of the most appropriate antigen. Current work seems to indicate that sections of parasite cuticles are adequate. A soluble antigen fluorescent antibody test has also been developed.

Ongoing evaluation of enzyme-linked immunosorbent assay indicates that this method is as sensitive as the IHAT and more sensitive than the IFAT. Preliminary research with the radioallergosorbent test to demonstrate IgE antibodies has yielded very sensitive results in the hands of some researchers and negative ones in the hands of others. Both methods still require more investigation, but their prospects appear excellent in regard to sensitivity.

Cutaneous reactions to *T. spiralis* antigens in patients are usually found after 4–6 weeks of infection. Some investigators ascribe particular diagnostic significance to the delayed-type reaction, whereas others rely more on the immediate type. Kozar (1971) found 74% positive reactions in 180 patients of trichinellosis who had presented the first symptoms within the preceding month and 76% positive results in 99 patients who had had the disease between 6 months and 20 years previously. Most reports, however, mention that skin tests identify only about 60% of the clinical cases. Among the acute patients with positive skin tests, 63% had immediate reactions, 16% delayed reactions, and 21% both. In the group with chronic infections, 24% of the reactions were of the immediate type, 43% of the delayed type, and 33% had both reactions in a sequence. Occasional Arthus reactions (intermediate hypersensibility) were also observed, which has also been reported to occur in rodents and swine.

From these results, it is clear that the immediate reaction is more useful to detect acute infection with clinical purposes, whereas the delayed reaction may be more useful to identify chronic infection in epidemiological studies. Not many critical studies have been done on the sensitivity and specificity of the skin tests, but, in a survey of 388 persons who had been skin tested for trichinellosis shortly before death, only 9 of 89 infected subjects had shown positive results, whereas 16 of 299 noninfected individuals had been positive. Recent observations indicate, nevertheless, that the specificity, sensitivity, and opportunity of the results of the skin tests are strongly dependent on the antigen employed (see below).

Problems in Immunodiagnosis of Trichinellosis

The immunodiagnosis of trichinellosis poses special problems because of the conflicting requirements of the particular worker. The clinician needs a test sensitive enough to detect the infection as soon as the symptoms appear (about the 12th day), but that does not react with residual antibodies from past infections or detect very light infections that are unlikely to produce disease. The epidemiologist wants a test that reveals infections of any age or intensity. The veterinary sanitarist prefers a test that identifies only infective larvae (alive and older than 16 days) at a density that constitutes risk of transmission to humans. All of them demand a simple method: the clinician does not see enough cases to justify elaborate procedures or paraphernalia; the epidemiologist deals with large numbers of samples; and the veterinarian handles mostly unwilling customers that need quick and inexpensive results.

Serological technology has reached a phase of development such that investigators can detect antibodies with almost any level of sensitivity that they may want. Unfortunately, considerably less encouragement has existed for the identification and purification of the antigens that *should* be used with our sophisticated technology. Some recent work with guinea pigs (Barriga, 1977) has demonstrated that different antigenic fractions of *T. spiralis* react at different times of infection and with different degrees of specificity in skin tests. Confirmation and extension of these findings in humans and swine is sorely needed.

With most serological tests, antibodies in patients are first found during the fourth week of clinical disease (about 40 days of infection on average), which is rather surprising because they can be detected within 2 weeks of infection in most experimental infections in animals. Part of this delay may be due to the relatively light infection of humans as compared with experimental animals; swine infected with 500 or more larvae became serologically positive 7 days later, whereas swine infected with 25 larvae did not show antibodies until day 62 of infection. It is also possible that the rather light human infection provides such an antigen:antibody ratio in the circulation that most of the antibody in the serum is combined with soluble antigen and does not react in vitro. Recent findings of antigen-antibody complexes in the circulation of patients with acute trichinellosis support this hypothesis. Finally, it is possible that the depression of the host immune response induced by *T. spiralis* for a few weeks after production of the new generation of parasites (see Chapter 7) may be partially responsible for the reduced formation of specific antibodies.

In any case, none of the available tests is completely satisfactory for the diagnosis of human or swine trichinellosis today. In humans, the skin tests are frequently used as a preliminary screening procedure and the bentonite agglutination test is employed for confirmation. When the results of this test do not conform with the clinical suspicion and the epidemiological antecedents, the performance of two other additional tests of a different type (complement fixation and indirect immunofluorescence, for example) is recommended. If positive results at low titers make the clinician suspect that the methods are detecting residual antibodies from past infections (which persist from 2 to 19 years after the original episode), repetition of the tests in 10 days will indicate whether the titers increased in that lapse. Such an observation suggests the presence of an acute, recent infection.

In swine trichinellosis, most methods are not sensitive or consistent enough to detect light infections that, nevertheless, are bound to spread the infection to humans. In an extensive evaluation, it was found that indirect immunofluorescence was more sensitive than trichinoscopy, but less sensitive than the digestion of pooled muscle samples. Since indirect hemagglutination is as sensitive as immunofluorescence or even more so, the IHAT may be also adequate to identify the infection in swine, although extensive studies seem to be lacking. The existence of automated equipment to perform the hemagglutination test nowadays makes it particularly appropriate to use in abattoirs.

Immunopathology

It is surprising how little is known about the pathophysiology of an infection that has been researched as much as trichinellosis has. The main reason may be that most of the information on trichinellosis has derived from studies in laboratory models, which are not nearly as affected by the disease as humans.

As reviewed by Larsh and Weatherly (1975), there is no doubt at present that *T. spiralis* causes an immunologically mediated inflammation of the small intestine, and the abundance of mast cells in this infiltrate suggests that vasoactive amines are produced locally. Most likely, these alterations are responsible for the digestive symptoms found in about a third of the patients of trichinellosis.

The cause of the edema is more obscure. Etiological roles have been attributed to the trauma caused by the larvae going through the capillaries, to foci of cell-mediated inflammation in the tissues, and to a general expression of Type I allergy. The recent finding of antigen-antibody complexes in the circulation of patients with acute trichinellosis indicates that an Arthus phenomenon (Type III allergy) may also be involved. Each one of these hypotheses can show arguments in its favor, but the persistence of the edema well beyond the 14th day of infection, when adults are expelled and no more larvae are found in the circulation (at least in laboratory rodents), suggests that this sign is produced by parasitic antigens rather than by the parasite itself.

The inflammation of the muscles, and occasionally of other tissues invaded by the wandering larvae, is certainly an immunological event that exhibits a characteristic secondary response. Nonspecific inflammation mediated by products released by the destroyed host's cells also may contribute to the general picture.

Undoubtedly, a considerable proportion of the symptoms of trichinellosis are due to an allergic status of the host to parasitic materials. It may not be wise to continue the search for parasitolytic drugs for the treatment of trichinellosis. The sudden release of a large amount of allergens from dead worms in a sensitized subject may certainly be dangerous and may require a previous preventive treatment of the patient.

The question of whether products of the invaded muscle are able to elicit immunity against self-components in the host does not seem to have been tackled experimentally. Some preliminary work by Kozar (1971), however, indicated that immunization with normal rat muscle reduced the susceptibility of mice to the intestinal developments of *T. spiralis* of rat origin. It is not known yet whether this represents adsorption of host's antigens by the parasite (in the fashion of *Schistosoma*) or sharing of components by both associates. I have found that mice with chronic trichinellosis respond with skin reactions to the injection of rat muscle extract.

Abundant work has been done lately on the regulation of the host immune response by *T. spiralis* infection (see discussion in Chapter 7).

THE NEMATODES OF THE BLOOD OR FILARIAE

When parasitologists speak of nematodes of the blood, they normally refer to the worms of the group Filaroidea, also called filariae or "filarial worms." Actually, most of the filariae important to human and veterinary medicine live in extraintestinal locations other than the blood, and despite the fact that the first-stage larvae of many species are found in the circulatory system, this is not the rule for all filariae. The early workers were apparently so impressed with the presence of a "worm" in the blood that they coined a term that survived past its usefulness.

The most important human filariae in the Americas are *Wuchereria bancrofti* and *Onchocerca volvulus*. Domestic animals harbor a large number of filarial worms, but these cause rather mild or sporadic diseases; *Dirofilaria immitis* of the dog, however, is of particular importance because of its wide distribution, high prevalence, and elevated pathogenicity. Table 3 presents some biological characteristics of the filariae mentioned in this chapter.

All filariae require a hematophagous arthropod vector that it is the intermediate host at the same time. First-stage larvae, customarily called *microfilariae*, are taken with the blood meal, develop to third-stage infective larvae usually in 1–3 weeks, and are deposited on the skin of a new definitive host during another meal. The infective larvae migrate to their normal location, moult to adulthood, and begin to lay microfilariae after a variable, but usually prolonged, prepatent period (about a year for *W. bancrofti* and *O. volvulus* and around 8 months for *D. immitis*).

Research on human filariases has been hampered by the presence of the infection mostly in areas deprived of appropriate medical facilities and by the lack of adequate animal models. For a long time, their control (when attempted) relied on reducing the population of vectors by the use of insecticides. The growing concern with environmental pollution at present has stimulated the search for alternative methods

Table 3. Some biological characteristics of the filarial worms mentioned in the text

Species	Natural host(s)	Location of		Vector	Pathogenicity[a]
		Adults	Microfilariae		
Brugia malayi	Humans, monkeys, felids	Lymphatics	Blood	Mosquitoes	+ +
B. pahangi	Carnivores, monkeys, others	Lymphatics	Blood	Mosquitoes	+
Dipetalonema perstans	Humans, apes	Body cavities	Blood	Biting gnats (*Culicoides*)	+?
D. reconditum	Dogs	Dermis	Blood	Fleas, lice	—
D. streptocerca	Humans, apes	Dermis	Dermis	Biting gnats (*Culicoides*)	+?
D. vitae	Hamsters	Dermis	Blood	Ticks	—
Dirofilaria immitis	Dogs, cats	Heart	Blood	Mosquitoes	+ + +
D. repens	Dogs, cats	Dermis	Blood, dermis	Mosquitoes	+?
Litomosoides carinii	Cotton rats, rodents	Body cavities	Blood	Mites	+?
Loa loa	Humans, monkeys?	Dermis, eye	Blood	Deer flies (*Chrysops*)	+ +
Macdonaldius oschei	South American snakes	Portal vein	Blood	Ticks	—
Mansonella ozzardi	Humans	Peritoneal cavity	Blood	Biting gnats (*Culicoides*)	+?
Onchocerca volvulus	Humans	Dermis	Dermis, eye	Black flies (*Simulium*)	+ + +
Setaria digitata	Cattle	Peritoneal cavity	Blood	Mosquitoes	+
Wuchereria bancrofti	Humans	Lymphatics	Blood	Mosquitoes	+ + +

[a] —, none known; +, weak; + +, moderate; + + +, severe.

of control. The recent inclusion of the filariases among the six tropical diseases that the World Health Organization considers the most important in the world has provided a further incentive for this work.

Our present knowledge of the immunology of filarial infections is quite imperfect and has many wide gaps, but the intense research activity that these two circumstances have generated is expected to provide many answers in the near future. The interested reader should consult the current scientific literature for the latest developments.

NATURAL RESISTANCE

In general, filarial worms are very specific parasites of their vertebrate hosts. *W. bancrofti* and *Mansonella ozzardi* affect only humans, and *O. volvulus* has been found naturally only in the monkey and the gorilla besides humans (chimpanzees accept experimental infections, however). The African species *Loa loa* infects humans and monkeys, but biological and pathological differences suggest that the anthropophilic and the zoophilic worms are at least different *parasite strains*. The Asiatic filaria *Brugia malayi* is found in humans, monkeys, and felids, but several indications also suggest that the infectivity for humans and lower mammals may depend on the strain of the parasite (Denham and McGreevy, 1977). The limited information available on human *Dipetalonema* species indicates that they also exist in apes. *D. immitis* is found in canids and felids, but in the latter the prevalence is much lower and patency is infrequent, which may indicate reduced susceptibility. A few dozen cases of *D. immitis* dirofilariasis have been reported in humans, but in virtually all cases a single worm, dead and encapsulated in the lung tissue, was found. The reasons for filarial specificity are unknown, but some scattered information has been collected.

One of the most critical phases of the filarial life cycle, and an obvious opportunity for natural resistance, is the invasion of a new host. The infective larvae are deposited on the skin of the selected host with a drop of fluid, and penetration through the puncture made by the vector while feeding must be achieved in a few minutes; otherwise, the larvae will dessicate and die. It therefore appears that the *microecological conditions* of the host's skin should have a considerable bearing on the ability of the parasite to survive long enough to initiate the invasion. The consistency of the infection of macaques injected with infective larvae of *Dirofilaria* species of carnivores, as opposed to the comparatively rare infection of humans in nature, suggests that skin penetration is an important barrier for human infection; probably the naked skin of humans favors the rapid dessication and death of the infective larvae.

The influence of *host's genetic factors* in susceptibility may be indicated by reported correlations between blood groups and filariasis prevalence and by relationships between race and severity of *W. bancrofti* filariasis. What these factors are and how they operate remain to be determined, however.

The *sex of the host* also appears to exert some influence on the natural resistance to the infection. Numerous surveys have demonstrated that *W. bancrofti* (especially the periodic variety in the South Pacific), *O. volvulus*, and *Dipetalonema perstans* are more prevalent in males, even when women are exposed to the infection as much as men. The same has been reported for *D. immitis* in dogs and for rodent filariae. That this difference is due to sexual hormones is suggested by the facts that it is not apparent in children and that it diminishes in populations of women older than 45 years of age.

Whatever the mechanisms of natural resistance are, they must operate during the invading generation of worms, because in several instances it has been demonstrated that microfilariae of domestic animals survive for some weeks when injected into laboratory animals. The infective larvae of *Setaria digitata* of the peritoneal cavity of bovines are unable to mature in sheep, goats, and horses, but they survive and behave as migratory larvae, causing extensive damage to the cerebrospinal system of these hosts.

ACQUIRED IMMUNITY

Infections by filarial nematodes result in the production of a variety of antibodies (Ogilvie and Worms, 1976). IgG and IgE antibodies have been identified by indirect means in humans, and IgM, IgG, and IgE antibodies have been demonstrated in dogs infected with *D. immitis*. These three classes have been also found in most laboratory models.

Cell-mediated immunity, manifested by delayed skin reactions, has been reported in humans by some authors, but others have been unable to demonstrate it. Recently, it was found that only the patients with the localized form of onchocercal dermatitis called "sowda" presented positive delayed skin tests, whereas those with the generalized form had only immediate-type reactions. Transformation of peripheral lymphocytes with extracts of *D. immitis* was negative in dogs infected with the homologous parasite, but inhibition of the leukocyte migration has been verified in human filariases.

The protective effect of the immunity, however, appears to be weak at best. Even in the presence of demonstrable immune responses, filarial infections are typically chronic events: *W. bancrofti* lives in the host for more than 10 years, *O. volvulus* for about 16 years, and *D. immitis* for about 7 years. Microfilariae of *M. ozzardi* or *D. perstans* injected in normal people or of *D. immitis* or *D. repens* injected in normal dogs persist for years in their circulation. Curiously, microfilariae of *W. bancrofti* or *L. loa* injected in noninfected humans disappear in less than a week (Nelson, 1966). This may indicate that the host's response is stronger against pathogenic filariae and is a further suggestion that immune reactions may play an important role in the production of pathology (see below).

Despite the general inefficiency of the effect of immunity on the current infection, there is some indirect evidence in humans and some experimental findings in lower animals that indicate some degree of protection against reinfections. The fact that adult parasites in humans and dogs do not accumulate as would be expected if there were continuous effective transmission, has been taken as a demonstration that older individuals in endemic areas are somewhat resistant to superimposed infections. This hypothesis was supported by the experiments of Bertram (1953), who found that *Litomosoides carinii* did not accumulate in the cotton rat (*Sigmodon hispidus*) over the parasitic burden produced by a single infection, when he established a system that allowed continuous transmission. This and other experiments demonstrated that the early infections produced a degree of resistance that retarded growth, delayed the moulting, and permanently stunted the parasites of a challenge. The finding of encapsulated masses of worms in various stages of destruction suggested that the immunity was lethal for at least some parasites (Otto, 1970). Recent experiments with *Brugia pahangi* in cats (Denham and McGreevy, 1977) showed that repeated infections over a prolonged period may produce almost complete resistance to an ulterior challenge. The fact that the prevalence of human filariasis in endemic areas levels off at about the third decade of life may suggest that immunological refractoriness also occurs in humans; this is not necessarily induced only by previous terminated infections with human filariae, but also possibly through multiple abortive infections with filarial worms of lower animals.

Several studies have shown lack of correlation between microfilaremia and adult parasite burdens, and it is known that the phenomenon of "occult filariasis," in which circulating microfilariae are absent although gravid females exist in the body, occurs in humans and in about 5% of dogs with *D. immitis*. These observations and the reports that microfilaremia is often absent in individuals with elephantiasis caused by the periodic strain of *W. bancrofti* (long-standing infections) or in individuals that suffered one episode of acute filariasis and left the endemic area (no further infections) suggested that prolonged antigenic stimulation or stimulation undisturbed by new infections resulted in immune responses that suppressed the presence of microfilariae in the circulation. In support of this thesis, experiments with *L. loa* in madrills showed that reinfections result in reduced microfilariemia and that repeated injections of microfilariae of *D. immitis* in noninfected dogs also produced decreasing microfilaremias. Some of the cats infected repeatedly with *B. pahangi* exhibited a precipitous termination of the microfilaremia that is reminiscent of the self-cure phenomenon described in *Haemonchus contortus* infections. The picture emerging from clinical and epidemiological observations and supported by experimental findings is that filarial infections produce immunological responses that little affect the current parasites and are only marginally protective in reinfections.

Recent work has demonstrated that at least some filariae are able to depress the host's immune responsiveness. The fact that immune protection becomes stronger with prolonged or repeated infections, however, may indicate that the main cause for filarial parasites to persist in the host is that they produce functional antigens insufficient in quality or quantity to elicit an effective response. This is not totally unexpected since parasites that achieve such an intimate contact with the host's tissues as filarial worms do would have had little chance for evolutionary survival if they were strongly immunogenic.

MECHANISMS OF ACQUIRED RESISTANCE

Very little is known about the mechanisms that affect the adult and preadult stages of filarial worms in the host. Vaccination of dogs with irradiated third-stage larvae of *D. immitis* produced almost total resistance to the development of the adults of a challenge, but only when this was given 3 or 4 months after the immunizing inoculation. The gap period before protection appears may indicate that the functional antigens are actually produced by later larval stages; the timing of the protection suggests that it is induced by late fourth-stage larvae (Knight, 1977). Denham and McGreevy (1977) have published results that indicate that cats undergoing numerous prior infections with *B. pahangi* develop an immune response lethal to the infective third-stage larvae of a challenge. Studies with *L. carinii* in rodents have also shown that the third-stage larvae are the best inducers of, and are most affected by, immune protection. The protection induced in cotton rats against *L. carinii* was produced by repeated infections, but not with dead parasites, which suggests that only metabolic antigens are functional.

Although all the developmental stages of the filariae appear to produce some degree of protection and to suffer its consequences to some extent, the most active (and the most studied) in these respects are the microfilariae. That the suppression of microfilaremia is an immunological event was demonstrated by the restoration of the microfilaremia when gravid *D. immitis* females were transferred from amicrofilaremic-infected dogs to noninfected dogs, and when amicrofilaremic hamsters with *Dipetalonema vitae* were treated with immunosuppressors. The suppression of the microfilaremia by injections of serum from dogs immunized with living microfilariae of *D. immitis* or *B. pahangi* indicates that the phenomenon is mediated by antibodies.

The exact function of the antibodies, however, has not been well identified. *Loa loa* of monkeys is destroyed in the spleen of the natural host, but there is no indication that the same happens in loaisis of humans, *L. carinii* infections of cotton rats, or *D. repens* infections of apes. However, splenic granulomas occur in *W. bancrofti* filariasis of humans, which might represent a defensive reaction against the parasite. Rats infected with *L. carinii* and monkeys injected with *W. bancrofti* produce antibodies that precipitate on the homologous microfilariae in vitro, but their significance in the suppression of the microfilaremia is unknown.

In vivo experiments with extracorporeal circulation in dogs have confirmed the prior in vitro observation that the serum of infected animals agglutinated the microfilariae. Possibly this aggregation of the parasites assists in their removal from the circulation. The existence of the phenomenon of occult filariasis indicates that the suppression of the circulating parasites must be very rapid or take place at the moment of larviposition.

Kobayakawa and collaborators reported recently from Japan that microfilariae of *D. immitis* were killed in vitro and in vivo by cells of the peritoneal exudate of normal guinea pigs when anti–*D. immitis* serum was added to the system. Since the supernatant fluid of the precipitation of the antiserum with ammonium sulfate was more active than the sedimented fraction (which contains the immunoglobulins), and the serum was active already 5 days after immunization, they felt that a substance other than an antibody (a lymphokine?) was involved. Tanner and Weiss (1978) subsequently found that mononuclear cells of the peritoneal exudate of normal hamsters adhered to microfilariae of *D. vitae* only when IgM antibodies of immune hamsters were added to the system. They felt that macrophages recognized the microfilariae–IgM antibody complexes and adhered to them as a first step in the trapping of the parasites.

The scant information available at present seems to indicate that the defense against filarial worms is mediated by humoral substances, mainly antibodies, possibly assisted by macrophages.

ARTIFICIAL PRODUCTION OF RESISTANCE

As with most nematodes, the limited experiments reported on induction of protection by inoculation of dead filariae or of their extracts have not succeeded. At least one experiment of vaccination of monkeys with supernatant fluid of cultures of *B. malayi* infective larvae did not produce resistance to a subsequent challenge. It appears, however, that this parasite does not release metabolic antigens in vitro unless it is briefly subjected to the host environment (in intraperitoneal diffusion chamber, for instance) before cultivation. Further investigation in this area is necessary.

The fact that repeated or prolonged infections or the inoculation of living microfilariae produce some degree of immune protection has stimulated a few attempts at vaccination with irradiated parasites. Inoculation of susceptible dogs with infective larvae of *D. immitis* irradiated with 20,000 rads produced almost total resistance to a homologous challenge; the protection took 3–4 months to develop and soon began to wane, however. Similar vaccination of dogs with *B. pahangi* gave lesser but still encouraging results. Vaccination of monkeys with irradiated *B. malayi* infective larvae also produced limited but significant protection to the homologous challenge. Attempts to vaccinate cats against *B. malayi* by the same procedures that were effective in dogs and in monkeys have been considerably less successful. This warns against extrapolating too readily the results gathered in one host species to other host species that might exhibit different reactivity. At any rate, the results so far obtained by vaccination with irradiated filarial parasites recommend the continuation of research along these lines.

The observations that onchocerciasis and *W. bancrofti* filariasis in humans are mutually exclusive and that *W. bancrofti* is less pathogenic in areas where the human population is likely to be intensively exposed to animal filariae have been interpreted as suggestions of heterologous immunity. Some preliminary evidence also indicates that dogs infected with *Dipetalonema reconditum* may have reduced susceptibility to infections with *D. immitis*.

Since there is evidence that humans react immunologically to the inoculation of animal filarial worms and that diverse species of filariae share antigens abundantly, it would not be too surprising to find that cross-protection actually occurs among filarial parasites.

IMMUNODIAGNOSIS

The examination of samples of blood or tissues is still the preferred method to diagnose filarial infections. These techniques, however, are inadequate during the prepatent period, in cases of occult filariasis, or in abortive (but symptomatic) infections of humans by animal filariae. Also, they are too time consuming for epidemiological surveys. For all these reasons, abundant effort has been put into devising reliable immunological procedures to identify these infections. Kagan (1963) produced a detailed review on this subject and Ambroise-Thomas (1974) wrote an update.

Two main circumstances that affect the immunological diagnosis of filariases must be kept in mind: filarial worms share important and numerous antigens, among themselves and with other helminths, which conspires against the specificity of the tests; and several systems of antibodies are stage specific (e.g., hemagglutinins and complement-fixing antibodies in canine dirofilariasis occur in connection with parasite moults and often disappear when microfilaremia is present), which demands the use of diverse immunological techniques at different periods of the infection. Also, the difficulties of obtaining antigens from the homologous parasites, particularly in the case of human filariases, has incited the wide use of heterologous antigenic preparations. *D. immitis* of dogs has been a favorite in this respect.

The immediate-type skin test has been widely used for the diagnosis of human filariasis, with amply divergent results: whereas some authors found a sensitivity from 70% to 90% with a specificity above 90% in humans using very diluted extracts of *D. immitis*, others reported many false-positive reactions. The use of a purified protein fraction has improved the test somewhat, but still it appears unsatisfactory. The antibodies responsible for this reaction must be elicited by the prolonged presence of microfilariae, since the reaction correlates with microfilaremia in adults but not in teenagers and the response is smaller in amicrofilaremic or treated patients. From the numerous and controversial reports in the literature, it seems that standardization of this test is urgently needed before making definitive statements in regard to its actual usefulness. A diagnostic procedure related to skin sensitivity is the Mazzotti test, in which the administration of diethylcarbamazine to patients with onchocerciasis often results in a cutaneous rash, presumably due to a reaction to antigens released by killed microfilariae.

The presence of delayed-type skin sensitivity in filariases has been affirmed by some investigators and denied by others. Recent reports indicate that only patients with the localized form of onchocercal dermatitis had cutaneous delayed hypersensitivity, although all forms of the infection gave immediate-type reactions. In one study, the inhibition of the leukocyte migration was positive in human patients serologically negative for filariae, and, in another, the lymphocyte transformation was negative in infected dogs with antibodies to *D. immitis*. It would be of interest to study whether there is a mutual exclusion between the antibody and the cell-mediated responses at some point in the evolution of the filarial infections.

Skin manifestations of the Arthus phenomenon (intermediate-tyr hypersensitivity) have been detected recently in all of 27 patients of onchocerciasis.

The complement-fixation test has also had mixed reviews by diverse authors. The presence of cross-reactions with normal serum and with serum of patients of hookworm disease, schistosomiasis, strongyloidiasis, ascariasis, paragonimiasis, and so forth, and the variable results of the same test with filariasis sera of different geographic origins, have brought discredit to this technique. The test has been very sensitive, however, in identifying tropical eosinophilia caused by filarial worms, and might be a useful tool if it were well standardized.

Precipitin tests have also gathered controversial reports, but the consensus seems to be that their sensitivity is too low and their cross-reactivity too great for practical use. In a recent study, the double diffusion test detected 41% of the cases and immunoelectrophoresis verified 73%. In another assay, 17 (71%) of 24 people from hyperendemic areas were positive by counterimmunoelectrophoresis. Reportedly, the patterns of the bands formed by immunoelectrophoresis with serum of human patients allow the differentiation among onchocerciasis, loaiasis, and wuchereriasis. A uroprecipitin test, in which the presence of antigen in the urine of the patient is detected by precipitation with a laboratory-raised antiserum, has been utilized with moderate success.

Agglutination tests with the antigen adsorbed to erythrocytes or to bentonite particles are routinely used in the U.S. Center for Disease Control, but their sensitivity varies from patient to patient and their cross-reactivity at low titer is extensive. In an evaluation, the indirect hemagglutination was positive in 73% of 37 patients and in 14% of almost 3000 controls. A more recent assay found only 57% sensitivity.

The indirect fluorescent antibody technique (IFAT), using adult worms, microfilariae, or soluble extracts as antigen, has given positive results in about 90% of the patients of diverse filariases, with very few cross-reactions when only titers above 20 are considered significant. A recent report (Grove and Davis, 1978) indicates that antiadult antibodies are found in all patients, whereas antimicrofilarial antibodies are present in only about 25% of them, and often are related to the existing pathology. Other authors have found that the IFAT detects only about 65% of the infections.

With the possible exception of the IFAT, there is no satisfactory technique for the immunological identification of filarial infections nowadays. The major problem appears to be the lack of purified antigenic preparations to provide the required sensitivity and specificity, this latter particularly to differentiate the most pathogenic from the less pathogenic species. It is hoped that they will be obtained in the near future as a result of the spur in research in filariases that has begun lately.

IMMUNOPATHOLOGY

Epidemiological, clinical, and histopathological considerations have convinced most investigators that the host's immune response is a major component in the pathogenesis of filariases. Although there is little solid proof of this, an impressive body of circumstantial evidence supports the presumption (Nelson, 1966).

The observation that severe symptoms of filariasis occur only in patients that are subjected to regular reinfections for many years in areas of endemia suggests that the disease is produced by the reactivity of the host rather than by the direct action of the parasite or its products. The facts that the specific immunity to filarial worms also develops very slowly and that the most serious damage (at least in cases of periodic *W. bancrofti* and *B. malayi* infections) coincides with the immunological suppression of the microfilaremia point to an association between the host's immune response and the parasite's pathogenicity.

During the course of all human filariases and of many filariases of domestic animals, it is common to observe peripheral eosinophilia and clinical manifestations of cutaneous or respiratory hypersensitivity, these latter often recurrent. The administration of the microfilaricidal drug diethylcarbamazine often aggravates or triggers the cutaneous symptoms in human filariases and occasionally produces an anaphylactic shock in dogs infected with *D. immitis*. This latter reaction has been reproduced by injection of microfilariae or of large doses of parasite extracts. All these phenomena are interpreted as manifestations of Type I hypersensitivity to bursts of antigens released by the parasites (during death, moulting, larviposition, etc.) in hosts sensitized by the prolonged current infection or by prior experiences with the worm. The consistent presence of eosinophilia, the clinical characteristics of these syndromes (which include enlargement of the lymph nodes), and a few reports of clinical improvement with inoculation of filarial extracts (desensitization?) favor the opinion that they are mediated by immune reactions rather than by the direct action of toxins.

This notion is also supported by the observation that those filariae that develop in abnormal hosts (e.g., *Macdonaldius oschei* of South American snakes in Asiatic snakes; *Setaria digitata* of cattle in horses, sheep, and goats; *B. malayi* of monkeys in cats as compared to *B. pahangi* of carnivores) are more pathogenic for these than for their habitual hosts. It seems easier to visualize these differences as dependent on the immune reactivity of the natural or abnormal host to parasite antigens than to assume such a regular variability of some hypothetical toxin, especially when the histopathology implies that an immunological reaction has taken place. Eventually, the presence of *Brugia* elephantiasis in humans, but not in carnivores, may be related to the effect of the bipedalism of the former on the mechanics of lymph circulation.

The histopathology of all filarial infections (at least in humans) appears to follow a similar trend: the early reactions to the parasites are edematous infiltrates containing eosinophils, plasma cells, and macrophages; neutrophils have been also reported by some authors. The same response is observed around collections of microfilariae that are presumably dead or dying. When the parasites remain stationary in the same location, the inflammation becomes chronic, adopting a granulomatous aspect first and developing into fibrous tissue later. Most of these advanced lesions appear around dead or dying worms, which has been taken as an indication that the response is triggered by the massive release of antigenic material.

The actual mechanism of the lesions is not known, but IgE, precipitating antibodies, and cell-mediated immunity have been demonstrated in filariases, so the possibilities are multiple. Recently, it has been reported that immediate and Arthus' hypersensitivity occur in all patients of onchocerciasis, but delayed hypersensitivity was only detected in those with the localized form of the disease. It is not known yet whether there is correspondence between a certain type of immunological reaction and a certain variety of clinical presentation, but this report and the finding of immune complexes in the kidneys of proteinuric dogs infected with *D. immitis* encourage further research.

Recent work with *B. pahangi* in cats (Denham and McGreevy, 1977) showed that the parasite in the lymphatic vessel produces a mild inflammation with accumulation of eosinophils, neutrophils, and plasma cells from the first week of infection. Toward the eighth week, the inflammation becomes chronic and the vessels are enormously distended; lymphoid follicles develop in the wall and lumen of the lymphatics, and granulomatous and fibrotic reactions appear in association with dead and disintegrating worms. The vessels become occluded and

218   Immunology of Parasitic Infections

their recuperation takes several years. Examination of the lymph nodes in such cats shows a very rapid cell-mediated–type response that begins to wane from the third month of infection onward and a more delayed antibody-type response that persists at high levels at least for 2 years. As in the case of human infections, these events are consistent with an immunological nature.

A few authors have defended the idea that damage by filarial worms is due mainly to toxins or to associate bacterial infections. Although the participation of these factors cannot be denied absolutely, all the available evidence favors the etiologic role of the host's immunity.

A syndrome of cough and dyspnea with eosinophilia and changing lung infiltration on x rays has been called "eosinophilic lung" or "tropical pulmonary eosinophilia." Because the infection subsided with diethylcarbamazine therapy, a filarial etiology was suspected; this was later confirmed by finding microfilariae in the lungs but not in the blood of patients. All indications are that this condition is allergic, but it has not been determined yet whether the causative filariae are human or lower animal parasites in origin. A clinically similar condition, often called Loeffler's syndrome, may be produced by *Ascaris* or *Strongyloides* migratory larvae in hypersensitive subjects.

It has been recently reported that at least some filarial worms produce a reduction of the immune reactivity of the host (see Chapter 7); the significance of this finding for the course of the homologous infection is unknown yet. The fact that specific antifilarial resistance increases with repeated infections suggests, nevertheless, that depression of the immunity is not the major mechanism responsible for the long-lasting presence of the filariae in their hosts.

SOURCES OF INFORMATION

Ambroise-Thomas, P. 1974. Immunological diagnosis of human filariasis: Present possibilities, difficulties, and limitations (a review). Acta Tropica 31:108–128.
Armour, J., and Bruce, R. G. 1974. Inhibited development in *Ostertagia ostertagi* infection—A diapause phenomenon in a nematode. Parasitology 69:161–174.
Banwell, J. G., and Schad, G. A. 1978. Hookworm. Clin. Gastroenterol. 7:129–156.
Barriga, O. O. 1977. Reactivity and specificity of *Trichinella spiralis* fractions in cutaneous and serological tests. J. Clin. Microbiol. 6:274–279.
Barriga, O. O. 1980. Evidence, nature and implications of the constitutive resistence to *Trichinella spiralis* in gallinaceous birds. Am. J. Vet. Res. (in press)

Befus, A. D., and Podesta, R. B. 1976. Intestine. In: C. R. Kennedy (ed.), Ecological Aspects of Parasitology, pp. 303–325. North-Holland Publishing Co., Amsterdam.

Benitez-Usher, C., Armour, J., and Urquhart, G. 1976. Studies on immunization of suckling calves with *Dictyocaulus*. Vet. Parasitol. 2:209–222.

Bertram, D. S. 1953. Labortory studies on filariasis in the cotton-rat. Trans. R. Soc. Trop. Med. Hyg. 47:85–106.

Butterworth, E. A. 1977. The eosinophilia and its role in immunity to helminth infections. Curr. Top. Microbiol. Immunol. 77:127–168.

Casarola, L., and Orlandi, M. 1977. Immune mechanisms in lambs infected with *Dictyocaulus filaria*. Ann. Fac. Med. Vet. (Pisa) 30:117–129.

Castellino, J. B. 1970. Immunological response in pigs to nematode infections with emphasis on serodiagnosis. Vet. Bull. 40:751–758.

Catty, D. 1976. Immunity and acquired resistance to trichinosis. In: S. Cohen and E. H. Sadun (eds.), Immunology of Parasitic Infections, pp. 359–379. Blackwell Scientific Publications, Oxford.

Clegg, J. A., and Smith, M. A. 1978. Prospects for the development of dead vaccines against helminths. Adv. Parasitol. 16:165–218.

Connan, R. M. 1976. Effect of lactation on the immune response to gastrointestinal nematodes. Vet. Rec. 99:476–477.

Denham, D. A., and McGreevy, P. B. 1977. Brugian filariasis: Epidemiological and experimental studies. Adv. Parasitol. 15:243–309.

Despommier, D. D., Campbell, W. C., and Blair, L. S. 1977. The *in vivo* and *in vitro* analysis of immunity to *Trichinella spiralis* in mice and rats. Parasitology 74:109–119.

Dineen, J. K. 1963. Immunological aspects of parasitism. Nature 197:268–269.

Dobson, C. 1972. Immune response to gastro-intestinal helminths. In: E. J. L. Soulsby (ed.), Immunity to Animal Parasites, pp. 191–222. Academic Press, Inc., New York.

Glickman, L., Schantz, P., Dombroske, R., and Cypess, R. 1978. Evaluation of serologic diagnostic tests for visceral larva migrans. Am. J. Trop. Med. Hyg. 27:492–498.

Goetzel, E. J., and Austen, K. F. 1977. Cellular characteristics of the eosinophil compatible with a dual role in host defense in parasitic infections. Am. J. Trop. Med. Hyg. 26(suppl.):142–150.

Grove, D. I., and Davis, R. S. 1978. Serological diagnosis of bancroftian and malayan filariasis. Am. J. Trop. Med. Hyg. 27:508–513.

Grove, D. I., Hamburger, J., and Warren, K. S. 1977. Kinetics of immunological responses, resistance to reinfection, and pathological reactions to infection with *Trichinella spiralis*. J. Infect. Dis. 136:562–570.

Ishizaka, T., et al. 1976. Immunoglobulin E synthesis in parasite infection. J. Allergy Clin. Immunol. 58:523–538.

Kagan, I. G. 1963. A review of immunological methods for the diagnosis of filariasis. J. Parasitol. 49:773–798.

Kagan, I. G. 1976. Serodiagnosis of trichinosis. In: S. Cohen and E. H. Sadun (eds.), Immunology of Parasitic Infections, pp. 143–151. Blackwell Scientific Publications, Oxford.

Kagan, I. G., and Norman, L. G. 1970. The serology of trichinosis. In: S. E. Gould (ed.), Trichinosis in Man and Animals, pp. 222–268. Charles C Thomas Publishers, Springfield, Ill.

Kazura, J. W., and Grove, D. I., 1978. Stage-specific antibody-dependent eosinophil-mediated destruction of *Trichinella spiralis*. Nature 274:588–590.

Kelly, J. D., Kenny, D. F., and Whitlock, H. V. 1977. The response to phytohemagglutinin of peripheral blood lymphocytes from dogs infected with *Ancylostoma caninum*. N. Z. Vet. J. 25:12–15

Khoury, P. B., and Soulsby, E. J. L. 1977. *Ascaris suum*: Immune response in the guinea pig. Exp. Parasitol. 41:141–159.

Knight, D. H. 1977. Heartworm heart disease. Adv. Vet. Sc. Comp. Med. 21:107–149.

Kozar, Z. 1971. Some current immunological aspects of trichinellosis. Wiad. Parazytol. 17:503–540.

Larsh, J. E., Jr. 1970. Immunology. In: S. E. Gould (ed.), Trichinosis in Man and Animals, pp. 129–143. Charles C Thomas, Springfield, Ill.

Larsh, J. E., and Weatherly, N. F. 1975. Cell-mediated immunity against certain parasitic worms. Adv. Parasitol. 13:183–222.

Love, R. J., Ogilvie, B. M., and McLaren, D. J. 1976. The immune mechanism which expels the intestinal stage of *Trichinella spiralis* from rats. Immunology 30:7–15.

Michel, J. F. 1974. Arrested development of nematodes and some related phenomena. Adv. Parasitol. 12:279–366.

Miller, T. A. 1971. Vaccination against the canine hookworm disease. Adv. Parasitol. 9:153–183.

Morgan, D. O., Parnell, I. W., and Rayski, C. 1951. The seasonal variations in the worm burden of Scottish hill sheep. J. Helminthol. 25:117–212.

Nelson, G. S. 1966. The pathology of filarial infections. Helm. Abstr. 35:311–336.

Nelson, G. S. 1970. Onchocerciasis. Adv. Parasitol. 8:173–224.

Ogilvie, B. M., and Jones, V. E. 1973. Immunity in the parasitic releationship between helminths and hosts. Prog. Allergy 17:93–144.

Ogilvie, B. M., Mackensie, C. D., and Love, R. J. 1977. Lymphocytes and eosinophils in the immune response of rats to initial and subsequent infections with *Nippostrongylus brasiliensis*. Am. J. Trop. Med. Hyg. 26:61–67.

Ogilvie, B. M., and Worms, M. J. 1976. Immunity to nematode parasites of man. In: S. Cohen and E. H. Sadun (eds.), Immunology of Parasitic Infections, pp. 380–407. Blackwell Scientific Publications, Oxford.

Otto, G. F. 1970. Insect-borne nematodes. In: G. J. Jackson, R. Herman, and I. Singer (eds.), Immunity to Parasitic Animals. Vol. II, pp. 913–980. Appleton-Century-Crofts, New York.

Poynter, D. 1966. Some tissue reactions to the nematode parasites of animals. Adv. Parasitol. 4:321–383.

Poynter, D. 1968. Parasitic bronchitis. Adv. Parasitol. 6:349–359.

Rose, J. H. 1973. Lungworms of the domestic pig and sheep. Adv. Parasitol. 11:559–599.

Sauerbrey, M. 1977. A precipitin test for the diagnosis of human abdominal angiostrongyliasis. Am. J. Trop. Med. Hyg. 26:1156–1158.

Schad, G. A. 1977. The role of arrested development in the regulation of nematode populations. In: G. Esch (ed.), Regulation of Parasite Populations, pp. 111–167. Academic Press, Inc., New York.

Sinclair, I. J. 1970. The relationship between circulating antibodies and immunity to helminth infections. Adv. Parasitol. 8:97–138.

Smith, J. W., and Wootten, R. 1978. *Anisakis* and anisakiasis. Adv. Parasitol. 16:93–163.

Stoll, N. R. 1929. Studies with the strongyloid nematode *Haemonchus contortus*. I. Acquired resistance of host under natural reinfection conditions out of doors. Am. J. Hyg. 10:384–418.

Stone, O. J., Newell, G. B., and Mullins, J. F. 1972. Cutaneous strongyloidiasis: *Larvae currens*. Arch. Dermatol. 106:734–736.

Taliaferro, W. H., and Sarles, M. P. 1939. The cellular reactions in the skin, lungs and intestine of normal and immune rats after infection with *Nippostrongylus muris*. J. Inf. Dis. 64:157–192.

Tanner, M., and Weiss, N. 1978. Studies on *Dipetalonema vitae* (Filaroidea). Acta Tropica 35:151–160.

Thorson, R. E. 1970. Direct infection nematodes. In: G. J. Jackson, R. Herman, and I. Singer (eds.), Immunity to Parasitic Animals. Vol. II, pp. 913–961. Appleton-Century-Crofts, New York.

Torres, P., and Barriga, O. O. 1975. Inter- and intra-specific antigenic relationships among some Ascaroidea. Acta Parasitol. Polon. 23:441–451.

Urquhart, G. M., and Armour, J. (eds.). 1973. Helminth Diseases of Cattle, Sheep and Horses in Europe. University of Glasgow, Glasgow.

Viens, P. 1977. La "larva migrans viscérale" à Montreal ou le somet de l'iceberg. Bordeaux Med. 10:697–698.

Wakelin, D. 1978. Immunity to intestinal parasites. Nature 273:617–620.

Warren, R. S. 1978. Dynamics of host responses to parasite antigens: Schistosomiasis and trichinosis. In: Colloque INSERM: Immunity in Parasitic Diseases, pp. 25–38. INSERM, Paris.

Chapter 5

Immune
Reactions to
Parasitic Platyhelminths

THE CESTODES OR TAPEWORMS

The cestodes that affect humans and domestic animals are characteristically inhabitants of the digestive tract of their definitive hosts, especially the small intestine, in their adult stage. The larval forms are dwellers in extraintestinal tissues of the intermediate hosts. Those of importance to medicine belong to the orders Pseudophyllidea and Cyclophyllidea.

The Pseudophyllidea (genera *Diphyllobothrium* or *Dibothriocephalus*, *Spirometra*, and others) are somewhat related to the digenetic trematodes. Bears, humans, dogs, and cats are their common definitive hosts. Two intermediate hosts are required to complete their life cycle: a copepod crustacean of the zooplankton (e.g., *Diaptomus*, *Daphnia*) for the development of the first larval stage or *procercoid*; and a vertebrate for the formation of the second larval stage or *plerocercoid* (or *sparganum*). The second intermediate host is a fish in the genus *Diphyllobothrium*, whereas, excepting fish, any vertebrate, including humans, may play that role in the genus *Spirometra* (Figure 12).

The Cyclophyllidea contain six families of medical importance, with *Taenidae* being the most prominent. The members of the *Taenidae* (genera *Taenia*, *Echinococcus*, and others) are transmitted by carnivorism in nature, with the adult cestode found in predator mammals and the larval stages in prey herbivores (Figure 13). Humans are the exclusive definitive host of *Taenia solium* and *Taenia saginata*, and the incidental intermediate host of *T. solium*, *Echinococcus granulosus*, and *Echinococcus multilocularis*. Other larval *Taenidae* are sporadically reported in humans.

Most of the remaining medically important Cyclophyllidea are parasites of the digestive tract of herbivores and utilize invertebrates, especially arthropods, as intermediate hosts. The genus *Hymenolepis*,

223

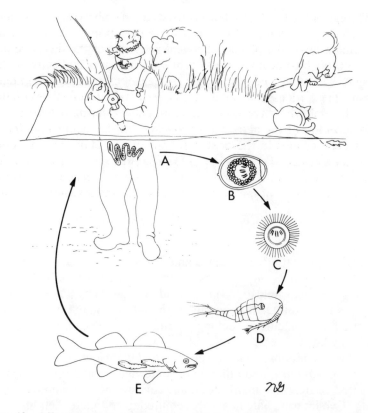

Figure 12.   Life cycle of a Pseudophyllidea cestode, *Diphyllobothrium latum*. The adult parasite lives in the small intestine of humans (**A**) and other piscivorous mammals. The eggs (**B**) that reach a fresh water environment release a swimming larva (*coracidium*, **C**) that is eaten by microcrustaceans of the plankton (*copepods*, **D**) and develop into a new larval stage (*procercoid*) in this first intermediate host. When an infected crustacean is ingested by an appropriate fish (**E**), the procercoid develops into a more advanced form (*plerocercoid*) and becomes ready to develop to the adult stage when consumed by the definitive host (**A**).

common in birds and rodents, has often been used as a convenient experimental model. *Hymenolepis nana* deviates somewhat from the common life pattern of the Cyclophyllidea in that it can use either arthropods or the intestinal villi of the definitive host as the environment to develop its larval stage (Table 4).

The process of infection, the knowledge of which is essential to understand the immunology to cestodes, has been studied for a few species. The general account that follows is based mainly on work done with *Taenidae* and *Hymenolepis*. For further details the reader should consult Smyth (1969) and Gemmell and Macnamara (1972).

The egg of the Cyclophyllidea consists of an embryo or *oncosphere* surrounded by a number of coverings. On ingestion of the egg by an intermediate host, the physicochemical activities of the digestive apparatus of this host destroy most of the egg envelopes and activate the oncosphere to break open the innermost egg membrane, and to initiate invasion. This process may entail active enzymatic mechanisms on the part of the embryo, since some secretion occurs during hatching.

Penetration of the activated oncosphere in the mammalian intestinal mucosa ensues in the next half-hour or so. Many investigators have demonstrated the presence of glands and secretions in the invading oncosphere and, although not all have succeeded in showing enzymatic alteration of the host's structures during invasion, it seems

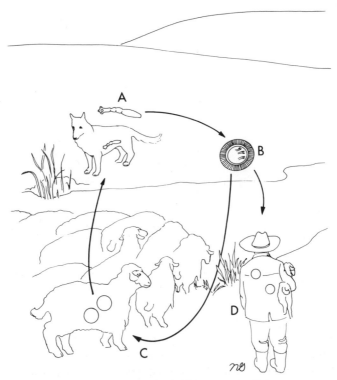

Figure 13.   Life cycle of a taeniid Cyclophyllidea cestode, *Echinococcus granulosus*. This tiny tapeworm (**A**) lives in the small intestine of dogs. Most often its eggs leave the host still in the gravid proglottid. Dessication releases the eggs (**B**), which contaminate pastures, water and, occasionally, the coat of the definitive host itself. Ingestion of contaminated grass or water by the intermediate hosts (**C**) allows the development of the larval parasite (*hydatid*), particularly in their liver and lungs. Consumption of these visceras by the definitive host results in the growth of new adult worms (**A**). Occasional ingestion of eggs by man causes human hydatidosis (**D**). Families of Cyclophyllidea other than *Taenidae* generally use invertebrates, especially arthropods, as intermediate hosts.

Table 4.   Hosts of the cestodes mentioned in the text

Species	Definitive hosts[a]	Intermediate hosts[a]
Diphyllobothrium (*Dibothriocephalus*) *latum*	Bears, humans, dogs, others	1st copepods; 2nd fish
Spirometra mansonoides	Cats	1st, copepods;
S. erinacei	Cats, dogs	2nd, vertebrates except fish
Taenia crassiceps	Foxes (dogs)	Rodents
T. hydatigena	Dogs	Sheep (others)
T. (*Multiceps*) *multiceps*	Dogs	Sheep (others)
T. ovis	Dogs	Sheep
T. pisiformis	Dogs (cats)	Rabbits
T. (*Taeniarhynchus*) *saginata*	Humans	Cattle
T. solium	Humans	Swine (humans, others)
T. (*Hydatigera*) *taeniformis*	Cats	Rats (mice)
Echinococcus granulosus	Dogs (foxes)	Sheep, others (humans)
E. multilocularis	Dogs, foxes (cats)	Rodents (humans, others)
Hymenolepis citelli	Ground squirrels (rats, mice)	Grain insects
H. diminuta	Rats (mice, humans)	Grain insects, roaches
H. microstoma	Mice (rats)	Grain insects
H. nana	Mice, humans	Mice, humans; grain insects
Dipylidium caninum	Dogs, cats (humans)	Fleas, lice
Davainea species	Poultry	Snails, slugs
Raillietina species	Poultry	Flies, others
Anoplocephala species	Horses	Mites
Moniezia expansa	Sheep (others)	Mites

[a] Some hosts considered abnormal but not unusual are in parentheses.

likely that penetration is assisted by the production of enzymes on the part of the embryo.

Once the oncosphere reaches a venule of appropriate caliber in the villus, it is taken by the bloodstream to its final location or to the neighborhood of that location, with the embryo reaching the ultimate site by active migration. A period of intense metabolic activity follows in which the embryonic cells of the oncosphere reorganize, multiply, form a cavity, and give origin to the larval scolex (*protoscolex*) and its surrounding membranes. Evidence of migration of the early larvae, before appearance of the protoscolex or hooks, suggests that enzymes to assist in the migration are produced by the parasite at this stage.

The outer layer of the cestode larvae is a biologically active membrane able to absorb nutrients and eliminate waste. The presence of a peripheral thick mucopolysaccharide membrane of parasite origin or of a dense connective tissue response by the host, as occurs in *E. granulosus*, does not appear to interfere in an important way with the permeability of the parasite structures.

For the life of the parasitic species to continue, it is necessary that the larval cestode be ingested by a definitive host. On ingestion of the larva, its cystic membranes are digested and the protoscolices become evaginated. Evagination is an active process on the part of the parasite, but needs to be stimulated by specific physicochemical conditions of the host's digestive tract.

The evaginated protoscolices attach to the host intestinal mucosa, at a site and with a depth that are mostly characteristic for the particular species of parasite, and begin to form a body or *strobila*. The report that neurosecretory material has been found in the scolex of *Hymenolepis diminuta* during primary strobilization, but not in the adult worm or in the larva, suggests that the formation of the strobila may be directed by a neurosecretory mechanism. Further strobilization, organogeny, and maturation will continue for the entire life of the adult cestode, while the last segments or *proglottids*, filled with eggs, become detached or explode regularly.

The morphology of the external layer of the cestode tegument is very similar to the absorptive-secretory surface of the host intestinal cells, with which it is in close contact. A number of enzymes have been detected in this tegument, but it is not known whether they are sesile or are actually released next to the host absorptive surface. In the latter case they might behave as important antigens.

NATURAL RESISTANCE

Natural resistance to cestodes has been recently reviewed in detail by Weinmann (1970) and summarized by Gemmell and Macnamara (1972) and by Gemmell (1976). Smyth (1968) and Yamashita (1968) have written with particular reference to *Echinococcus*.

Adult Cestodes

Cestodes, especially the adult stages, have a remarkable specificity of host: whole groups of cestodes are restricted to particular orders of mammals or birds, and many individual species of tapeworms will develop only in the appropriate species of definitive host. This specificity may result from several parasite- or host-related factors.

Genetic Restrictions Specificity may be a result of genetic restrictions imposed by the habits of the parasites. Many cestodes self-fertilize frequently (e.g., *T. saginata* and *T. solium*) and some larvae multiply vegetatively (e.g., *E. granulosus*); these processes of inbreeding tend to form genetically uniform populations that may lack the genetic plasticity to adapt to environments slightly different from the intestine of the usual definitive hosts. Herd (1977) has reported a remarkable resistance to *E. granulosus* in some dogs that might have a genetic basis. Extension of these studies may indicate the existence of subpopulations with diverse susceptibility within the same host species.

Host Stimuli   The first problem faced by a larval cestode on ingestion is to find the adequate host stimuli for evagination. In the few cases in which something is known, these factors may vary from a single enzyme (e.g., pepsin) to a complex combination of digestive secretions for various tapeworms. In most cases, bile is required to initiate or to accelerate the process of evagination. Smyth (1969) found that deoxycholic acid, or herbivore bile, which is rich in it, lyses the protoscolices of *Echinococcus*, whereas carnivore bile stimulates the rate of evagination. It is possible that bile composition represents a biochemical factor determining specificity by assisting in the development or by killing the parasite. There are isolated reports that indicate that other biochemical factors, such as pH, gas concentrations, and temperature, are important for the establishment of the cestode in the definitive host, but there seem to be no systematic studies in this area.

Structural or Chemical Conditions   Indirect evidence indicates that the adult cestode requires rather restricted structural or chemical conditions for establishment and development. It has been shown for several species that the developing tapeworms colonize a site of the intestine for a few days on infection, and move, anteriorly or posteriorly, to the typical habitat of the adult later on. It is not clear, however, whether the changing morphology and biochemistry of the worm during development demand diverse structural or physiological environments at different times. It does seem certain that the parasite will not persist in a host that fails to provide the necessary conditions.

The influence of chemical factors is evidenced by the profound effect of the host diet on the parasites: starvation or restriction of carbohydrate intake (or of fatty acids to a lesser extent) interfere with the establishment, growth, maturation, and longevity of some tapeworms. Variations in the intake of proteins and vitamins, on the other hand, have negligible effect on the cestodes, perhaps because the parasites can obtain these nutrients directly from the tissues of the host. In some cestodes, a circadian rhythm has been detected by which the worms migrate along the alimentary canal in correspondence with the meals of the host. Evidently, only vertebrate species whose diets coincide with the requirements of a particular cestode will be regarded as "appropriate" hosts by it.

Histological observations suggest that the morphology of the intestine also plays a role in the natural resistance to adult tapeworms. It appears reasonable that a small scolex will necessitate a different substrate for attachment than a larger scolex: actually, the protoscolices of *E. granulosus* establish themselves deeply within the intestinal crypts of the dog during the early stages of the infection, but move out toward the lumen as the scolex and the suckers grow. Differences in the microanatomy of the intestinal mucosa among carnivores might have some influence on the exclusive susceptibility of the dog to this parasite. Furthermore, the depth that the scolex reaches in the definitive host varies in different cestodes: although it has a rather superficial adherence in the human *T. saginata*, it buries very deeply in *Davainea* and *Raillietina* of chickens and in *Anoplocephala* of horses. Again, only the host that offers an adequate structural substrate will be appropriate for the development of the tapeworm.

Experiments in vitro have also indicated the importance of a suitable physical substrate to direct the physiology of the tapeworms: protoscolices of *E. granulosus* form strobila in a biphasic medium with coagulated protein, but not in a monophasic fluid medium or when the solid substrate is agar.

Some morphological characteristics of the intestine may actually act through physiological mechanisms. In the 1940s it was demonstrated that the resistance of young mice to infection with *H. nana* was due to the shortness of their small intestine and that a correlation existed between intestinal emptying time and number of parasites that became established.

Although abundant work is still needed in this area, it appears to be well demonstrated that the natural susceptibility to tapeworm infections depends partially on the appropriate combination of physicochemical conditions in the alimentary tract of the definitive host. Nevertheless, adult cestodes may tolerate some latitude in the physicochemical conditions of their habitats, since eventually they will become established in some abnormal hosts (e.g., *H. nana* in hamsters and grey squirrels). That the situation here is less than ideal is denoted by the fact that the growth rate of the tapeworm is usually depressed. The proper set of conditions, however, may occur at an intestinal location different from that in the habitual host, since the site of establishment of the parasite in the intestine varies from one host species to another.

Presence of Other Digestive Organisms Apart from the inherent physicochemical characteristics of the host's intestine, the presence of other digestive organisms may influence the susceptibility to cestode infections. Apparently the normal intestinal flora of guinea pigs is deleterious for *H. nana*, because the parasite grows into mature adults in germ-free animals but develops poorly in conventional guinea pigs. Similarly, elimination of bacitracin-sensitive flora in rats promotes the growth of *H. diminuta*. Simultaneous infections of mice with *H. nana* and with *Nippostrongylus brasiliensis* or *Trichinella spiralis* also reduces the susceptibility to the tapeworm. The fact that no modifications of the susceptibility were detected when the nematode infection preceded the tapeworm infection by several days indicates that the phenomenon was not an expression of acquired cross-immunity. These phenomena may or may not be related to the *crowding effect*, frequently observed in tapeworm infections, in which the size of individual worms is inversely related to the number of specimens in the intestine of the host. The rate of maturation of the tapeworms is not usually affected, however. Some research done on the crowding effect has demonstrated that competition for carbohydrates plays an important role in its production.

Ecological Opportunity   To a limited extent, lack of ecological opportunity for infection may have some influence on host specificity in some cases. Human infections with the rat tapeworm *H. diminuta* or the dog tapeworm *Dipylidium caninum* are not rare events when the appropriate conditions for transmission occur.

Infection by Single Tapeworms   A yet-unexplained phenomenon is the frequency with which some tapeworms (*T. saginata*, *T. solium*, *Diphyllobothrium latum*) are found in the human intestine as single specimens. It has been variously hypothesized that the infection of the intermediate hosts would not be prevalent enough for humans to ingest more than one larva during the life span of the worm in most cases; or that the existing parasite would induce strong resistance against reinfections; or that the tapeworm would secrete substances inhibitory to the development of other specimens; or that competition for some essential threshold nutrient would allow the growth of only one tapeworm.

It seems that in areas such as the Middle East and North Africa, from which up to 37% of human infection by *T. saginata* has been reported, and with a parasite whose life span is estimated to be up to 25 years, there must be plenty of opportunity for multiple primary infections or for reinfection to occur. Nevertheless, most infections in these areas are still with "solitary" worms. Substances that will inhibit development have not been detected in other tapeworms; neither the degree of acquired immunity or competition for nutrients that will explain the absolute absence of accompanying specimens has been observed.

A possibility is that the host can tolerate only a threshold amount of parasite biomass before the immunological or nonimmunological mechanisms of defense begin to operate. This hypothesis is supported by the finding that the efficacy of the immunity to *H. diminuta* reinfections in mice is proportional to the size of the infective dose and by the existence of the crowding effect. In the case of cestodes as monstrously large as the human "solitary" tapeworms, a single specimen might provide enough biomass to trigger an effective immune response or to deplete a threshold nutrient. In any case, this notion requires investigation.

Larval Cestodes

Much less is known about the natural resistance to the larval stages of tapeworms than to the adult forms. As a general rule, the larvae appear to enjoy a broader specificity than the corresponding adults. Many cestodes that are restricted to the intestine of a single species of definitive host (e.g., *E. granulosus*, *T. solium*, *Spirometra mansonoides*) possess larvae that will grow in a wide array of intermediate hosts.

A large part of the susceptibility to the infection with larvae seems to be related to the presence of appropriate *host's stimuli* for oncospheral activation at the intestinal level. Many larval cestodes will develop in unsuitable hosts when the oncospheres are artificially activated and injected parenterally. To some extent, this situation might be expected, because the differences in the intestinal environment among species of mammals must be more marked than the variations in their *milieux interieurs*.

It would seem that the nonspecific inflammation caused by the early migratory larvae should have some detrimental effect on their ability to survive in the host. Inoculation of *H. nana* cysticercoids in preformed acute or chronic inflammatory sites, however, did not result in any apparent alteration of their development (Weinmann, 1970).

Differences in the development of the larvae of *Taenia taeniformis* in different strains of rats or of *E. granulosus* in sheep indicate the influence of the *host's genetic make-up*. It is not clear at this moment, however, whether this should be regarded as a manifestation of natural resistance or as a diverse ability to mount a response of acquired immunity.

ACQUIRED IMMUNITY

Adult Cestodes

The superficiality of the contact between the adult cestodes and the tissues of the host led to the belief in the past that tapeworms did not induce immunity in the definitive host. This notion was supported by the clinical observation that, left untreated, the intestinal cestodiases of humans persisted for years. Early research with a number of tape-worm species pointed in the same direction. As recently as 1977, Nuti reported that the levels of IgM, IgG, and IgA were not changed in 19 patients with *T. saginata*. The expulsion of some short-lived worms like *H. nana*, *Moniezia expansa*, or *E. granulosus* was attributed to the normal life span of the parasite, or to the crowding effect. *Hymenolepis nana* presented a special situation: since the definitive hosts also harbor the larval stages of the worm, which are intimately associated with the host tissues inside the intestinal villi, they may express a strong intestinal immunity that is elicited by the larvae rather than by the adult worms.

Studies initiated in the 1960s have shown, however, that prolonged or repeated infections with various species of lumen-dwelling *Hymenolepis* or with *E. granulosus* will produce some degree of resistance to the current infection or to reinfections of the definitive host with the homologous parasite. In the particular case of *H. nana*, whose adults live less than a month in normal mice, it was found that their life span was considerably longer in vitro or in x-irradiated hosts. This appears to be a satisfactory argument to assume that the immunity of the host sets a limit to the persistence of the tapeworm.

The use of sensitive, modern immunological methods has demonstrated the presence of specific antibodies at least in intestinal infections by *D. latum* and by *T. solium* in humans, by *E. granulosus* in dogs, by *H. diminuta* in rats and mice, and by *H. microstoma* in mice. The levels of IgG and IgE, but not of IgM or IgA, were elevated in some patients with diphyllobothriasis, and the concentrations of IgG, IgA, and IgE were augmented in mice with *Hymenolepis microstoma*. Host IgM, IgG, IgA (presumably antibodies), and complement were demonstrated on the tegument of *H. diminuta* of mice and rats, and IgE antibodies have been detected in dogs with *E. granulosus*.

Although our own studies have indicated that *E. granulosus* in dogs may elicit cell-mediated immunity, additional work is still needed to determine whether this is a general phenomenon among the cestodes.

It seems well demonstrated at present that many, if not all, cestodes are able to induce some degree of immune response during their intestinal stages. The intensity of this response, and its practical importance, therefore, may be influenced by the quality of the parasitic antigens and by the extent of the host-parasite contact in each particular case. The latter must depend on the intimacy of the contact, the biomass of the parasite population, and the longevity of the infection. All these factors will favor the passage of antigens to the host to elicit a stronger immune response. Also, a tapeworm implanted deeply in the tissues of the host may be more accessible to the immune mechanisms.

Larval Cestodes

Because of the intimate association between the parasite and the host tissues in the case of larval cestodes, a rather strong immune response would be expected in these instances. In fact, clinical observations as well as experimental infections have demonstrated a variety of immune responses with diverse species of cestodes. Because of its medical importance, a great deal of the effort in research has been dedicated to human hydatidosis.

A number of studies that began in the 1930s have shown resistance to reinfection with the larval stages of *T. taeniformis* in rats, *Taenia pisiformis* in rabbits, *H. nana* in mice, *T. saginata* in cattle, and *Taenia hydatigena* in sheep. The reports about *E. granulosus* infection in sheep are somewhat controversial, but it appears that this species does not elicit a strong protective immunity, or that it has effective mechanisms to elude it (see below and Chapter 7). Clinical observations suggest that the same occurs in humans.

At least in the cases of *T. taeniformis*, *H. nana*, and *T. saginata*, the immune resistance persists long after the larval parasites are removed or have died, which is rather unusual in helminth infections. It would be of interest to study whether this long-lasting stimulation of the immune system of the host is related to the high content of carbohydrate characteristic of the cestodes, or to the complex organic-inorganic composition of their calcareous granules. It is not known at present whether those carbohydrates may behave as antigens and remain in the body of the host for long periods, like the polysaccharides of pneumococcus, or the complex structure of the granules may act as an immunological adjuvant.

Although the production of antibodies appears to be the rule in infections of mammals by larval cestodes, only recently have the classes of antibodies involved begun to be investigated. IgM, IgG, and IgA antibodies have been detected in human hydatidosis, and the frequency with which immediate-type skin tests and symptoms of anaphylaxis are found in these patients suggest very strongly the presence of IgE antibodies. Reaginic antibodies have been demonstrated indirectly in several larval cestodiases. IgM and IgG (presumably antibodies) have been found on the protoscolices of *E. multilocularis* infecting laboratory animals. Similar findings indicate the presence of IgM and IgG antibodies in mice harboring the larva of *T. taeniformis*.

The routine production of skin reactions of the delayed type has verified the presence of cell-mediated immunity in hydatidosis and other infections by larval tapeworms since the turn of the century. These findings have been recently extended by the reports of specific lymphocyte transformation in human patients with cysticercosis and in rabbits with *T. pisiformis* larvae, and of macrophage activation in mice with multilocular hydatidosis.

MECHANISMS OF ACQUIRED RESISTANCE

Adult Cestodes

Since it was believed for a long time that intestinal tapeworms did not elicit any measurable immunity, little work has been done in this area until very recently. In the late 1950s it was reported that serum from artificially immunized rabbits reduced the infectivity of the cysticercoids of *H. nana*, stimulated the motility of the adult worm, and produced precipitates on its surface. Later work could not confirm these results using serum from infected mice.

Recent work has demonstrated that immunization of mice with eggs or with a whole or purified extract of adults of *H. microstoma* induces protection against the intestinal infection. The protection transferred to susceptible mice was greater with injections of serum than with inoculations of cells, which suggests a preferential role for the circulating antibodies. However, other work showed that, although the rejection of *H. nana* from the intestine of mice may involve specific antibodies, the expulsion of *H. diminuta* was not affected by the suppression of the humoral response or by the injection of immune serum. Incubation of *H. diminuta* cysticercoids with immune serum did not reduce their infectivity either.

The scarce reports on the participation of circulating antibodies in protection against infections by adult cestodes seem controversial at the moment, but several factors must be taken into account for their proper interpretation. First, adult cestodes are weakly immunogenic in natural infections and it may take several infections (or even parenteral immunizations) before the appropriate antibodies reach a concentration in the serum that could transfer measurable resistance or produce visible reactions. Second, protection against reinfections by adult tapeworms may be mainly a local immune response at the intestinal level, in which case transfer of serum would not be expected to be successful. In fact, resistance to the infection of suckling rats with the larval stage of *T. taeniformis* has been related to the IgA content of the colostrum of the mother. On the other hand, the litters of bitches infected with *E. granulosus* have been demonstrated to be susceptible to homologous infection, which negates a strong or sustained effect of colostral antibodies in this species. Third, despite the presence of antibodies, the parasite may possess effective mechanisms to elude the deleterious effect of the host immune response. Recent investigations indicate that cestodes can produce substances able to activate the complement system, which might be used as a way to divert the immunity before it affects the parasite (see below and Chapter 7). Further research on the role of the humoral response against infections by adult cestodes appears necessary.

Recent work has demonstrated that protective immunity to adult *H. diminuta* and to adult or larval *H. nana* does not occur in the absence of thymus, but the ability to produce this response is restored by thymic implants. Although this does not constitute an incontrovertible proof of the role of cell-mediated immunity in resistance to tapeworms, it certainly encourages further investigations in that direction.

Some effort has been devoted to the identification of the stage of development of the adult tapeworm at which protective immune mechanisms are expressed. Incubation of cysticercoids of *H. nana*, spargana of *Spirometra*, or protoscolices of *Echinococcus* in sera of animals immunized or infected with the homologous parasite produces precipitates around the respective forms of the worms. In one experiment, the serum of mice resistant to *H. nana* did not induce any effect on the cysticercoids, but the intestinal secretions were lethal to the larvae.

Studies with *H. diminuta* (a normal parasite of rats) in mice demonstrated that the cysticercoids grow normally at the beginning of the infection, but, on about the 10th day, growth ceases, destrobilation occurs, and death ensues. When transferred to their normal rat host, however, the worms completed their development. Infections of mice (the normal host) with *H. microstoma* produced demonstrable antibodies but not measurable inhibition of the development of the worms. The difference between the results of these two experiments might be attributable to the fact that the antigens of the cestode may be more potent for an abnormal host than for the host to which it has been associated for millions of years (see "Imitation of Host Antigens" in Chapter 7). If this is the case, the basic variation between both models could be more quantitative than qualitative. On the other hand, immunosuppression of the host has been shown to result in more, larger, or longer-living worms in the cases of *H. nana*, *H. microstoma*, and *E. granulosus*.

The evidence presented above indicates that the deleterious effect of the host immune response on adult cestodes may act at three levels: on the early ingested larvae (presumably only in hosts who have developed specific immunity previously), on the developing cestode, and on the established adult worm.

In relation to the immunological mechanism of damage itself, there is ample proof that the tegument of the adult cestodes can be destroyed by the action of complement. Some recent communications have indicated that adult specimens of *E. granulosus* are able to activate the complement system in the absence of known antibodies, which probably corresponds to triggering of the alternative pathway. Ultimately, complement activation directly by the parasite may be a mechanism of protection against the host immune response (see below and Chapter 7). It is likely that other immunological mechanisms also play a role in the acquired resistance to infections by adult cestodes, but the respective investigations still remain to be done.

No work appears to have been done on the effect of cell-mediated immunity in intestinal cestodiasis.

Larval Cestodes

Because of its medical and economic importance, acquired resistance to larval tapeworms has commanded greater interest among the investigators. A great deal of work has been done with *E. granulosus*, which constitutes an important infection of humans in some countries, and with *T. taeniformis*, which is a convenient laboratory model transmitted between cats and rats or mice.

Role of Antibodies   The participation of antibodies in the protection against larval cestodiasis had already been demonstrated in the 1930s by experiments of transfer of resistance with the serum of infected hosts. At present, it is possible to confer resistance by this method at least against *T. taeniformis* in rats, *T. pisiformis* in rabbits, *T. hydatigena*, *Taenia ovis*, and *E. granulosus* in sheep, and *H. nana* in mice. Intestinal secretions of infected hosts transferred resistance to infections by eggs of *H. nana* and *T. taeniformis*. Protective immunity to *T. taeniformis* in rats, to *H. nana* in mice, to *T. hydatigena* and *T. ovis* in sheep, and to *T. saginata* in cattle has also been transferred by ingestion of colostrum. Although colostral transfer is not accepted nowadays as an absolute proof of antibody activity, recent reports have demonstrated that the protection from the infection by *T. taeniformis* larvae transferred with colostrum, intestinal secretions, or serum is related to the IgA content of the colostrum and intestinal secretions or the IgG content of the serum. The protective properties of these preparations could also be abolished by adsorption with activated oncospheres. At present, this constitutes evidence of antibody activity. Furthermore, inoculation of serum from mice infected with *T. taeniformis* protects athymic mice against the homologous infection.

The literature reports failures and successes in the transfer of resistance to *T. saginata* larvae with serum or with colostrum. This conflict may be due to the stage of development of the cysticerci in the donor: some studies have shown that, whereas the serum taken early in the infection by *T. pisiformis* or *T. taeniformis* prevented the infection of the recipients, serum taken several weeks later allowed the establishment of the parasites but had a deleterious effect on the growing larvae (see below). At present, there is little room to doubt that antibodies play a preponderant role in the protection against the larval stages of the cestodes.

Since vasoactive amines are lethal to the early embryo in vitro (although inconsistently), there was a suggestion that IgE antibodies might be protective through the release of mediators of anaphylaxis. However, inoculation of inhibitors of Type I allergy did not reduce the passive protection to *T. taeniformis* larvae afforded by inoculation with immune serum in rats. This report tends to negate a role for IgE antibodies in the resistance to tapeworm larvae.

On the other hand, Baron and Tanner (1977) have demonstrated that specifically and nonspecifically activated macrophages are protoscolicidal for *E. multilocularis*, and that this activity is enhanced by incubation in immune serum. Thus, there is current evidence that antibodies act against larval cestodes at least in two ways: by activating the complement system (see below) and by facilitating phagocytosis.

Role of complement   Early studies had already indicated that the humoral response acted on the parasite mainly by activation of the complement system. Recent investigations have supported those reports by establishing that infection by *T. taeniformis* or *E. multilocularis* larvae is much more severe in decomplemented animals than in normal hosts. Studies with protoscolices of *E. granulosus* have shown that they are able to activate the complement system even in the absence of antibodies, or in the serum of guinea pigs genetically deficient in the factor C4 of complement. In all probability this activation occurs through the alternative pathway. On the other hand, the reports about *E. multilocularis* indicate that this parasite is unable to activate complement in the absence of antibodies, which suggests that the activation in this case is by the classic pathway.

The surprising finding was made a few years ago that antibodies and complement factors are occasionally present together inside healthy looking larvae of *E. granulosus* or *T. taeniformis*. There is current evidence that several cestodes actually can produce substances that will deplete complement in their immediate vicinity, possibly avoiding direct damage in this way. This phenomenon is discussed further in Chapter 7.

Role of Cell-mediated Immunity   The participation of cell-mediated immunity has been little investigated. Since the high susceptibility of athymic mice to infections with larvae of *T. taeniformis* is abolished by transfer of immune serum, it would appear that antibodies are enough to explain the observed resistance. Nevertheless, this area deserves investigation since recent reports claim that transfer of peritoneal cells from rats infected with *T. taeniformis* into noninfected rats reduced by 50% the susceptibility of the latter to a challenge with eggs, as compared to the controls.

Developmental Stages Affected   Much effort has gone into the identification of the developmental stages of the cestode larvae that are affected by the immune response. Several reports have demonstrated directly or indirectly a harmful activity on the oncospheres in the alimentary tract. Serum from immune rabbits forms precipitates around the activated oncospheres of *T. pisiformis*, and the oncospheres of *H. nana*, *T. hydatigena*, *T. ovis*, and *E. granulosus* do not readily penetrate the intestinal wall of immune intermediate hosts. Cattle, sheep, and rabbits immune to oral infection with *T. saginata*, *E. granulosus*, and *T. pisiformis*, respectively, do allow development of the larvae when infected by parenteral inoculation of activated oncospheres, a procedure that eludes the local immunity in the intestine. In the case of *H. nana*, it has been shown that the intestinal secretions of immune hosts transfer more protection than their serum. With *T. taeniformis* it has been found that, although the protective immunity of the intestinal secretions is mostly related to the IgA content, the resistance transmitted by serum is instead correlated to its IgG content.

Oncospheres are readily destroyed by complement in the presence of antibodies. It is possible, additionally, that the simple delay of the invasion of the intestinal wall by the activity of antibodies (by neutralization of penetration enzymes, perhaps) is enough to prevent the infection, since Gemmell (1976) has observed that oncospheres are rapidly digested in artificial intestinal fluid.

Fewer experiments have been performed with the developing larvae in the host's tissues, but it has been observed that the inoculation of immune serum in normal rabbits or rats causes an accelerated tissue response and the premature death of the early developing larvae of *T. pisiformis* or *T. taeniformis* in the liver.

The successful use of protoscolices as antigens in diagnostic tests for larval cestodiases and the harmful effect of homologous immune serum on the motility of *E. granulosus* protoscolices and on the tegument of *T. taeniformis* and *H. nana* larvae have suggested that immune responses against the well-established larvae indeed occur. That these reactions are actually against the mature larvae rather than against cross-reactive antigens produced by other stages of development has been confirmed by inducing similar responses in animals that received only the mature larvae by surgical transfer. The frequent observation that an increasing proportion of *E. granulosus* larvae does not develop protoscolices in host species that appear to be progressively less and less adequate (sheep, cattle, pig, humans) suggests that this sterilization might be an immunologically mediated phenomenon.

There is no proof of this, however, and, even if immunological, this effect might equally well be the result of mechanisms operating during the early development of the larva.

It is surprising, however, that most infections by larval tapeworms elicit some degree of protection against reinfections, but only occasionally does the immune response appear to be effective against the coexisting mature larvae. The activation of the complement directly by the parasite might have survival value (see above and Chapter 7). In addition, Kwa and Liew (1978) have shown that mature *T. taeniformis* larvae survive normally when transferred to immune rats, but die quickly if they are trypsinized before transplantation. This finding suggests that the larvae are covered by some proteinaceous material, presumably host antibodies, that effectively neutralizes the host immunity.

It seems well proven at this time that the intermediate host commonly responds immunologically to three stages in the development of larval cestodes: the activated oncosphere, the early developing larva, and the mature larva. It is likely that the deleterious part of each response is directed against antigens peculiar to each particular stage. The response seems to be effective mainly on the first and possibly the second stage, however, with little action on the mature larva. This latter form appears to possess special mechanisms of survival in the immune host; at least one of them is reminiscent of the concomitant immunity described for *Schistosoma*.

ARTIFICIAL PRODUCTION OF RESISTANCE

Against Adult Cestodes

Little work has been done on the immunological prophylaxis of cestode infections in their definitive hosts, and those early studies using homogenates of worms were rather discouraging. The inoculation of extracts of adult *H. diminuta* in rats or the larval extracts of *Spirometra erinacei* in cats failed to protect against homologous challenges. The injection of homogenates of adult worms, protoscolices, or germinal membranes of hydatid cysts in dogs produced only weak resistance to *E. granulosus* infection. Only recently a report appeared claiming that considerable protection against the infection of mice with *H. microstoma* is elicited by administration of extracts of the adult worm.

The use of living stages of the parasites has not given clearly superior results. The implantation of adult *H. diminuta* in rats did not protect against reinfections, and the injection of activated oncospheres or protoscolices of *E. granulosus* in dogs produced only weak resistance to homologous challenge. The inoculation of oncospheres of *H.*

nana in rats has conferred variable protection to the subsequent infection with *H. diminuta*. Injection of eggs of *H. microstoma* in mice, on the contrary, has been reported to produce resistance to the reinfection, although less than the inoculation of extracts of the adult worm.

An attempt to vaccinate dogs with excretory/secretory (ES) products of adult *E. granulosus* yielded encouraging results, but there were later indications that the resistance presumably elicited by the procedure may have been an expression of natural resistance to the infection in some dogs rather than of acquired immunity (Herd, 1977).

In all fairness, the experiments above should be regarded as heroic and the partial success of some of them is rather admirable. Now that information is accumulating that the immune response to digestive parasites appears to be largely a phenomenon triggered and expressed locally, the parenteral inoculation of parasites does not seem appropriate to produce effective immunity against intestinal cestodes. Additionally, even today there is an almost absolute lack of information about functional antigens in adult tapeworms and about their sharing by the diverse developmental stages, which makes the election of the appropriate immunizing form particularly uncertain. Considerably more knowledge will have to be gathered and research done before deciding whether immunization against the digestive stages of cestodes is or is not feasible.

Against Larval Cestodes

The larval cestodes seem to be a partial exception to the notion that parasite homogenates do not elicit protective immunity. Neither inoculation of hydatid fluid nor inoculation of dead oncospheres of *T. hydatigena* produced resistance to the respective infections in sheep, but some degree of resistance to the infection by larvae was induced with larval extracts of *T. taeniformis* in rats and of *E. granulosus* and *T. multiceps* in sheep, and with extracts of adult *T. pisiformis*, *T. taeniformis*, and *H. nana* in rabbits, rats, and mice, respectively. In addition, the injection of homogenates of adult *S. mansonoides* caused the encapsulation of the respective spargana in monkeys.

The relative success of the immunization against larval forms with parasitic extracts apparently stimulated a number of attempts to improve those results by the use of living, metabolizing stages of tapeworms. Injection of eggs of *T. taeniformis* in rats or of *T. hydatigena* in sheep produced a strong immunity to reinfections. However, the protective immunity appears to be directed to the very early stages of development, because the serum of rats immunized in this way will protect newly infected rats only during the first 4 days of their infection.

Activated oncospheres of *E. granulosus*, *T. ovis*, or *T. hydatigena* injected in sheep protected them from the homologous infection. Oncospheres of *T. hydatigena* even protected calves against infection with eggs of *T. saginata*, although the inoculation of oncospheres of this latter to immunize cattle against the homologous parasite has given controversial results. Implantation of larvae of *Taenia crassiceps* or infection with eggs of *H. nana* in mice protected against the infection with eggs of the corresponding parasite. Curiously, the infection of mice with cysticercoids of *Hymenolepis citelli* (a parasite of the ground squirrel) induced strong protection against the challenge with eggs of *H. nana* or with homologous cysticercoids, but the resistance to the former persisted much longer than to the latter. Infection with cysticercoids of *H. diminuta* in mice produced absolute refractoriness to the reinfection with homologous cysticercoids, very strong resistance to eggs of *H. nana*, and somewhat less strong resistance to cysticercoids of *H. nana*. The more effective rejection of the challenge with eggs of *H. nana* rather than with its cysticercoids may be related to the fact that the eggs will develop into tissue-dwelling larvae, which must elicit a stronger secondary response and be more accessible to the immune mechanisms than the lumen-dwelling parasites derived from the cysticercoid infection.

Implantation of adult *H. nana* in mice did not result in measurable immunity to egg challenge.

Since the immune response to larval cestodes rarely is lethal for the concurrent parasites, it would be pointless to vaccinate with stages that lead to the development of larvae. In order to avoid the formation of larvae while preserving some sustained production of metabolic antigens, some attempts to use oncospheres sterilized by irradiation have been made. Inoculation of such oncospheres of *T. taeniformis* in rats produced strong resistance, but tests with oral or intraruminal infection with irradiated eggs of *T. saginata* in cattle were somewhat less successful. This is a rational procedure that may deserve further study, but advances in the in vitro culture of parasites have permitted the production of ES preparations that are easier to analyze and standardize.

Inoculation of ES products of larvae or of activated oncospheres of *T. taeniformis* in rats, *T. saginata* in cattle, and *T. ovis* in sheep has produced strong resistance to the reinfection that is transferred to the offspring with the colostrum but, at least in cattle and sheep, does not interfere with the subsequent vaccination of the youngsters. However, the colostral antibody appears to be directed against the very early stages of development, because it reduces the number of larvae that develop but does not affect the survival of those that had already attained establishment (Rickard et al., 1977).

Recent reports from Australia have claimed that infection of sheep with *T. hydatigena* eggs prior to infection with *Fasciola hepatica* metacercariae confers a strong resistance to the latter, when these infections occur 12 or more weeks apart. This is a curious case of cross-protection between remotely related organisms that may have practical implications if confirmed.

Although considerable work remains to be done, we can envision that protection of humans and domestic animals from cestodiases through vaccination may become feasible in the predictable future.

IMMUNODIAGNOSIS

Since infections with adult cestodes rarely produce evident symptoms before patency, and the determination of the presence of eggs or proglottids in the feces of the host is attained by rather simple procedures, little interest has existed in developing immunodiagnostic methods for intestinal cestodes. Nevertheless, Capron's group recently reported the diagnosis of prepatent infections with *T. saginata* in man by immunoelectrophoresis.

Infections by larval tapeworms, on the contrary, are as a rule inapparent until they become dramatically symptomatic. When they do, the symptoms are usually ambiguous and likely to be confused with those caused by a number of other conditions. Except in very few cases (e.g., ocular cysticercosis), the direct identification of the parasite requires major surgery or necropsy, which obviously is undesirable or impossible in human patients.

Several immunological techniques have been devised for the indirect identification of larval cestodes, most of them dealing with human hydatidosis. It has also been suggested that the use of these methods may improve the efficiency and reduce the cost of the inspection for porcine and bovine cysticercosis in abattoirs.

The immunological diagnosis of hydatidosis has been recently reviewed by Kagan (1976) and Varela-Diaz and Contorti (1974), and that of bovine cysticercosis by Geerts et al. (1977). A group of experts has written a paper about cysticercosis (World Health Organization, 1976) and Rydzewski et al. (1975) have provided new information on human cysticercosis.

One of the oldest and most widely known tests for hydatidosis is the skin test introduced by Casoni in 1911. This is an immediate-type reaction (Type I) that, less frequently, is followed by a reaction of delayed hypersensitivity (Type IV). Using native hydatid fluid as the antigenic preparation, the sensitivity of the procedure appears to be inversely related to its specificity: in one study, 87% of 75 cases of proven hydatidosis were detected but 15% of noninfected controls were also positive; in another series in which only 75% of 40 proven cases were positive, the proportion of false-positive reactions was only about 8%. Reduction of the protein content of the hydatid fluid (to about 15 μg of N_2/ml in humans) improves its specificity markedly, as does boiling for 15 min. Measurement of the area of reaction in the skin and comparison with noninfected controls also contributes to improving specificity and sensitivity: an area over 2 cm^2 was frequently associated with actual infection in one study. A lipoprotein believed to be the major reacting antigen in this test has been recently purified. Because of the simplicity of this test and its convenience for the diagnosis of individual patients as well as for epidemiological studies, it may be very useful if it is ever adequately standardized.

Skin reactivity for diagnostic purposes has been investigated less in other larval cestodiasis, but work with human and swine cysticercosis from T. solium, cattle cysticercosis from T. saginata, and sheep cysticercosis from T. multiceps has revealed the presence of abundant nonspecific results. It is possible that the nonspecific reactions are actually processes of increased vascular permeability mediated by the complement-activating products found in many larval cestodes, which may be present in the antigenic preparation for diagnosis.

The complement-fixation test (CFT) has been used for the diagnosis of human hydatidosis for about 70 years already, but, despite that, most of the available information about it is widely controversial. The literature indicates a degree of sensitivity that fluctuates from 36% to 100% and a nonspecificity that varies from 0.4% to 28% in different studies. Unfortunately, cross-reactivity with the serum of cancer patients is not uncommon, despite the fact that differentiation between neoplasia and hydatidosis is often of particular interest for the clinician. Early reports indicated that the CFT became negative in about 6 months in the absence of the parasite, which was very convenient for evaluating the success of the treatment, but a recent communication failed to confirm this fact. In cattle cysticercosis, on the contrary, it has been reported that the CFT is more sensitive in detecting calcified than viable larvae. There are some indications that the CFT may be very sensitive and specific when the appropriate antigen and technique, adequately standardized, are used. In the meanwhile, other tests are preferred in practice.

Different variations of the indirect hemagglutination test (IHAT) have been used in the last 20 years or so in human hydatidosis. Reportedly, this method exhibited a sensitivity of 52% to 93% in the hands of various investigators. It can yield very high titers (up to 200,000) in patients, but cross-reactions are rather frequent at low dilutions. Cross-reactivity with serum of cancer patients is very rare, nevertheless. Some authors consider that only reactions at titers either equal to or greater than 200 are specific for hydatidosis; others recommend testing every new batch of antigen with negative and positive sera to determine the diagnostic titer for each particular lot. Curiously, sera of patients with pulmonary larvae are less reactive in the IHAT than sera of patients with hepatic hydatids. The same has been reported for the skin test and immunoelectrophoresis. This may be a reflection of some special association between the larva and the lung tissue that affects the parasite metabolism, and therefore its production of antigens, since I have found that protoscolices of lung cysts survive for a shorter time in vitro than protoscolices of liver cysts (Barriga, 1971). The IHAT is the standard test used currently in the U.S.A. for the diagnosis of human cysticercosis.

An agglutination test with the antigen absorbed on latex particles has given results similar or slightly superior to the indirect hemagglutination test. This method is preferred by some workers over hemagglutination for the diagnosis of hydatidosis in individual patients or in epidemiological studies because it is simpler and quicker. In cattle cysticercosis, it is particularly sensitive in the diagnosis of early infection.

The indirect fluorescent antibody technique (IFAT) has consistently proven to be somewhat more sensitive than the IHAT in several trials with human hydatidosis. Since IgM antibodies do not persist long in the absence of antigen release, this method may be utilized to investigate the success of the treatment by using appropriate anti-IgM fluorescent serum. In human cysticercosis, this technique has proven to be less sensitive but more specific than the hemagglutination test. In cattle cysticercosis, some investigators have reported satisfactory results and others have found that the cross-reactivity is so extensive as to make the test useless. A modification of the IFAT, employing a soluble antigen absorbed on paper, has been found to be able to diagnose 82% of cases of hydatidosis with 4% of nonspecific reactions.

The tests of precipitation in gel have traditionally been used for the analysis of antigenic preparations, but it has been possible to identify up to 80% and 90% of cases of human hydatidosis by double diffusion and by counterimmunoelectrophoresis, respectively, in various studies. In 1967, Capron and collaborators found that one electrophoretic precipitation line (band 5) was specific for *Echinococcus* in man, and was absent in sera cross-reactive with hydatid fluid (Figure 14). Since then, immunoelectrophoresis has grown to be the diagnostic technique of choice for hydatidosis for many researchers. The sensitivity of this technique is high (about 85%) and its specificity is excellent. It is reported that this test becomes negative a few months after removal of the larva, which makes it appropriate to investigate

A

B

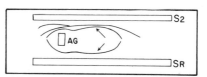

Figure 14.   Diagnosis of human hydatidosis by immunoelectrophoresis. Plate A shows band 5 (**arrows**), characteristic of human infection by *Echinococcus* larvae, in the sera of 2 patients (S_1 and S_2) reacted against hydatid fluid (**AG**). Since S_2 developed an atypical band 5, this serum was reacted again in plate B, in parallel with a reference serum (S_R) that was positive for the antigen of band 5; this time S_2 yielded a characteristic precipitation line, but its identity would have been secured otherwise by its near fusion with the corresponding band of the reference serum.

whether some parasite has remained in the operated patient. It is likely that the recent purification of the specific antigen of band 5 will bring important advances in the immunodiagnosis of hydatidosis.

Preliminary studies on the use of enzyme-linked immunosorbent assay in the diagnosis of human hydatidosis and human cysticercosis and of the radioallergosorbent test for the identification of human hydatidosis have yielded encouraging results, but further assays are necessary before advocating their practical use.

Differences in the results obtained with various techniques using the same sera and antigenic preparations suggest that some antigens may react in some tests but not in others. Considering this fact, some investigators with long experience recommend the simultaneous use of several tests (especially IHAT, IFAT, and immunoelectrophoresis) as a way of improving the sensitivity and specificity of the diagnosis of human hydatidosis. The same may be true for cysticercosis, but we do not have enough information available yet.

Since small hydatids in the thickness of the liver may be impossible to detect during the surgical treatment, considerable interest exists in some method that will indicate whether the patient still harbors parasites after surgery. This will also be an early warning of the production of secondary hydatidosis consequent to surgery. The investigation of some specific immune response that wanes soon after the parasite is eliminated would be particularly appropriate for these purposes. The literature indicates that the complement-fixation test would fulfill this requirement, but Todorov et al. (1976) could not find support for this belief. Determination of the presence of band 5 by immunoelectrophoresis or of IgM antibodies by immunofluorescence may be used to this end, but the fact that both tests are essentially qualitative limits their usefulness.

ANTIGENS

There is a profusion of studies of antigens in cestodes, but few of them appear to be directed to the identification and isolation of the antigens that are important for diagnosis or protection. Kagan and Agosin (1968) have written a detailed review of the antigens of *Echinococcus* with some references to other genera.

Although it is rather unexpected that parasites as specialized as the cestodes have retained antigens common to other animal groups, the fact is that extensive cross-reactivity seems to be the rule in tapeworms. *E. granulosus*, for instance, shares antigens with *E. multilocularis*, *T. solium*, *T. saginata*, *T. taeniformis*, *T. pisiformis*, and *M. expansa*. The antigen of band 5, originally thought to be exclusive to *E. granulosus*, has also been found in extracts of *E. multilocularis* and *T. hydatigena*. The cross-reactivity goes far beyond the usual systematic boundaries, since anti–*T. saginata* serum reacts with extracts of the trematodes *F. hepatica*, *Dicrocoelium dendriticum*, and *Schistosoma mansoni* and of the nematode *Onchocerca volvulus*.

The electrophoretic study of hydatid fluid from sheep cysts in 1965 showed the presence of nine bands of parasite origin and 10 bands that corresponded to serum proteins of the host (Kagan, 1967). More recent work has demonstrated the presence of host materials inside other larval cestodes. The current thinking is that these substances are absorbed rather than synthesized by the parasite.

The ubiquity of the antigens of cestodes poses particular problems in devising specific immunodiagnostic tests, but it may be beneficial to the development of immunoprophylactic procedures, as suggested by claims of protection of sheep against the highly pathogenic *F. hepatica* afforded by a previous infection with the more innocuous *T. hydatigena*.

Extensive cross-reactivity among the different developmental stages of the same species has also been found: the antiserum obtained by immunization with extracts of adult *H. nana*, for instance, reacts with the eggs, the cysticercoids, and the adults of the homologous species. However, the fact that resistance to challenges is much stronger when the host is immunized with the specific stages against which the protective response operates suggests that the functional antigens are rather specific to each stage. This is important to consider for successful immunoprophylaxis.

The same need not be necessarily true in the case of immunodiagnostic tests, since even a nonfunctional antigen that is specific for a particular tapeworm and elicits a strong response will be adequate for purposes of identifying the infection. The latitude allowable in the purification of diagnostic antigens, therefore, may be considerably greater than that required in attaining immunoprotecting materials.

Work with hydatidosis has demonstrated that the source of the antigenic preparation (hydatid fluid, usually) may influence to a large extent the results of the immunodiagnostic tests, in regard to sensitivit˙ as well as specificity. The appropriateness of the hydatid fluid for diagnostic assays seems to be related to the host species and to the fertility and state of development of the parasite. Human and ovine cysts are reported to yield the best results, and bovine cysts the worst. Similarly, fully mature hydatids with a large number of protoscolices give more and better antigens.

The increase in specificity of the hydatid fluid by incubation at 100°C for 15 min appears to be related to the removal of host components that precipitate with this treatment, whereas many of the parasite antigens are heat stable. Unfortunately, this is not true for the antigen of band 5, so that boiled hydatid fluid is not adequate for immunoelectrophoresis and visualization of band 5.

The differences among the results of the various tests for hydatidosis seem to be connected to a considerable extent to the composition of the antigenic preparations and to the ability of each test to express the reaction between a particular antigen and its corresponding antibodies. Complement-fixation tests with the same sera but with antigens prepared with laminated membrane, with protoscolices, or with hydatid fluid all gave divergent results. A polysaccharide fraction of the fluid was reactive in the complement-fixation and the latex agglutination tests, but not in the indirect hemagglutination test. A purified lipoprotein was very active in skin tests, but yielded negative results by indirect hemagglutination and immunoelectrophoresis with a number of positive sera. With the indirect immunofluorescent antibody technique it has been determined that larval scolices fluoresce more than scolices of the adult worm, and that most of the fluorescence is located in the hooklets.

An ingenious technique of purification of the antigen of band 5 allowed Capron's group to produce a monospecific serum with which it was demonstrated that the respective antigen was located in the germinal layer and in the interior of the protoscolices (Yarzabal et al., 1976). This antigen is a lipoprotein of 69,000 daltons, stable at pH 4.0 to 7.0 and to 56°C, but labile to 70°C in an acidic environment or to 100°C. It is more concentrated in human and sheep cysts than in pig and cattle larvae, and more concentrated in hepatic than in pulmonary cysts. It reacts well in indirect hemagglutination tests.

Some isolated work done on the antigenic composition of other tapeworms seems to indicate that the findings in hydatidosis are representative of the phenomena that occur in cestode infections in general. Despite the revolutionary advances in the techniques of purification of biomacromolecules in the last decade or so, the characterization and isolation of cestode antigens has lagged sadly behind for the most part. With the current availability of immunodiagnostic tests of exquisite sensitivity and the indications that the immunoprophylaxis of tapeworm infections is feasible, it appears that the most urgent endeavor in this area should be the attainment of appropriate antigenic preparations.

IMMUNOPATHOLOGY

Skin-sensitizing antibodies have been indirectly demonstrated in infections by several species of tapeworms, and the presence of IgE antibodies has been verified directly on a few occasions. Their participation in the pathology of the disease is mostly conjectural, however.

The presence of adult cestodes in a host is mainly silent, producing only few symptoms in a rather small proportion of human patients and remaining asymptomatic in most domestic animals. Some of the symptoms, nevertheless, such as epigastric pain and sporadic diarrhea, nausea, and pruritus, may be attributed to the activity of anaphylactic antibodies. The occasional existence of eosinophilia and lymphocytosis in these patients and the recent report that the intestinal secretions of rats infected with *H. nana* contained 2 times more histamine than the secretions of noninfected rats support this hypothesis. Adequate confirmation is necessary, nevertheless.

Numerous experimental infections of laboratory animals with spargana have demonstrated that the early migration of the larvae causes infiltration of the host tissues with eosinophils, lymphocytes, and plasma cells. Experimental infections in humans showed the production of local urticaria, edema, and erythema with systemic eosinophilia. All these changes are characteristic of reaginic allergy, and the immunological nature of the lesions in this particular case is supported by the fact that they were largely prevented in monkeys by prior vaccination with specific antigens.

The Cyclophyllidea larvae of mammals normally cause more tissue damage, and the corresponding symptoms, during the early migration to their final location and after their death in the tissues. The viable, well-established larva usually leads a quiet life except when its size interferes with the function of the surrounding tissues (e.g., *E. granulosus, Taenia multiceps*). The tissue response to the larvae is a cellular infiltrate in which eosinophils, macrophages, lymphocytes, and plasma cells are present in variable proportions. This morphological composition is typical of immunological events and, at least in the early migratory stages, must be mediated by antibodies because the local inflammation has been enhanced in rabbits infected with *T. pisiformis* and in rats infected with *T. taeniformis*, by inoculation of specific immune serum.

For a long time it has been accepted that the temporal distribution of the tissue reactions in relation to the stage of development of the larva was due to the amounts of antigens produced by the parasite at each time, and to the corresponding immune reactions of the host. It was assumed that the accelerated metabolism of the early migratory larva released a large amount of antigenic substances in torn tissues so that they could be taken readily by the host's lymphoid system. Death of the parasite would again mean sudden liberation of much antigen because of the disintegration of its membranes. The well-established parasite, on the contrary, with a slower metabolism, would be relatively isolated from the immune apparatus, so that the host could deal better with moderate amounts of antigens, released gradually.

Although this notion is still tenable, with some restrictions, the new findings with respect to nonimmunological activation of the complement system by the parasite and the possible concealment of the larva by adsorption of host material permit an alternative explanation. The reduced reactivity to the mature larva may be the expression of the development of self-protective mechanisms on the part of the cestode to elude the immunity of the host. Conceivably, these mechanisms will take some time before becoming fully operative and will be lost on death, so that early and dead parasites would be better stimulants of and more susceptible to the immune reaction of the host. In any case, fascinating possibilities for research in this area lay open, and the control of some of the symptoms of cysticercosis by immunological means might be feasible in the future.

The accidental rupture of hydatids in the host frequently produces generalized urticaria and pruritus, and occasionally causes a systemic anaphylaxis that may be fatal. This syndrome is attributed to the sudden absorption of massive amounts of antigen through the serose membranes of a host previously sensitized by the presence of the parasite. Support for this theory has been provided by the recent reproduction of the condition by intravenous injection of hydatid fluid in infected sheep. Noninfected sheep did not respond in the same way, which rules out the activity of a toxin.

Certain symptoms of chronic hydatidosis, such as dyspnea, generalized pruritus, localized pains, and loss of weight, have been traditionally imputed to an allergic state of the host, and the "biological treatment" of Calcagno, which amounts to a procedure of hyposensitization with hydatid fluid, allegedly restored the well-being of many patients. Nevertheless, a critical review done recently on the published results of this treatment failed to disclose any beneficial effect. It must be said, however, that hyposensitization procedures with complex antigenic preparations such as hydatid fluid rarely yield satisfactory results (see Chapter 6). It would be interesting to assay this method with purified antigens when more information about the allergens of *E. granulosus* is available.

Clinicians have sporadically expressed the hope that chemotherapeutic developments might in the future replace the surgical treatment of human hydatidosis and cysticercosis. The evidence available at present that damage by larval cestodes appears to be largely an immunologically mediated event makes one wonder whether it would be advisable to kill the parasite while it is in the host, and risk the massive release of sensitizing antigens. Perhaps the rational use of immunological procedures would be able to achieve one while preventing the other.

Recent work reviewed by Hammerberg and Williams (1978) has demonstrated that *T. taeniformis* and other cestodes are able to activate the complement system by nonimmunological means. Inoculation of extracts of *T. taeniformis* larvae generated anaphylatoxic activity in the serum, changed vascular permeability, reduced the levels of circulating complement, and inhibited the complement-mediated inflammatory responses in the skin. These findings are still too new to guess all their implications, but it is likely that this mechanism plays some role in the course of the homologous infection, and maybe in the pathology caused by concurrent pathogens.

There seem to be no works on the relationship between cestode infections and diseases by autoimmunity, but this may be a fruitful field of investigations. Since the membranes of the parasite are permeable to host macromolecules, these might be changed by the metabolism of the parasite inside the larva, and acquire new antigenic reactivities that might be detectable by the host as non-self material. The limited work done on the depression of the host's immune response by cestodes is reviewed in Chapter 7.

THE DIGENETIC TREMATODES
OR FLUKES

The trematodes are usually divided into three major groups (Monogenea, Aspidobothrea, and Digenea) with uncertain relationships. Of these groups, only the Digenea are relevant to medicine. The Digenea share with the Cestoda several evolutionary characteristics, of which the possession of an absorptive-secretory tegument and the existence of a life cycle that includes definitive and intermediate hosts are important to this discussion.

As in the cestodes, the tegument of the digenea exerts a number of metabolic functions that may play important roles in the immunology of the respective infection. In the case of the schistosomes, it has been verified that materials of their surface membrane are constantly being shed, contributing to the formation of circulating antigen-antibody complexes that may have a pathogenic activity; this membrane also adsorbs and synthetizes host materials that hide the parasite from the host's immune system. Little is known about the participation of the tegument in the immunity to other trematodes, but any actively metabolizing membrane in intimate contact with the host tissues must have plenty of opportunities to stimulate or to modulate immune responses.

The life cycle of most trematodes (Figure 15) demands an aquatic environment: the eggs passed by the definitive host usually mature in water and release a ciliate larva (*miracidium*) that swims around in search of a specific snail for a few hours. Once this intermediate host is located (with the assistance of chemicals produced by the mollusk), the miracidium penetrates the snail by a combination of mechanical and enzymatic activities. The "wrong" species of snail usually destroys the invading parasite, whereas the "right" mollusk allows its development through several stages (*sporocyst, rediae, cercariae*). Normally some of these developmental stages multiply inside the snail (a process called *pedogenesis* or *polyembriony*, depending on whether one considers the mother stage a juvenile or an embryo), so that one miracidium may give origin to several hundred cercariae. The cercariae escape from the snail's tissues and commonly encyst to metacercariae in or on a second intermediate host or on aquatic plants.

This phase of the trematode life cycle can experience some modifications (related to the time of formation of the miracidia, the manner of penetration in the snail, the number and generations of intramolluscan stages, the number of intermediate hosts, etc.) according to the species being considered. The most relevant to our present interest, however, is the way in which the new flukes invade the definitive host.

In the schistosomes, the cercariae are attracted nonspecifically by lipids on the skin of the host and penetrate actively, so that this has no participation in the process of host selection until the juvenile parasites are actually migrating in its tissues. At this time, it has been found that the natural resistance exhibited by the host correlates well with the degree of inflammation elicited by the parasite. It is not clear,

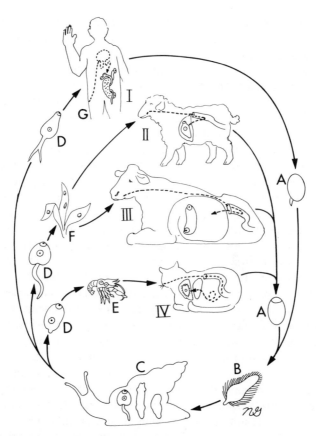

Figure 15.   Life cycle of the trematodes considered in the text: *Schistosoma mansoni* (**I**), *Fasciola hepatica* (**II**), paramphistomes (**III**), and *Paragonimus kellicotti* (**IV**). The eggs of the parasites (**A**) are discharged with the feces of the host and, in an aquatic environment, they release a swimming larva (*miracidium*, **B**). The miracidium penetrates its corresponding specific snail (**C**) and multiplies in it for two or three larval generations; the details of this multiplication may change according to the species of trematode. A new swimming stage (*cercaria*, **D**) leaves the snail after an appropriate lapse and encysts to metacercaria in a second intermediate host (**E**) or on aquatic plants or in the water (**F**), or enters the definitive host directly (**G**). If encysted, the metacercaria is subsequently ingested by the definitive host and develops to the adult stage. A period of migration in the definitive host (represented by a dashed line) generally precedes the establishment of the parasite in the adult habitat.

however, whether this inflammation is brought about by the living parasite and is lethal to it, or whether it appears after the parasite is already fatally affected by other mechanisms and begins to degradate. At any rate, it appears that any mammal that enters into contact with contaminated waters will be an appropriate target for the initial attack by the cercariae of schistosomes.

All the other medically important trematodes enter the definitive host by the digestive route, and the young flukes are released from the metacercarial cyst by a process not too different from those that we saw for hatching and exsheathment in the nematodes. Studies with *Fasciola hepatica* have shown that the outer wall of the cyst is removed by the mechanical actions of the teeth or the gut, but the inner wall is perforated by the fluke after activation by physicochemical conditions of the gastrointestinal tube of the host. For *F. hepatica*, these conditions are appropriate temperature, redox potential, CO_2 levels, and bile concentration. The necessary stimuli are less complex for other species. Some evidence indicates that release from the cyst is assisted by an excysting secretion of yet-unknown composition. In this group of trematodes, it seems logical to assume that the ecological opportunity to acquire the metacercariae (which is related to the diet of the host) and the physicochemical conditions of the gastrointestinal tract must play important roles in the susceptibility and natural resistance of a vertebrate species to a particular species of trematode.

Once the young fluke penetrates the definitive host, it migrates rather extensively: sometimes it appears to progress steadily toward the final habitat of the adult parasite, but at other times it seems to wander aimlessly in the body before reaching its final location. Most of the natural resistance of a host is expressed during this migratory phase: the parasites may be killed, stunted, or rendered sterile. Since this phase lasts for several weeks, in many cases it is not absolutely clear whether the antiparasite activities are a manifestation of natural resistance or the result of a developing acquired immunity.

Most trematodes are much stricter in the selection of their snail host than of their vertebrate host; often only a single genus of mollusk (and a single species in a determined geographic area) will be adequate for a given species of trematode to grow, whereas several species of vertebrates will be appropriate to sustain the development of the same fluke to adulthood. Based on this different restricted specificity, many investigators believe that Digenea began its parasitic life as parasites of mollusks and that the vertebrate host is a rather late acquisition in the evolution of the group. Some work has suggested that, on repeated invasions by miracidia, snails can develop reactions that in some ways resemble the vertebrate immune responses.

The systematic classification of the Digenea is somewhat complex and not particularly helpful for medical purposes. Customarily in medical parasitology, the trematodes have been categorized according to the location of the adult fluke in the vertebrate body. Thus, it is usual to talk about blood flukes, liver flukes, intestinal flukes, and lung flukes.

Because of the distribution of the snails during geological times, most trematodes of medical importance occur in Asia and Africa. We will review the most important species that occur naturally in the Americas: *Schistosoma mansoni* among the blood flukes; *Fasciola hepatica* among the hepatic trematodes; *Paramphistomum* species among the intestinal flukes, and *Paragonimus* species among those that affect the lungs.

THE SCHISTOSOMES

The schistosomes are a group of medically important flukes characterized by the possession of several peculiar features: their eggs lack an operculum; their cercariae (with bifurcated tails and no pharynxes) actively penetrate the skin of the host; they are unisexual; and the adults live in the bloodstream. A number of species that affect humans and domestic (and wild) animals exist, but the best known and the most researched are the human schistosomes: *Schistosoma mansoni, Schistosoma haematobium,* and *Schistosoma japonicum.*

Various estimations have claimed that between 125 (Wright, 1972) and 200 (Warren, 1973) million people are infected with *Schistosoma* at present, with possibly 1% of them showing clinical manifestations. Huge and increasing amounts of research funds have been poured on schistosomiasis in the last 15 years or so: as a consequence, our knowledge of the fine details of schistosome infections exceeds that of any other parasitic disease. Unfortunately, all this information has not been organized yet in a practical context for the effective control of the infection in humans. As an expression of the generous funding, publications on the immunology of schistosomiasis are numerous and new information is being produced daily. Some recent reviews have been written by Smithers and Terry (1976) and Warren (1978a,b). They give lists of important references.

The life cycle of the schistosomes outside the vertebrate host corresponds approximately to the general model for the trematodes. Once the cercariae leave their specific snail they seek and penetrate the skin of a definitive host with the assistance of enzymatic secretions. On penetration, the cercaria undergoes morphological and biochemical changes to become a schistosomulum. The invasion of the epidermis by the cercaria takes about 15 minutes, and the resulting schistosomulum migrates in the dermis until penetrating a blood or lymph vessel that will take it to the lungs, where they are found about the fourth day. After a period of growth, the young parasites migrate to the liver, where they become sexually mature, mate, and continue toward their final habitat (mesenteric veins for *S. mansoni* and *S. japonicum*; vesical plexus for *S. haematobium*, and nasal vessels for *Schistosoma nasalis* of cattle, goats, and horses).

The eggs laid in the vessels develop an embryo able to produce enzymes that permeate outside the shell and open the way through the tissues and into the lumen of the respective organ (intestine, urinary bladder, or nose). More than half the eggs produced, however, remain in the tissues of the host. Prepatency is about 4–5 weeks for *S. japonicum*, 6 weeks for *S. mansoni*, and 10–12 weeks for *S. haematobium*. The life span of the flukes in humans may reach up to 20 or more years.

Natural Resistance

This subject has been discussed in some detail by Lewert (1970) and by Smithers and Terry (1976). Schistosomes, in general, do not exhibit a very strict specificity for the definitive host: *S. japonicum* grows in humans, other primates, many ungulates, and numerous rodents; *S. mansoni* can develop in primates, rodents, insectivores, marsupials, and cattle; *S. haematobium* can infect primates and some rodents.

Peculiar restrictions of hosts, probably with genetic bases, are known to occur in some strains: the Formosan variety of *S. japonicum* grows to maturity in rodents and carnivores but only rarely in humans or other primates; and Puerto Rican strains of *S. mansoni* do not reach patency in guinea pigs or dogs, whereas South American strains do. It is not completely clear, however, whether these examples are instances of natural resistance or cases of particularly effective acquired immunity produced by parasites strongly immunogenic or that lack appropriate mechanisms to circumvent the host's response.

Some studies have shown that the sex of the host may influence its susceptibility to the parasite, but further work is necessary to determine exactly the extent and mechanisms of this influence.

A number of workers have demonstrated that skin structure is a major factor in the susceptibility to the infection: the cercariae appear to be attracted to the skin by lipids on its surface, and initiate invasion independent of the host species. However, death of the invading parasites ensues during their epidermal migration, in direct proportion to the resistance of the host: up to 50% of the cercariae of *S. mansoni* die in rats, about 30% in mice, and only 10% in hamsters. The mortality is higher when the cercariae or the host are older.

The lower rate of survival of cercariae 8 or 24 hr old, as compared to cercariae 2 hr old, appears to be related to their glycogen content which decreases with time. Apparently, the mechanism of host resistance in these cases is purely passive: the difficulty of going through the epidermis brings about the exhaustion of the sources of energy of the parasite and results in death. Incubation of the host snail in glucose solutions prior to the escape of the cercariae increases the glycogen content and the survival of the cercariae during skin invasion.

The mediating factor of natural resistance at the skin level must be the susceptibility of the skin of different host species to depolymerization by the cercarial enzymes. The less susceptible the skin is, the longer it will take the cercariae to complete their journey to the vessels, and the more chances of exhaustion will exist. Studies with *S. mansoni* have demonstrated that the mortality of cercariae invading the skin of 1-month-old mice is over 3 times greater than that of similar cercariae invading the skin of 2-day-old animals. This difference is attributed to the higher degree of polymerization of the skin of older animals.

Acquired Immunity

Infection of humans with schistosomes results in the production of antibodies of the IgM, IgG, and IgE classes, and in manifestations of cell-mediated immunity. Obvious ethical restrictions have prevented the direct demonstration of acquired resistance in the human host. On a few occasions, the infection of volunteers previously infected (naturally or artificially) has failed to establish the parasite, but lack of proper controls diminishes the validity of these results. Epidemiological observations in endemic areas have shown that schistosomiasis is a disease of the young, and that the passage of eggs decreases with the age of the population examined. This characteristic may indicate production of acquired immunity as well as age immunity, or reduced exposure of the older population to the parasite. Later studies showed that people exposed to a primary infection as adults did not exhibit particular resistance to it, which tends to dismiss age immunity as an important factor. The participation of the exposure to the infection, however, is more difficult to dismiss, since there is a strong correlation between rates of infection and activities in the water. Warren (1973) has presented a lucid discussion of the immunological and ecological factors that may control the prevalence of human schistosomiasis. At any rate, based on findings in laboratory animals, most investigators accept that humans develop immune resistance against schistosome infections. The exact importance of immune resistance in the control of the condition under natural circumstances remains to be evaluated, however.

Animal models have been amply employed to determine the immunology of schistosomiasis, the most popular being mice, rats, and various species of lower primates. Studies with laboratory animals have shown that the immune resistance to schistosomiasis is effective against new infections but not against the persisting infection that elicited the immunity. This phenomenon was called *concomitant immunity* since it was reminiscent of the corresponding phenomenon that occurs in cases of tumors. Some observations suggest that concomitant immunity also occurs in humans: repeated counts of the number of eggs passed by a group of children during a period of 3 years showed a remarkable consistency of the results; the children passing more eggs during the first year continued to pass the highest number at the end of the observation period, and those passing fewer eggs continued to have low egg counts. Since exposure to the infection was unchanged during the study, only two logical explanations were considered: either the rate of acquisition of new parasites was equivalent to the rate of loss of old worms, or the eggs came from long-lived schistosomes in circumstances in which new infections were unsuccessful. All experimental evidence favors the latter possibility.

Numerous experiments in animal models have demonstrated that the pathology produced by schistosomiasis is mediated by immune phenomena. Although strict proof that this is also the case in humans is lacking, the similarity of the pathology in man and animals and the consistency of the pathophysiological mechanisms in diverse species of laboratory animals leave little doubt.

Since no correlation has been found between resistance and other manifestations of humoral or cell-mediated immunity, protection must be evaluated by recovery of the parasites of a challenge infection. This may be done by collection of the adult worms from their final habitat or from the lungs during the passage of the schistosomula through this organ. This latter method does not indicate whether the parasites have been killed or simply delayed in their migration or whether they will finally reach their adult stage, but does allow the attainment of results 4 or 5 days after the challenge instead of waiting several weeks. Taking into account its limitations, lung recovery is a useful technique.

When mice are challenged at weekly intervals after a primary infection with *S. mansoni*, reduced numbers of schistosomula are found in their lungs from the challenges given between the first and fourth week of the immunizing infection and from the seventh week onward. Counting of the worms that reach adulthood, however, indicates that the resistance of the first to fourth weeks only delays the migration, and does not decrease the number of parasites that reach the adult stage. The resistance expressed from the seventh week onward, on the contrary, reduces the final parasitic load. This second wave of resistance reaches its peak between the 12th and 16th week, when 75–90% of the parasites of the challenge are eliminated. The remaining worms persist for long periods, which makes the mouse model particularly

appropriate to study chronic schistosomiasis. The early wave of resistance may be transferred to normal mice with IgG antibodies. The
late parasiticidal wave can be transferred with serum to animals deficient in T cells but that have eosinophils, and can be abolished by
treatment with antieosinophil serum (see below).

Weekly challenges with *S. mansoni* in previously infected rats
show that immunity begins to develop by the third week of the primary
infection, reaches a peak in the sixth week and wanes totally after the
11th week. Rats are unusual hosts in the sense that they destroy virtually all schistosomes 4 to 6 weeks after the infection (which makes
them adequate models to study the acute phase of the infection). Evidently, the resistance elicited by schistosomes requires the presence
of the parasite to remain effective; otherwise, it wanes rapidly.

Lower primates, especially rhesus monkeys, have been used in
the hope that they will represent a better experimental approximation
to the actual situation in humans. Rhesus monkeys are readily infected
with, and develop a strong resistance to, *S. mansoni* and *S. japonicum*.
The resistance produced in this species appears to be related to the
degree of antigenic stimulation: infection with 600 cercariae of *S. mansoni* causes two-thirds of the parasites to die after the 12th week of
infection, and causes the number of eggs per worm passed in the feces
to decrease suddenly; however, infection with 100 cercariae does not
induce these changes—the egg output instead increases after the 12th
week, to taper down later on. In both cases, the infection terminates
spontaneously after 1 or 2 years.

Considerable variation in the susceptibility of several species of
primates to schistosomiasis has been detected: rhesus monkeys appear
to be particularly prone to develop strong immunity against *S. mansoni*,
whereas *Cercopithecus* does not evidence any protection after an infection. Nowadays, baboons (*Papio hamadryas*) are believed to be the
closest model to the human infection. In this species, an infection with
S. mansoni persists for years, but it protects the host against new challenges after 16–18 weeks of the primary infection.

Mechanisms of Acquired Resistance

The developmental stage of the parasite that induces immune protection
has been identified through a series of ingenious experiments. Transfer
of 80 pairs of adult *S. mansoni* to schistosome-naive monkeys resulted
in strong resistance of these to the challenge with cercariae 8–14 weeks
later. This experiment demonstrates that the migrating schistosomula
are not critical for the production of protection. Dead worms did not
induce resistance, but transversally sectioned parasites, unable to pro-

duce eggs, did. This shows that metabolic products of the adult worms that are not related to egg production act as functional antigens.

Inoculation of schistosome eggs produces humoral and cell-mediated responses, the latter being the major pathologic mechanism of the disease (see below), but no evidence of immune protection. These results have been elegantly confirmed by infections with parasites of a single sex that elicit strong resistance to challenges despite the absence of oviposition.

Immunization with cercariae irradiated in such a way that the resulting schistosomula die in the liver 2–4 weeks after infection (or even earlier) has been reported on several occasions to produce some protection in mice and in monkeys. However, repeated exposures are necessary and it may take several months after the last inoculation for an effective immunity to be detected. These characteristics of the schistosomulum immunization make one wonder whether the observed protection is really elicited by the young flukes or by a few stunted adults that might have developed, not necessarily in their normal habitat. Irradiated cercariae, however, are as immunogenic as normal cercariae in rats. Nevertheless, the rat model is evidently different from the human infection, since rats normally eliminate the parasites before patency.

The reduction of the number of schistosomula recovered from the lungs of immune animals, and other arguments that are discussed in relation to concomitant immunity, indicate that the immune mechanisms that kill the parasites in immune hosts operate in the early phase of the infection, possibly within the first week. Until recently, the intimate mechanisms of immune protection against schistosome infections remained a mystery. Transfer of partial protection with serum or with cells had been reported on several occasions, but this had not explained the totality of the acquired resistance produced by natural infections in most instances. Findings in the last few years (reviewed by Capron et al., 1978) have contributed greatly to our current understanding of the problem.

A few years ago it was demonstrated that infected animals produced an IgG antibody that was lytic for the surface membranes of the schistosomum in vitro, in the presence of complement. Induction of this antibody in rats or mice, however, did not confer protection in vivo, and the corresponding titers remained high in rats infected more than 12 weeks previously, despite the fact that they were completely susceptible to the infection again. The reasons for the lack of activity of this antibody in vivo are not known yet; it is tempting to speculate that host materials that coat the parasite (see below and chapter 7) prevent the action of the antibody, but this will not explain the lack of resistance to the very early infection, since the young schistosomula are not covered by host antigens during the first days of infection.

In 1975, it was reported that schistosomula were killed by incubation in a mixture of normal leukocytes and of heat-inactivated serum of infected individuals. The phenomenon was verified in infections of humans, baboons, and rats. Later work demonstrated that an IgG antibody (possibly in the form of antigen-antibody complexes) reacted with the parasites to produce an opsonizing effect that facilitated the attachment of eosinophils to the schistosomula. The eosinophil-parasite association resulted in lesions to the worm and in degranulation of the eosinophilic cell. It was also found that the lytic effect of the eosinophils was enhanced by the presence of mast cells, possibly because of the eosinophil chemotactic factor produced by these cells. Intense work is currently being done to identify the immediate mechanism by which the effector eosinophils damage the parasite, but the findings at this moment are rather controversial. The typical histology of the lesions of acute schistosomiasis in the immune host and the abolishment of the resistance by administration of antieosinophil serum suggest very strongly that this phenomenon also occurs in vivo. Some preliminary work has provided indications that the schistosomula may be able to activate complement nonimmunologically by the alternative pathway, and some investigations still in progress have suggested that products of the complement activation may be opsonizing for the eosinophils. If these findings are confirmed, it could be further hypothesized that eosinophilic leukocytes opsonized by nonimmunologically activated complement might have an important participation in the natural resistance to *Schistosoma* infections. Conversely, eosinophils opsonized by specific IgG antibodies or by products of immunologically activated complement would be important in the mechanisms of acquired resistance.

About the same time, it was communicated that normal macrophages incubated with unheated serum of immune rats were able to adhere to and kill schistocomula. The same phenomenon has been demonstrated with serum of infected people and infected baboons. Further studies showed that IgE antibody-antigen complexes in the serum bound and activated normal macrophages in a first step, and stimulated the adherence of the cells to the parasite and the production of damage in a second phase. Although no strict proof that this mechanism operates in vivo has been provided yet, the existence of circulating antigen-antibody complexes in schistosomiasis is well known and makes the event feasible in vivo.

Experiments in rats have shown that the IgG-eosinophil system is particularly active from the fourth to the sixth week of infection, whereas the IgE-macrophage system is effective from the seventh to the 12th week. Whether these mechanisms are also sequential in the production of protection in other host species remains to be investigated. Since the IgG antibody involved in the IgG-eosinophil system in rats is an anaphylactic antibody and close correspondence has been found between protection and IgE antibody levels in rats, Capron thinks that anaphylactic antibodies are essential to the expression of immunity against schistosomes, through an indirect effect that favors the association between the effector cells and the target.

A disturbing fact is the persistence of adult parasites in hosts that are fully resistant to reinfections. This phenomenon has been verified in all the common animal models with all three human schistosomes, and epidemiological observations strongly suggest that it also occurs in humans. By association with a similar phenomenon known in tumor immunology, it was called *concomitant immunity*. Evidently, the immune protective mechanisms are fully effective against the early migrating schistosomula but fail against the adult, established parasites. A decade ago, Smithers and collaborators demonstrated that the developing parasites adsorbed host materials on their surface, thus antigenically disguising their presence in the host organism. This subject is elaborated in Chapter 7.

Artificial Production of Resistance

Past attempts to vaccinate laboratory animals with the dead parasite, its extracts, or its metabolic products have produced modest results at best. A recent experiment of vaccination of mice with *S. mansoni* cercarial antigens showed that, although the animals exhibited histological reactions to the challenge closely resembling those of immune animals, no protection occurred. Apparently the cercarial material lacks some constituent that is essential for the production of effective immunity. In another assay, secreted antigens of *S. mansoni* were repeatedly inoculated subcutaneously and intraperitoneally with adjuvants in mice previous to a challenge, but 60% as many worms developed as in the controls when they were infected later.

Recent reports have communicated that the infection of mice, domestic ruminants, and rhesus monkeys with highly irradiated schistosome cercariae produces a considerable level of protection against reinfections. Since the same technique has not been effective in baboons or chimpanzees, its value in the control of the human infection is doubtful, but it may be adequate to protect domestic animals in enzootic areas.

Inoculation of BCG in mice has been reported to produce considerable, although inconsistent, protection against a subsequent infection with *S. mansoni*. Its mechanism of action remains conjectural, although BCG is a well-known activator of macrophages.

Since immune protection against schistosomiasis takes a rather long time to develop and antigen sharing among trematodes is extensive, attempts have been made to produce resistance by infection with heterologous species of schistosomes that, it is hoped, will persist for some time in the abnormal host but will not reach maturity. An earlier trial of vaccination of rhesus monkeys against the Japanese anthrophilic strain of *S. japonicum* by infection with the Formosan zoophilic strain has been repeated and reevaluated recently. Unfortunately, the Formosan strain induced only partial cross-protection, but matured and produced granulomas. Since a large proportion of the pathology of schistosomiasis is due to the tissue reaction to trapped eggs, and this may be a cumulative phenomenon, immunizing infections that progress to the egg-laying stage are inadequate for the prevention of the human infection. If the egg production is low, they might be appropriate to vaccinate domestic animals, since their shorter life span may prevent the development of extensive damage. In support of this idea, the infection of mice or rhesus monkeys with bovine *Schistosoma bovis* or *Schistosoma mattheei* produced very low parasitic loads but strong protection against subsequent challenges with *S. mansoni*. Infections of cattle and sheep with *S. bovis*, *Schistosoma turkestanicum*, or *S. haematobium* reduced the adults of a later heterologous challenge by 8–42% and the tissue eggs by 15–91%. In a recent experiment, infection of mice with *Fasciola hepatica* reduced a subsequent infection with *S. mansoni* by 56%; inoculation of *F. hepatica* antigens in mice or hamsters produced 25–49% protection. Although common antigens are demonstrable in vitro between *S. mansoni* and the nematodes *Ascaris suum* and *Trichinella spiralis*, infection with these latter does not confer immunity against the schistosome. Undoubtedly the latest advances in the study of the mechanisms of immune protection in schistosomiasis will assist in the location of the functional antigens and will indicate which are the most adequate methods to stimulate the reactions most relevant to protection. In all probability, the next years will bring very significant progresses in the procedures of immunization against schistosomiasis (see "Antigens" below).

Immunodiagnosis

Althouth the clinical diagnosis of schistosomiasis is commonly based on the finding of characteristic eggs, a profusion of immunological methods has been proposed for use in those clinical cases in which egg excretion level is low or has not begun yet, or in epidemiological surveys. Some methods, such as the leukocyte-mediated histamine release technique, are sensitive and specific, but appear to belong more to the research laboratory than to a diagnostic center.

The immediate-type skin test becomes positive in infected people between 4 and 8 weeks after the infection and continues to yield positive results for years, even after successful treatment. The sensitivity of this test depends largely on the antigen utilized and on other factors such as age and sex of the patient and site of inoculation (more sensitive with extracts of adult worms, in adults than in children, in males than in females, and in the upper back instead of the forearm). The specificity is variable: cross-reactions with other trematode infections are common, and positive results in cases of cercarial dermatitis are obtained when cercarial antigens are used.

The delayed-type skin test has been shown to be too insensitive for diagnostic purposes. There are some indications, however, that the presence of this reaction may be related to repeated exposures to the infection in the past and to the existing degree of pathology.

The complement-fixation test is considered by some workers to be one of the most reliable serological techniques for the identification of schistosomiasis. A recent study using the standard technique recommended by the U.S. Public Health Service with antigens from adult worms found 100% specificity, but only 69% sensitivity. The antibodies persisted at the same titer (16) in one patient for $1\frac{1}{2}$ years after treatment, but were below 8 in patients who had been treated 6 or more years previously.

Several types of precipitation tests have also been used in the diagnosis of schistosomiasis. In a study with 73 individuals passing schistosome eggs and 49 noninfected subjects, the double diffusion test (DDT) showed about 85% specificity with either adult or cercarial antigens, but, whereas the sensitivity was 67% with the former, it was only 48% with the latter; circumoval precipitation (COP) with fresh eggs showed about 95% sensitivity and specificity in the same survey. The sensitivity of COP was much less dependent on the intensity of the infection than that of the DDT.

The finding of schistosome antigen in the urine of patients of *S. japonicum* prompted the assay of a uroprecipitin test for the demonstration of the infection: this technique appears to give more consistent results in infections with *S. japonicum* than with *S. mansoni*. Two antigens (and specific antibodies) have been found recently in the milk of a proportion of infected women.

A number of agglutination tests have also been developed, using either particles sensitized with antigen, or cercariae. Equivocal results obtained with the former and the need for a constant supply of living parasites for the latter have reduced their popularity. Some otherwise appropriate tests that utilize living parasites (e.g., the "cercarienhül-lenreaktion," the miracidial immobilization test) have followed the same fate.

The indirect fluorescent antibody test (IFAT), with particulate or soluble antigens, has become a favorite among the traditional serological methods. Its sensitivity in a recent study, however, was only 73% when used with sections of adult worms and 70% when used with fixed cercariae; its specificity was 86% in the first case and 65% in the latter (Ruiz-Tiben et al., 1979). Paying particular attention to a polysaccharide antigen of the cecum of the adult fluke, the sensitivity of the IFAT in 49 patients was 100% when IgM antibodies were investigated, and 86% when IgG antibodies were detected. Contrary to what might be expected, the lesser activity of the IgG antibodies was found exclusively in the chronic patients. In both cases the specificity was 97%.

In a recent comparison among the DDT, COP, enzyme-linked immunosorbent assay (ELISA), and radioimmunoassay (RIA), it was reported that the respective specificities were 83%, 96%, 92%, and 79%, and their sensitivities were 41%, 95%, 75%, and 95%, respectively. COP and RIA were almost equally sensitive for all age groups or intensities of infection; the sensitivity of the DDT and ELISA was inversely related to the age of the patient and directly related to the intensity of the infection. Further studies with ELISA have shown that acute patients have higher antibody titers against cercarial antigen than against adult worm antigen, and that the contrary is true for chronic patients. A similar phenomenon had been reported with the IFAT, which correlates well with the life cycle of the parasite in the host.

Antigens

Some of the most recent advances in this area have been reviewed by Smithers and Terry (1976) and Bout et al. (1978). Recent studies with bidimensional immunoelectrophoresis have shown that extracts of *S. mansoni* possess at least 60 different antigen systems. This complexity has been discouraging in the attempt to purify and characterize these components, but the ingenious application of modern immunological techniques has made possible important progress in the last years. The main goals in the study of schistosome antigens have been to obtain materials to use in sensitive and specific diagnostic tests, to produce immune resistance to the infection, and to investigate their biological significance for the parasite and the host.

The ample array of antigenic substances in schistosomes has been shown to include stage-, species-, and genus-specific antigens, as well as extensive cross-reactive materials. Some 15 years ago, it was found that, of 21 antigens demonstrable in extracts of adult *S. mansoni*, 11 were shared with the eggs, 14 with the cercariae, and 12 were also present in the excretions and secretions of the mature worms; 19 were common to *S. haematobium* and 10 to *S. japonicum*. It was also found that flukes grown in mice or in hamsters shared four or five antigens, respectively, with the host; two of the antigens common to *S. mansoni* and to hamsters were also present in humans. Cross-reactivity with materials of the snail intermediate host, with other trematodes, and with nematodes have also been reported.

The biological significance of these findings is discussed in Chapter 7. Their practical importance for the diagnosis is dual: on the one hand, the extensive cross-reactivity must be responsible for the numerous nonspecific reactions often found in schistosome immunodiagnosis; on the other hand, it permits workers to obtain diagnostic antigens from the most accessible stages (the adult) or even from species that grow particularly well in laboratory animals, and to use them in the diagnosis of the human condition. The functional antigens of schistosomes have not been definitively identified yet, but there are some suggestions that they may be shared by the schistosomulum and the adult fluke. If this is so, vaccination of the host with selected antigens obtained from the adult worm may arrest the infection at an early stage, before the most serious pathology occurs.

A few schistosome antigens have been purified and/or characterized lately. Two allergens that react specifically with anti–*S. mansoni* IgE antibodies were identified in 1977; they were present in adult worms as well as in cercariae. One of them was subsequently purified and had a molecular weight of 15,000 to 20,000. Since allergens must be protective antigens in view of the recent discoveries of lethal antischistosoma activity of the IgE-macrophage association (see above), this finding may have very important repercussions for the immunoprophylaxis of the infection.

The use of affinity chromatography with antischistosoma drugs has permitted the purification of the antigens that act as targets of the drug in the worm. These discoveries have facilitated the study of the mechanism of action of the drugs and allowed the use of specific antigens for immunization. Active immunization of rats with drug-target antigens and passive transfer of antiemetine-target serum in mice have led to significant protection against the infection. Actually, the drug-target antigen produces antibodies that are specifically directed against the parasitic structures susceptible to the drug! Four of these antigens had enzymatic activity and were located in the cecum of the adult worm; the other two had no known enzymatic function and were located in the parenchyma of the fluke.

A genus-specific antigen, called "antigen 1" for its position in the immunophoretogram, has also been identified. It appears to be a polysaccharide of 600,000 daltons formed by polymerization of molecules of 10,000 daltons. It is located in the cecum of the adult worm and causes the formation of early but persistent antibodies in infected people. Another genus-specific antigen ("antigen 4") was found to be a malate dehydrogenase also located in the cecum of the adult fluke. This antigen has been detected in circulating antigen-antibody complexes in humans. The corresponding antibodies appear early in the human infection. Immunization of mice with it produced significant reduction of the worm burden of a challenge.

Since antigens on the surface of the worm are likely to be the target of protective immune reactions and may play an important role in the refractoriness of the adult fluke to the host's immunity, some work has been done in this direction. By isotopic labeling it was found that the half-life of the schistosoma surface membrane is only 30 hr. This means that the fluke is constantly shedding surface antigens. This characteristic may facilitate the production of circulating antigen-antibody complexes with the consequent effect on the pathology of the infection, and may allow the worm to circumvent the host immunity in a more effective manner. The latter is elaborated in Chapter 7.

Recently, it has become possible to identify three major serological antigens (MSA) in schistosome eggs; MSA_1 appears only in mature eggs, seems to be a constituent of the hatching fluid, and is the major component that elicits granuloma-forming reactions.

Work on schistosoma antigens is very actively progressing at present, and exciting news is expected in the near future.

Immunopathology

The participation of immunity in the pathology of schistosomiasis has been the subject of a recent review by Warren (1978b). The penetration of cercariae in the skin of the host, even if they die shortly after invasion, constitutes sufficient stimulation to elicit production of skin-sensitizing antibodies and cell-mediated immunity. Subsequent attacks will produce the corresponding allergic manifestations in the skin. These reactions to cercarial invasion are not strictly specific, but their immunological nature has been verified histologically in humans and experimentally in laboratory animals.

Occasionally, individuals undergoing a primary schistosome infection develop an acute syndrome with fever, eosinophilia, urticaria, and enlargement of the lymphoid organs, shortly after the beginning of oviposition. There is some evidence that this condition may be related to the presence of specific antigen-antibody complexes in the circulation. Glomerulonephritis has been frequently reported in infections with *S. mansoni*, and specific antigen-antibody complexes, including IgM and IgG, have been recovered from the kidneys of infected individuals. Although strict causal relationships are still lacking, these findings suggest that the schistosomiasis-associated glomerulonephritis may be a result of the parasitic infection.

The major pathogenic mechanism in schistosomiasis, however, is the sequence of vascular obstruction, granuloma formation, fibrosis, and vessel recanalization that occurs in the host's tissues as a reaction to the trapped eggs. The finding that the granulomatous response was specific and occurred with more intensity on reinfections revealed its immunological nature. Experiments of cell and serum transfer, of differential suppression of humoral and cell-mediated responses, and of production of lymphokines directly by isolated granulomas proved that it was a manifestation of cell-mediated immunity (this appears not to be true of *S. japonicum,* however). Recently, the egg antigen responsible for the sensitization of the host and for the subsequent granulomatous reaction has been identified.

Autoantibodies against liver, smooth muscle, nuclear material, and mitochondria have been reported in a proportion of patients of schistosomiasis. Their clinical significance has not been determined yet.

FASCIOLA HEPATICA

Fasciola hepatica is the major liver fluke affecting ruminants throughout the world. Primarily a parasite of sheep and goats, it readily infects a number of other ruminants, rodents, and lagomorphs, and it is also found in pigs, horses, and occasionally carnivores. Human infections, once thought to be rare, are not too infrequent as judged by work done mainly in Chile and in France. A British-Peruvian-French team working in the Peruvian Altiplano found the infection by fecal exam in 9% of 1001 school-age children. Reviews on the immunology of fascioliasis have been written recently by Platzer (1970) and Smithers (1976). After a period of comparative neglect, the studies in this area have increased considerably in the last years.

The premammalian portion of the life cycle of *F. hepatica* is typical of the trematodes: a period of multiplication occurs in *Limnaea* snails, and the resulting cercariae encyst to metacercariae on aquatic plants or in the water. On ingestion by the definitive host, the young fluke goes through the intestinal wall into the peritoneal cavity and invades the liver parenchyma within the next 2 to 4 days. A period of growth and wandering in this organ follows until the worms reach the bile duct (about 40 days after the infection in sheep, 49 in cattle, 45 in rabbits, 28 in rats, and 24 in mice). Sexual maturation and oviposition are attained 2 to 3 weeks later.

Natural Resistance

Sheep and goats, lagomorphs, and murine rodents are the animals most susceptible to the infection and to the disease. The tissue reaction in these species is severe, but most of the parasites (larval or adult) survive and do not become isolated from the host tissues. Cattle, humans, guinea pigs, and horses respond to the late-migrating forms or to the adult parasites with a fibrotic tissue reaction that kills or isolates many worms. Pigs, hamsters, and carnivores react sooner and destroy most of the flukes during the liver migration. These differences have been attributed to nonspecific tissue reactions rather than to immunological events, but this opinion should probably be revised now.

The relative infrequency of the human infection may be due to ecological circumstances: practically the only common source of infection for humans is watercress grown in contaminated marshes, and this delicacy is not appreciated by everyone.

Infection of rats with doses of 1, 5, 10, 15, or 20 metacercariae resulted in the maturation of a similar proportion of adult worms in all cases, which negates the existence of nonimmune mechanisms of control of the parasitic load such as crowding effect or competitive inhibition. Experiments with rabbits showed similar results. A number of studies with sheep have demonstrated that 43% of 70, 40% of 100, 39% of 200, and 37% of 600 metacercariae administered reached adulthood. It appears then that a fairly constant proportion of the infective dose grows to maturity, with independence of the size of the dose. This is an unusual occurrence in parasitism; its causes in fascioliasis are unknown. These results must be taken cautiously, however, since the individual variability is great: of 7 sheep infected with 100 metacercariae each, 8 adult flukes were recovered from one and 52 from another; of 10 sheep receiving 200 cysts each, the recovery ranged from 24% to 59% (Dawes and Hughes, 1964).

Experiments in rats have suggested that this host develops age resistance, although some negative results have also been reported.

Acquired Immunity

Examination of the infection in sheep, cattle, humans, and laboratory animals has consistently demonstrated the production of a humoral immune response in fascioliasis. Precipitating, agglutinating, complement-fixing, and skin-sensitizing antibodies (IgE and IgG) are readily demonstrated in these hosts.

The presence of cell-mediated immunity in fascioliasis has been indirectly demonstrated by transfer of protection with lymphoid cells in cases in which the serum of the same donors was ineffective for those purposes. This has been verified recently in rats by production of delayed skin tests and by histological examination, and by in vitro correlates in other laboratory animals.

Sheep being the natural host of *F. hepatica*, it is rather surprising that neither field observations or laboratory experiments have succeeded in demonstrating any important degree of acquired resistance to current infections or reinfections. There are reports of sheep infections lasting for 11 years without evidence of immunological rejection, and the infection of older animals that have undergone previous infection is, if anything, more severe than a primary infection. Sinclair (1975) showed experimentally that one prolonged infection or five short ones did not reduce the number of worms or the pathology produced by a later challenge. However, he did detect a quicker eosinophilic response and a higher hyperglobulinemia on reinfection, which was reminiscent of a secondary immune response. Several other investigations in sheep and goats have yielded essentially the same results. Infection of sheep treated with corticosteroids resulted in increased pathology and larger flukes, but the number of worms recovered was not significantly greater than in control animals: these effects were attributed to the nonspecific antiinflammatory activity of the drug rather than to an immunosuppressive action, although there seem to be no strong arguments for preferring one alternative over the other. The lack of protection is even more surprising because specific antibodies and secondary responses have been consistently verified in infected sheep.

It appears evident that sheep do mount an immune response against *F. hepatica*, but this is ineffective in affording protection. Since acquired immune protection has been well proven in the case of fascioliasis in other host species, this phenomenon must be connected with the particular reactivity of sheep to the parasite or its products rather than with some characteristic inherent in the worm. At any rate, this might well be an example of a parasite eluding the effects of the host's immunity (see Chapter 7).

Unlike sheep, cattle produce an effective protective immune response (Kendall et al., 1978). Naturally infected animals often eliminate the parasites (by expulsion or destruction) in 1 or 2 years in the absence of reinfection. Experimentally, it has been found that an infection with 1000 metacercariae terminated artificially after 8 months produces enough resistance to reduce by 82% the number of flukes that develop following a challenge administered 3 or 22 weeks after the end of the primary infection. In this case, the few parasites recovered were much smaller and the host showed less precipitating antibody than the controls. This latter result indicates a negative correlation between resistance and presence of precipitating antibodies, which might suggest that these antibodies are not relevant to protection. Eventually, the reduction of antibodies might be due to immunoelimination by combination with antigens released by the broken parasites.

Since sheep do not develop resistance and cattle are expensive to work with, a number of laboratory models using rabbits, rats, and mice have been devised to study the mechanisms of immunity to *F. hepatica*. All these hosts have been shown to respond to a primary infection with production of immunoresistance to reinfections.

Mechanisms of Acquired Resistance

A few ingenious experiments have been developed to identify the site where the resistance is expressed. Seventeen-day-old worms have been transferred from the livers of immune or normal mice into the peritoneal cavity of normal mice: in the former case the parasites died, whereas in the latter they continued the infection until attaining maturity. This has been taken as proof that the migration in the liver of the immune donors had damaged the fluke in a way that was incompatible with its continued existence.

In another experiment, significantly fewer worms were recovered 2 days after infection from immune mice than from normal mice. This finding indicates the presence of protective mechanisms in the intestinal wall or in the peritoneal cavity. The facts that adult *F. hepatica* survive for long periods in the bile ducts of the host and that they will live for at least 7 weeks when transferred to the peritoneal cavity of normal rats or rabbits indicate that the ductal environment is not detrimental for the parasite, despite the recent verification of antibodies in it. Therefore, there is evidence that *F. hepatica* is affected by the host immune response during its early period of invasion through the intestine and into the peritoneal cavity and later during its migrations in the liver. Once the bile ducts are reached, the parasites appear to be protected even from the boosting effect of a reinfection.

Some related work has been done to identify the stages of the parasite that elicit immunity: subcutaneous implantation of metacercariae, 4-week-old juveniles, adult flukes, or eggs in rats showed that all but the adults produced protection against a challenge with 30 metacercariae 2 weeks later. These results agree well with the findings reported above and pose the question of whether the lack of immune response to the worms in the bile ducts is related to the stage of the parasite, the site in the host, or both.

In regard to the immune mechanisms of protection, a number of early experiments had failed to transfer immunity by inoculations of serum of resistant hosts into susceptible animals. Recent studies have shown that serum of mice immunized with *F. hepatica* excretory/secretory (ES) products or of mice with a 25-day-old infection reduced the number of worms that developed and the host mortality when injected into susceptible mice. Sera of animals immunized with somatic materials of the flukes or undergoing a 100-day-old infection did not produce protection in the recipients. These findings indicate that only antibodies to metabolic products are protective, and that the corresponding functional antigens are produced only early in the infection.

Work with isogeneic rats and with a pair of monozygous twin cattle demonstrated that lymphocytes of animals infected for 8 or more weeks transferred from 67% to 100% protection, whereas the cells of animals with 4-week-old infections produced less than 20% protection in the recipients. Serum transfers protected 25% at the most (Corda et al., 1971).

Taking these two studies as a whole, the indication is that immune resistance to fascioliasis is a biphasic phenomenon: a humoral protective response to parasite metabolites is produced early in the course of the infection and wanes later on; at that time, it is replaced by a protective cell-mediated response, which is stimulated by the more advanced parasitic stages.

In other experiments, it has been observed that inoculation of large amounts of serum from immune animals into susceptible rats appears to cause the destruction of the parasites before they invade the liver, whereas transfer of immune cells accelerates and intensifies liver infiltration and results in dead parasites in this organ. A possible explanation for this duality is that antibodies are particularly effective during intestinal penetration or peritoneal migration, whereas effector T cells are more efficacious in the hepatic parenchyma. Even if this speculation is true, it is not known whether the different effectiveness of both branches of immunity is related to the antigens produced by the parasite at different stages in its life cycle or to the site where the reaction takes place.

Artificial Production of Resistance

Dawes and Hughes (1964) have produced a detailed review of the early attempts to vaccinate against *F. hepatica* infection. Clegg and Smith (1978) and Smithers (1976) have commented on this topic in more recent works.

Many assays of vaccination with somatic extracts of the parasite have failed or have resulted in negligible levels of protection against reinfections. Injections of ES products collected from worms incubated in vitro have generally given rather high levels of protection, although some inconsistant results have been reported. These latter may be due to improper handling of functional antigens, which might be particularly unstable in *F. hepatica*: injection of ES products collected over a 4-hr period did not elicit protection in mice, but, when the incubation medium was harvested periodically and injected immediately, the protection to a challenge was very high.

Proof that the immune response to the functional antigens released by the parasites is sytemic rather than local was obtained by inducing resistance with the subcutaneous implantation of metacercariae, 4-week-old juveniles, adults, or eggs of *F. hepatica* in rats. No resistance was produced by the adults in this experiment, and they have been shown to be considerably poorer stimulators of protective immunity than the migratory forms in others. In a recent study, a single adult *F. hepatica* implanted subcutaneously in rats induced production of cell-mediated immunity and of precipitating antibodies that were already evident at the end of the first week; on challenge, however, the number of worms that developed was two-thirds of that in the controls, and the pathology produced was similar to that in the controls. It appears that the quality or quantity of the functional antigens released by the adult parasites, even when they are at a location other than the bile ducts, is rather deficient.

Since irradiation of the metacercariae with 3000 r interrupts the development of the young flukes during the second week of infection, the vaccination with these forms has been attempted in the hope of providing a fairly prolonged source of functional antigens with reduced pathology. Although encouraging results have been reported in rats, failures have been communicated in mice and rabbits. Even if this method proves to be effective in domestic ruminants in the future, technological developments for the production of an adequate number of metacercariae will be necessary.

None of the techniques of vaccination assayed have given significant protection in sheep. As already mentioned, this seems to be a peculiar characteristic of the reactivity of sheep to *F. hepatica*. Strong resistance to *F. hepatica* challenges has been reported, however, by a previous infection with *Taenia hydatigena* (see section on "The Cestodes or Tapeworms," above).

Immunodiagnosis

A large part of the pathology of fascioliasis is due to the liver damage produced by the migratory juveniles; since the infection is not patent at this time, the identification of the condition must rely on indirect means. Besides, the current methods of detecting eggs in the feces have a low efficiency.

Benex et al. (1973) studied the serological response of sheep experimentally infected with *F. hepatica* by tests of complement fixation, latex agglutination, hemagglutination, fluorescent antibodies, immunoelectrophoresis, and double diffusion in agar with a delipidized antigen of adult worms. They found that all tests become positive during the second or third week of infection, peaked at the sixth to eighth week, and persisted at least for 15 weeks. Successful treatment lowered the titers, but, with the exception of the precipitating tests that became negative by the ninth week, specific antibodies were still detectable until the 15th week. Gundlach, in Poland, found similar results in cattle: hemagglutinins appeared on the second or third week of infection and precipitins on the fourth or fifth; they peaked during the seventh to 11th week and persisted at least until the 18th week. IgE antibodies appeared on the fourth week and increased from the 20th week of infection.

In a study with 67 children who were passing *F. hepatica* eggs, 10.5% were positive by complement fixation, 42% by immunofluorescence, 51% by immunoelectrophoresis, and 82% by skin test (Stork et al., 1973). The results of this particular investigation may have been affected, however, by the fact that several months elapsed between the attainment of the serum samples and the performance of the serological tests.

The immediate-type skin test with low concentration of antigen (30 μg of N_2/ml) appears to be a sensitive and specific method to diagnose fascioliasis. It diagnosed 82% of the infections in a group of 67 children passing *F. hepatica* eggs, but only 60% of 109 children who were positive to the fecal exam or to other immunological tests. This work, however, was done by injecting only 0.05 ml of an antigenic solution containing 6 μg of N_2/ml; it is possible that a larger dose might have improved its sensitivity. A recent report from Capron's group in France about the isolation of an *F. hepatica* species-specific allergen will probably improve the skin test in the future. At least in cattle, this test became positive about the third week of infection and still persisted on the twentieth week; after the sixteenth week the typical immediate reaction was followed by local manifestations of an Arthus phenomenon. Although delayed-type skin reactions occur in rats implanted subcutaneously with an adult worm, they have not been used for diagnostic purposes.

The complement fixation test (CFT) lacks in sensitivity and specificity: of 109 children positive to fecal or other immunological exams, only 14% yielded positive CFT. In cattle, this test was positive in only

50% of 226 infected animals and in 29% of 69 noninfected controls. Many researchers feel that the CFT does not have a place in the diagnosis of fascioliasis nowadays.

The reports about precipitation tests are discordant. A report from Germany states that precipitation in agar was negative in 284 animals belonging to an infected cattle herd, but in Poland, 53% of 226 infected cows were positive by the same test. Results comparable to these latter ones were obtained in sheep in Holland. Immunoelectrophoresis yielded positive results in 51% of 67 children passing eggs and in 48% of 109 children positive to fecal or immunological exams. Hillyer and Capron (1976) reported that all of 37 patients showed positive counterimmunoelectrophoresis results with a crude extract of the parasite, but there were extensive cross-reactions with patients of hydatidosis, trichinellosis, cysticercosis, and amebic hepatitis. Purification of the antigenic preparation improved the specificity without diminishing the sensitivity. In sheep, this exam identified all infections in 49 animals harboring 4 or more flukes and 75% of those caused by 1 to 3 worms. Extensive cross-reactivity has also been reported in precipitation tests in fluid phase with crude extracts.

In a comparative study with the CFT and the precipitation in fluid phase, the indirect hemagglutination test was found to exceed the others in sensitivity and specificity. This technique identified the infection in all 25 sheep with more than 10 flukes, in 92% of 24 with 4 to 10 parasites and in 66% of 32 with 1 to 3 worms. Latex agglutination in cattle was positive in 47% of 69 animals passing eggs and in 20% of the noninfected controls.

The indirect immunofluorescent antibody test (IFAT) in sheep appears positive about the 20th day of infection and becomes negative before the 150th day of successful treatment. A trial in cattle showed 87% sensitivity with only 4% nonspecificity; the titers declined 4 weeks after treatment and were virtually absent 4 weeks later. Of 67 children passing eggs, the IFAT identified the infection in 42% and of 109 children diagnosed by several procedures, it revealed the infection in 49%: in this study, however, only titers greater than 40 were considered positive. The same test in cattle diagnosed all infections when a titer of 8 was taken as significant: unfortunately the nonspecificity appears to increase in direct proportion to the sensitivity; 32% of the noninfected controls were also positive.

The limited studies reported in regard to enzyme-linked immunosorbent assay indicate that the respective antibodies appear in cattle about the fifth week of infection and exhibit a sensitivity of 93%, with a remarkable specificity.

Apart from the newest tests, which still require further evaluation, it is evident that no test is satisfactory for the diagnosis of fascioliasis nowadays; perhaps the skin test is the least misleading at present. The major problem seems to be a consistent direct correlation between sensitivity and nonspecificity: Van Tiggele and Over (1976) found that gel diffusion in sheep had an overall sensitivity of 49% with 5% nonspecificity; the indirect hemagglutination test identified 84% of all infections, but was positive in 27% of sheep harboring no parasites; and counterimmunoelectrophoresis demonstrated 90% of all infections, but gave positive results in 30% of animals with no evidence of infection.

Clearly this inconvenience depends on the complex antigenic composition of the trematodes, which, besides having species- and stage-specific antigens, share antigenic substances extensively with other invertebrates. The purification of species-specific antigens and its use in some of the modern serologic tests should allow virtually any degree of sensitivity that the investigator may want, without the interference of cross-reactions. Fortunately, rapid progress is being made in this area at present: a specific allergen appropriate for use in skin tests has already been identified, and fractions particularly reactive in precipitation reactions, indirect hemagglutination tests, manifestations of immediate allergy, and in vitro correlates of cell-mediated immunity have been located.

Different studies have shown that the major antigens relevant to the diagnosis of and protection against fascioliasis are located in the developing intestinal and genital structures, and in the surface coat of the parasite. Undoubtedly the coming years will bring important advances in the immunology of *F. hepatica* infections.

Immunopathology

Two major problems related to the immunopathology of fascioliasis have begun to be studied only recently. One is the possibility that the parasite causes damage by inducing immune reactions to the host's own materials (not unexpected from an organism that provokes extensive tissue destruction); the other is the identification of the mechanisms that allow the persistence of the worms in hosts that manifest evidence of active immunity.

In regard to the former, recent work has found that sheep infected with *F. hepatica* develop complement-fixing and hemagglutinating antibodies that react with liver, kidney, and lung extracts of noninfected sheep, cattle, or people. The lack of organ or species specificity of these antibodies suggests that they are directed against some fundamental tissue component that is common to all mammals. The fact that they increase even more after a secondary infection confirms that they respond to antigenic stimulation. Their significance in the pathology of the infection has not been assessed yet.

On the other hand, it has been reported that injection of *F. hepatica* somatic antigens in mice causes depression of the packed erythrocyte volume and that erythrocytes of infected mice are agglutinated by antimouse serum during the anemic phase of the infection. These findings suggest that fascioliasis induces the production of autoantibodies against the host's erythrocytes that might result from some antigen sharing between the parasite and the red cells, or from the adsorption of worm antigens on the erythrocytes. The extension of this phenomenon to other hosts and its participation in the characteristic anemia of the infection are unknown at present.

The remarkable ability of *F. hepatica* to invade and survive in sheep that show evidence of specific immunity has been a puzzle to parasitologists for a long time. Investigation of the presence of adsorbed host material on the worms, to disguise their presence in the fashion of *Schistosoma*, has yielded negative results. Some authors have speculated that *F. hepatica* migrates constantly so "today's" response takes place where the parasite was yesterday. This may well be true in the case of cell-mediated immunity, in which the process of recruitment of cells may take over 24 hr, but it is difficult to accept in the case of antibodies that must be circulating and whose reaction with the proper antigen is virtually instantaneous. Besides, the exhibition of this notable capacity to physically escape the site of the immunological reaction in sheep but not in other hosts species would also require an adequate explanation.

A recent paper reported that metabolic products of *F. hepatica* reduce by about 70% the viability of rat lymphocytes in vitro and partially inhibit the adherence of peritoneal cells to the parasites when incubated together in immune serum. Since the experiments were done in rats, which exhibit adequate immune resistance to fascioliasis, it is not known whether these findings are relevant to the elusion of immunity by the worm in sheep. It must be concluded that the reason why *F. hepatica* persists in sheep despite immunity is not known.

THE PARAMPHISTOMES

There is still some controversy about the taxonomy of the paramphistomes and about the proper designation of its members. The most common species in the U.S. is *Paramphistomum microbothroides*, which may be a synonym of *Cotylophoron cotylophoron*. Its life cycle outside the mammalian host is similar to that of *Fasciola hepatica*, although it uses different and diverse snails as intermediate hosts. This is rather exceptional among the trematodes, since they usually exhibit a high degree of specificity for the intermediate host.

The definitive hosts are domestic and some wild ruminants. The young flukes live in the anterior portion of the small intestine for a couple of months and then migrate to the rumen, where they attain sexual maturity during the third or fourth month of infection. Paramphistomiasis is regarded almost as an asymptomatic infection in the U.S., but the larval stages are responsible for serious outbreaks of disease, with 30% or more lethality, in several parts of the world.

The immunology of paramphistome infections has been little researched. Horak (1971) has written a review that includes extensive references to his doctoral thesis on the subject.

Immune Response and Acquired Resistance

The infection causes the formation of skin-sensitizing, precipitating, and complement-fixing antibodies. Field observations have suggested that an infection elicits immune protection against subsequent infections, which is supported by the results of experiments of artificial immunization. Infection of sheep or goats with 40,000 metacercariae (native or irradiated) or of cattle with 40,000 to 100,000 metacercariae protected them against a subsequent massive challenge. This acquired resistance is expressed by the expulsion of a large proportion of the newly excysted flukes of the challenge infection, and by the delay of migration and the inhibition of growth of those few parasites that achieved establishment. When the challenge is massive, part of the preexisting immunizing infection is also eliminated.

The immune protection may persist for up to 3 years in sheep and for at least 1 year in cattle, but it necessitates the continuous presence of adult parasites, even if only a few, in the resistant host; otherwise, it wanes rapidly. It also depends on the size of the immunizing dose: too few metacercariae do not produce detectable immunity, but too many will kill the animals. For protective immunity to develop, the immunizing parasites must go through their larval stages; the implanation of adults in the rumen of susceptible hosts does not elicit protection against reinfections. It appears that the larvae are essential to initiate the production of immunity, but the adults suffice to maintain it. It also appears that sheep under 1 year of age are not competent to develop protection.

The actual mechanisms of immune protection are not known, but a recent report from the U.S.S.R. claims that the daily injection of 80–120 ml of ox hyperimmune serum for 3 days improved the appetite and weight gain of infected cattle.

Immunodiagnosis

Because the pathology of paramphistomiasis occurs mainly during the prepatent period, some interest has existed in devising immunodiagnostic tests. Complement-fixing antibodies are produced during the infection, but the serum of infected sheep has proved to be anticomplementary, which makes the technique impractical. Precipitation tests on adult parasites, or in gels with parasite extracts, also demonstrate antibodies, but their use has not become popular.

The immediate-type skin test, as used by Horak (1971), was too nonspecific since it gave positive reactions in animals infected with *Fasciola*, *Schistosoma*, or nematodes. Although insufficient information is provided, it appears that the concentration of antigen used in these tests was too high by today's standards. Employment of purified fractions of parasite extracts with 10 μg of protein per ml has considerably improved the sensitivity and specificity of this technique.

PARAGONIMUS SPECIES

The immunology to *Paragonimus* infections has not been extensively investigated. Relatively recent reviews that contain immunological information have been written by Yokogawa (1969) and by Sogandares and Seed (1973). There is no complete agreement among the taxonomists yet about the proper systematics of this genus, but the species most widely known are *Paragonimus westermanii* of the Far East and *Paragonimus kellicotti* of North America. A number of other species exist.

The members of this genus are not too specific; *P. kellicotti* appears to be originally a parasite of mink that readily infects cats, dogs, and wild carnivores, and occasionally other mammals (pigs, goats, humans). In Japan, Habe recently infected a number of vertebrate species with metacercariae of *P. westermanii* and recovered 96% of the parasites from cats, 82% from dogs, about 60% from rats, hamsters, and mice, 52% from wild boars, 34% from rabbits, 23% from guinea pigs, and 0.6% from chicken. Although the infective doses were not uniform, these results suggest different degrees of natural susceptibility among these hosts.

The life cycle of *Paragonimus* outside the vertebrate host is typical for the trematodes, with a period of multiplication in the snail *Pomatiopsis lapidaria*, followed by encystment to metacercaria in several crayfish in the case of *P. kellicotti*. On ingestion by a cat, the young fluke excysts, penetrates into the peritoneal cavity from the duodenum in less than 2 hr, and wanders in the liver or the peritoneum for 2 or 3 weeks. Then it goes through the diaphragm to the lungs, settles (usually in pairs) in the neighborhood of a bronchiole, and becomes encapsulated in 2 or 3 weeks. Eggs begin to appear in the bronchial secretions (and in the feces when ingested) 6 weeks after the infection.

Immune Response

Study of the cat infection has shown the presence of complement-fixir antibodies from the 30th to the 100th day of infection. The infection of rhesus monkeys with two African species of *Paragonimus* produced complement-fixing and hemagglutinating antibodies beginning 14 to 21 days after the infection; the former tended to disappear after 6–8 months, but the latter persisted. Precipitating antibodies have been demonstrated in human and cat infections. Increased IgE and specific IgE antibodies have been found in the pleural exudate and in the serum of infected people.

Immune Protection

The rarity of heavy human infections has been taken as a demonstration of acquired immunity to the reinfection. This assumption, although unfounded at the time, has nevertheless been supported by recent work on rats: an immunosuppressive combination of corticosteroids administered to rats since the first day of an infection with *P. miyazakii* increased their susceptibility to the infection and increased the size and the number of eggs laid by the parasites, but depressed the number of precipitating antibodies revealed by diffusion in agar. The administration of prednisol only, however, did not modify the proportion of flukes that developed.

Although it seems clear that acquired immune protection exists in paragonimiasis, its efficiency and mechanisms remain unknown.

Immunodiagnosis

The immunodiagnostic techniques are useful in paragonimiasis for the diagnosis of extrapulmonary forms of the infection, for the differential diagnosis with other pulmonary lesions, and for evaluation of the results of the therapy.

An immediate-type skin test with purified antigens is highly sensitive and specific, but is inadequate for clinical use since it remains positive for years after the death of the parasite. It is very appropriate for epidemiological studies, however. For some reason, cats do not respond to it.

Precipitation tests in fluid phase and in semisolid media, as well as a uroprecipitin test (which reveals antigen in the urine with potent antisera prepared in the laboratory), have been devised but do not seem to be popular.

The complement-fixing antibodies disappear in humans 3–9 months after successful therapy, so the corresponding technique has been favored to assess the results of the treatment. Investigators with ample experience trust this method enough to treat suspected patients that exhibit positive tests even if eggs of the parasite are not detected.

A recent comparison between the complement-fixation test (CFT) and the indirect hemagglutination test (IHAT) in 74 patients infected with *P. uterobilateralis* in Africa (Oelerich, and Volkmer, 1976) revealed 71 (96%) with IHAT titers greater than 80 and 61 (82%) with CFT titers higher than 10. The same test with 37 patients of *P. africanus* showed 87% positive by IHAT and 81% positive by CFT. Although the IHAT appears to be more sensitive than the CFT, experiments in monkeys suggest that the former antibodies persist longer than those revealed by the CFT, so that they do not indicate the result of therapy as readily.

Recent trials with the indirect fluorescent antibody technique, using particles of delipidized worms as antigen, diagnosed the infection in all of 25 patients and were positive (at titer 16 or higher) in only 1 of 80 individuals with other parasitoses. Of 8 cured cases, 6 gave negative results and 2 were doubtful (positive at titer 8). In addition, the results were the same when serum recovered from dried blood samples was used instead of fresh blood. This method may prove to be the technique of choice in the future.

Immunopathology

Choi and Lee reported recently that serum of rabbits infected with *Paragonimus* reacted against extracts of normal tissues. This work requires confirmation and expansion to determine whether the infection may induce the formation of antibodies against the host's own materials.

SOURCES OF INFORMATION

Baron, R. W., and Tanner, C. E. 1977. *Echinococcus multilocularis* in the mouse: The *in vitro* protoscolicidal activity of peritoneal macrophages. Int. J. Parasitol. 7:489–495.

Barriga, O. O. 1971. Sobrevida de escólices de *Echinococcus granulosus* en solución salina y en líquido hidatídico a diferentes temperaturas. Bol. Chile. Parasitol. 26:80–84.

Benex, J., Guilhon, J., and Bernabe, R. 1973. Étude comparative des diverses méthods de diagnostic immunologique de la fasciolose hépato-biliaire expérimentale du mouton et influence du traitement sur le persistance des anticorps. Bull. Soc. Pathol. Exotique 66:116–128.

Bout, D., Carlier, Y., Dessaint, J. P., and Capron, A. 1978. Characterization and purification of *S. mansoni* antigens. In: Immunity to Parasitic Diseases. Colloque INSERM-INRA, pp. 71–85. Editions INSERM, Paris.

Capron, A., Dessaint, J. P., and Capron, M. 1978. Effector mechanisms in immunity to schistosomes. In: Immunity to Parasitic Diseases. Colloque INSERM-INRA, pp. 217–230. Editions INSERM, Paris.

Clegg, J. A., and Smith, M. A. 1978. Prospects for the development of dead vaccines against helminths. Adv. Parasitol. 16:165–218.

Corda, J., Roberts, A. R. J., and Urquhart, G. M. 1971. Transfer of immunity to *Fasciola hepatica* infection by lymphoid cells. Res. Vet. Sci. 12:292–295.

Dawes, B., and Hughes, D. L. 1964. Fascioliasis: The invasive stages of *Fasciola hepatica* in mammalian hosts. Adv. Parasitol. 2:97–168.

Geerts, S., Kumar, V., and Vercruysse, Jr. J. 1977. *In vivo* diagnosis of bovine cysticercosis. Vet. Bull. 47:653–664.

Gemmell, M. A. 1976. Immunology and regulation of the cestode zoonoses. In: S. Cohen and E. H. Sadun (eds.), Immunology of Parasitic Infections, pp. 333–358. Blackwell Scientific Publications, Oxford.

Gemmell, M. A., and Macnamara, F. N. 1972. Immune response to tissue parasites. II. Cestodes. In: E. J. L. Soulsby (ed.), Immunity to Animal Parasites, pp. 235–272. Academic Press, Inc., New York.

Hammerberg, B., and Williams, J. F. 1978. Interaction between *Taenia taeniformis* and the complement system. J. Immunol. 120:1033–1038.

Herd, R. D. 1977. Resistance of dogs to *Echinococcus granulosus*. Int. J. Parasitol. 7:135–138.

Hillyer, G. V., and Capron, A. 1976. Immunodiagnosis of human fascioliasis by counterelectrophoresis. J. Parasitol. 62:1011–1013.

Horak, I. G. 1971. Paramphistomiasis of domestic ruminants. Adv. Parasitol. 9:33–72.

Kagan, I. G. 1967. Characterization of parasitic antigens. In: Immunological Aspects of Parasitic Infections, pp. 25–42, 138–145. Pan-American Health Organization, Washington, D.C.

Kagan, I. G. 1976. Serodiagnosis of hydatid disease. In: S. Cohen and E. H. Sadun (eds.), Immunology of Parasitic Infections, pp. 130–142. Blackwell Scientific Publications, Oxford.

Kagan, I. G., and Agosin, M. 1968. *Echinococcus* antigens. Bull. W.H.O. 39:13–24.

Kendall, S. B., et al. 1978. Resistance to *Fasciola hepatica* in cattle. J. Comp. Pathol. 88:115–122.

Kwa, B. A., and Liew, F. Y. 1978. Studies on the mechanisms of long-term survival of *Taenia taeniformis* in rats. J. Helminthol. 52:1–6.

Lewert, R. M. 1970. Schistosomes. In: G. J. Jackson, R. Herman, and I. Singer (eds.), Immunity to Parasitic Animals. Vol. II, pp. 981–1008. Appleton-Century-Crofts, New York.

Oelerich, S., and Volkmer, K. J. 1976. Antikorper- und Immunoglobulinbestimmungen bei Patienten mit afrikanischer Paragonimiasis. Tropenmed. Parasitol. 27:44–49.

Platzer, E. G. 1970. Trematodes of the liver and lung. In: G. J. Jackson, R. Herman, and I. Singer (eds.), Immunity to Parasitic Animals. Vol. II, pp. 1009–1019. Appleton-Century-Crofts, New York.

Rickard, M. D., Adolph, A. J., and Arundel, J. H. 1977. Vaccination of calves against *Taenia saginata* infection using antigens collected during *in vitro* cultivation of larvae. Res. Vet. Sci. 23:365–367.

Ruiz-Tiben, E., et al. 1979. Intensity of infection with *Schistosoma mansoni*: Its relationship to the sensitivity and specificity of serological tests. Am. J. Trop. Med. Hyg. 28:230–236.

Rydzewski, A. K., Chrisholm, E. S., and Kagan, I. G. 1975. Comparison of serological tests for human cysticercosis by indirect hemagglutination, indirect immunofluorescent antibody and agar gel precipitin tests. J. Parasitol. 61:154–155.

Sinclair, U. B. 1975. The resistance of sheep to *Fasciola hepatica*: Studies on the pathophysiology of challenge infections. Res. Vet. Sci. 19:296–303.

Smithers, S. R. 1976. Immunity to trematode infections. In: S. Cohen and E. H. Sadun (eds.), Immunology of Parasitic Infections, pp. 296–332. Blackwell Scientific Publications, Oxford.

Smithers, S. R., and Terry, R. J. 1976. The immunology of schistosomiasis. Adv. Parasitol. 14:399–422.

Smyth, J. D. 1968. *In vitro* studies and host-specificity in *Echinococcus*. Bull. W.H.O. 39:5–12.

Smyth, J. D. 1969. The Physiology of Cestodes. Oliver and Boyd, Edinburgh.

Sogandares, F., and Seed, J. R. 1973. American paragonimiasis. Curr. Top. Comp. Pathobiol. 2:1–56.

Stork, M. G., et al. 1973. An investigation of endemic fascioliasis in Peruvian village children. J. Trop. Med. Hyg. 76:231–235, 238.

Todorov, T., Stojanov, G., Rohov, L., Rasev, R., and Alova, N. 1976. Antibody persistence after surgical treatment of echinococcosis. Bull. W.H.O. 53:407–415.

Van Tiggele, L. J., and Over, H. F. 1976. Serological diagnosis of fascioliasis. Vet. Parasitol. 1:239–248.

Varela-Diaz, V., and Contorti, E. A. 1974. Hidatidosis Humana: Técnicas para el Diagnostico Immunológico. Centro Panamericano de Zoonosis, Buenos Aires.

Warren, K. S. 1973. Regulation of the prevalence and intensity of schistosomiasis in man: Immunology or ecology? J. Infect. Dis. 127:595–609.

Warren, K. S. 1978a. Dynamics of host responses to parasite antigens. In: Immunity in Parasitic Diseases. Colloque INSERM-INRA, pp. 25–38. Editions INSERM, Paris.

Warren, R. S. 1978b. The pathology, pathobiology and pathogenesis of schistosomiasis. Nature 273:609–612.

Weinmann, C. J. 1970. Cestodes and Acanthocephala. In: G. J. Jackson, R. Herman, and I. Singer (eds.), Immunity to Parasitic Animals. Vol. II, pp. 1021–1059. Appleton-Century-Crofts, New York.

World Health Organization. 1976. Research needs in taeniasis-cysticercosis (Memorandum). Bull. W.H.O. 53:67–63.

Wright, W. H. 1972. A consideration of the economic impact of schistosomiasis. Bull. W.H.O. 47:559–566.

Yamashita, J. Y. 1968. Natural resistance to echinococcosis and the biological factors responsible. Bull. W.H.O. 39:121–122.

Yarzabal, L., Dupas, H., Bout, D., and Capron, A. 1976. *Echinococcus granulosus*: Distribution of hydatid fluid antigens in tissues of the larval stage. Exp. Parasitol. 40:391–396.

Yokogawa, M. 1969. *Paragonimus* and paragonimiasis. Adv. Parasitol. 7:375–388.

Chapter 6
Immune
Reactions to Arthropods

Medical arthropodology has traditionally emphasized the role that arthropods play in the transmission of infections, often at the expense of other facets of the host-parasite relationship that are also of great interest. Besides their undeniable and overwhelming importance as vectors, the arthropods can affect the tissues of their hosts by inoculation of irritant or antigenic products (*hematophagous arthropods*), or by the injection of venoms or toxins (*poisonous arthropods*), or they can behave as etiologic agents of disease by themselves (*invasive arthropods*).

In every case, the secretions or detritus of the arthropods act as antigens that often may cause defensive or pathologic reactions on the part of the host. Many workers also believe that the transmission of numerous arthropod-borne infections may be critically affected by tissue modifications in the site of the bite, caused by the host immune response to it. The tissue changes may facilitate or hinder the feeding process of the arthropod (and subsequently the inoculation of microorganisms), or may influence the virulence mechanisms (e.g., diffusion, multiplication) of the injected pathogens.

For the purposes of review, I have divided the arthropods into hematophagous, invasive, and poisonous organisms. It is convenient to keep in mind, however, that this classification is rarely absolute; rather, there is a considerable degree of overlap among the three categories.

Excellent reviews on aspects of the immunology to arthropods have been recently written by Feingold et al. (1968), Benjamini and Feingold (1970), Gaafar (1972), and Nelson et al. (1977).

THE HEMATOPHAGOUS ARTHROPODS

The hematophagous arthropods constitute a tremendously heterogeneous group of invertebrates that are grouped together mainly for teaching purposes (Figure 16). Among them, we can find insects that visit the host only sporadically (e.g., mosquitoes, biting flies, kissing-bugs), insects that live permanently in the immediate neighborhood of the host (fleas, bed-bugs) and arthropods that remain constantly or for long periods on the host (lice, hard-ticks, ked or louse-fly of sheep). Their host specificity is also widely variable. Some show little discrimination in the election of their source of meals (mosquitoes, kissing-bugs), others exhibit some preference for specific hosts (ticks, fleas), and still others are strict parasites of one single vertebrate species (lice, ked or louse-fly of sheep). Their common characteristic, however, is that they penetrate the skin of the host with their buccal parts to obtain blood, during which process they inoculate the host with diverse substances with antigenic properties.

Natural Resistance

Because of the heterogeneity of this group, it is practically impossible to establish general principles of natural resistance that are applicable consistently to all members of the group. Besides, the respective studies have been scarce and not completely conclusive.

Recently Friend and Smith (1977) have reviewed the *stimuli* that influence the food-seeking behavior, the host selection, and the feeding process in hematophagous arthropods. In general, visual stimuli, especially movement and color of the host, and olfactory signals seem to be important in the location of hosts by flying insects. Odor, temperature gradient, and CO_2 gradient (the latter in some cases, at least) appear to be the most relevant stimuli for apterous arthropods. The exploration of the skin prior to the actual suction of blood is directed mainly by the temperature gradient and by the contact with the host, whereas the location of the food and the actual suction of it depend on the osmotic and chemical composition of the host's fluids, especially on the presence of nucleotides, which are particularly abundant in erythrocytes and platelets. The size of the hemoglobin crystals formed in the arthropod gut subsequent to a meal will also determine which mammalian species is an adequate source of blood. I do not know of any instance in which all the stimuli have been put together to explain a particular host-parasite association among the hematophagous arthropods, but it seems self-evident that only those vertebrates that offer the sequence of most appropriate stimuli will be attacked, exclusively or preferentially, by the corresponding arthropod.

Field observations and a few experiments in the laboratory have indicated that the *behavior of the host* also influences its susceptibility to the attack by arthropods, especially by those that remain on the host for rather prolonged periods. The mechanical reactions to

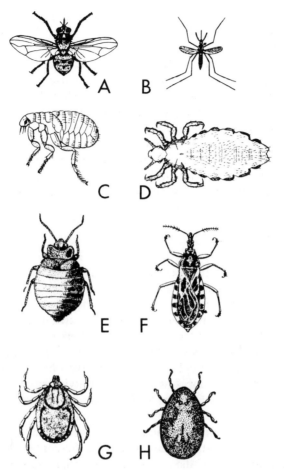

Figure 16. Some hematophagous arthropods. **A,** *Stomoxys calcitrans,* the stable fly; **B,** *Aedes aegypti,* the yellow-fever mosquito; **C,** *Ctenocephalides* species, the dog and cat fleas; **D,** *Pediculus humanus,* the sucking louse of humans; **E,** *Cimex lectularius,* the bed-bug; **F,** *Triatoma* species, a kissing bug, one of the vectors of *Trypanosoma cruzi;* **G,** *Dermacentor* species, a common hard tick in the U.S.; **H,** *Argas* species, a soft tick of birds. Reprinted by permission from *Arthropods, Reptiles, Birds, and Mammals of Public Health Significance.* Publication No. 1955. Public Health Service, Washington, D.C. (1967).

the sight, sound, or contact with the flying arthropods (fleeing, hiding, concealing the susceptible parts of the body) impedes the successful attack, and grooming contributes to eliminating a considerable proportion of permanent ectoparasites. Amputation of the forelimbs in mice or of the beak in birds results in the development of louse or mite infestations that are much more severe than in complete animals. The influence of the social behavior of man (crowding, personal hygiene) on the prevalence of louse infestations and on scabies has been known for centuries.

On a theoretical basis, and considering the predilections of the distribution on the host exhibited by many ectoparasites, some authors have speculated that *characteristics of the host's skin* may facilitate or hamper the attack by certain arthropods. This notion seems reasonable, and some examples tend to support it, e.g., long-haired cattle frequently harbor more lice than short-haired animals. Other observations, however, limit its validity considerably: epidemiological studies have found no significant difference in the prevalence of *Dirofilaria immitis* for either long-haired or short-haired dogs, which indicates that this characteristic does not disturb the attack by the mosquito vectors. Cebu cattle are more resistant to ticks than conventional breeds, but no correlation has been found between this resistance and the thickness of the respective skins. Studies with mice susceptible and resistant to lice and with sheep susceptible and resistant to the louse-fly have also failed to demonstrate a relationship between susceptibility and histological characteristics of the host's tegument.

In a large number of host species it has been possible to show that a proportion of animals in the population (herd, flock, etc.) is less susceptible to the attack than the rest of the individuals in that population. Certain observations have suggested that this difference may depend on *genetic factors*. Experimental studies with *Bos indicus* and *Bos taurus* have revealed that the former exhibits a degree of resistance to ticks that is expressed by a prolongation of the life cycle of the arthropod and a decrease in the weight and fecundity of the adult parasites. Breeding experiments showed that this characteristic is inherited mainly as a dominant characteristic. It is not known, at present, however, whether this resistance is related to behavioral pecularities of Cebu cattle with regard to the attack by the arthropod, or to their physiological ability to react to the bite with inflammatory reactions that, in turn, may be nonspecific or immunologic. A certain parallelism between the resistance of *B. indicus* to ticks and to mosquitoes suggests that the mechanism of resistance may be nonspecific rather than immunologic. Experiments with mice susceptible to and resistant to lice have demonstrated that the latter exhibit quicker mechanical and inflammatory responses to the insects' bites than do the susceptible mice.

Acquired Immunity

Allergic Reactions   The oral secretions of numerous arthropods contain substances that are irritating, toxic, or have histaminelike activity. These compounds may cause rapid tissue alterations, even in hosts that have not been subjected to attack by that particular species of arthropod previously. Besides this nonspecific damage, the antigens universally present in the oral secretions often trigger diverse immunological reactions in the hosts.

Although little researched, it is possible that the quality of the host reaction is influenced by the feeding characteristics of the arthropod. Some arthropods (*solenophages*), such as lice, fleas, *Xenopsilla*, and piercing dipterans, penetrate directly the lymph or blood vessels of the host to suck their meal. Others (*telmophages*), such as most acarids, mallophages, and biting dipterans, tear the host tissues apart and take solubilized tissue materials along with the oozed fluids. It can be speculated that the antigens of the oral secretions go especially to the regional lymph nodes and to the spleen in the first case, whereas they are mainly deposited in the thickness of the skin in the second instance. Experience has shown that antigens that reach the spleen often stimulate a humoral response preferentially, whereas antigens that remain in the skin are particularly prone to produce cell-mediated immunity. It would be interesting to study whether these differences actually occur in relationship to the peculiar mode of feeding of the arthropods.

Occasionally, the bite of hematophagous arthropods produces allergic reactions that are expressed systemically, but in most cases the manifestations of the allergy remain restricted to the skin.

Benjamini and coworkers demonstrated the usual sequence of the cutaneous allergy to insect bites by following the alterations produced in guinea pigs subjected to periodic attack by fleas. During the first 4 days of daily bites, the skin remained normal to the external or histological exam (Stage I). Beginning on the fifth day, the rodents responded to the bites with local erythema and papules that appeared 24 hr after the bites; the dermis showed a diffuse infiltration of mononuclear cells at the histological exam (Stage II). From the ninth day, the bites produced local edema, erythema, and pruritus within 20 min, with abundant eosinophils in the dermis; this was followed by the 24-hr reaction (Stage III). After the 60th day, only the 20-min reaction occurred (Stage IV), and toward the 90th day the bites did not cause any reaction detectable by external or histological examination (Stage V).

This different reactivity has been interpreted in the following manner: Stage I corresponds to the lag period required to mount an immune reaction; Stage II is an expression of cell-mediated immunity; Stage III represents a sequence of reaginic hypersensitivity (Type I allergy) and delayed hypersensitivity (Type IV allergy); Stage IV is characterized by the abatement of cell-mediated immunity, with persistence of the reaginic reactivity; and Stage V represents a period of lack of cutaneous reactivity to the antigens inoculated by the arthropod.

Because the antigens of the oral secretions of arthropods have not been isolated as pure macromolecular species yet, the strict verification of the immune nature of these reactions has not been possible. However, in the light of numerous indirect proofs, very few investigators doubt that they are really immunological processes. The lag period required for the production of lesions and the histological aspect of these constituted already strong arguments in this respect. Lately, it has been possible to transfer immediate-type hypersensitivity from an allergic patient to a normal subject by inoculation of serum. Similarly, delayed hypersensitivity has been transferred between laboratory animals by injection of lymphoid cells. These experiments constitute satisfactory proof of the immune nature of the respective responses. Additionally, both expressions of tissue damage have been suppressed by the administration of antihistamics and of immunodepressor agents, respectively.

There is no agreement yet, however, on whether the nonreactivity of the Stage V is due to tolerance to the antigens developed by the host, or to the production of circulating, blocking antibodies that prevent local reactions. The opinions of the specialists are also divided in regard to whether a single antigen is responsible for the different reactions at various times or whether each reaction is an independent process elicited by a different antigen.

Numerous observations in humans and other animals suggest that these five stages are the usual sequence of cutaneous reactions to the repeated injection of small doses of antigens. An identical sequence has been verified in cases of bites by bed-bugs (*Cimex* species), sandflies (*Phlebotomus* species), mosquitoes (*Aedes* species), lice (*Phthyrus* and *Pediculus*), or other arthropods, and as a consequence of the periodic inoculation of ovalbumin or heterologous sera. At any rate, the transition from one stage to the next is gradual and the persistence of each stage may vary greatly. In humans, the hypersensitivity to the bite of an arthropod species may last for years without reaching the stage of cutaneous nonreactivity.

In general, the allergic manifestations of arthropod bites are more common and intense in humans than in domestic animals. Among the latter, the carnivores and the equines appear to be the most susceptible. On occasion, the local reactions may reach exaggerated proportions, with production of bullae or sloughs. It is suspected that the lesions in these cases are produced by an Arthus reaction resulting from the formation of precipitating, complement-fixing antibodies. In strongly sensitized subjects, the arthropod bite may trigger systemic reactions such as generalized papular eruptions, asthma, or even anaphylactic shock.

The cutaneous reactions to arthropod bites may also influence the efficacy of the transmission of arthropod-borne infections. In the case of *Pasteurella pestis*, it has been demonstrated that the bacilli inoculated by the vector flea lack a capsule and are therefore susceptible to the phagocytic activity of the neutrophils. However, if these bacilli penetrate macrophages in the neighborhood of the site of injection, they will multiply in the host cells to leave them as encapsulated organisms, indifferent now to the actions of the frustrated neutrophils. In the case of avian *Plasmodium*, it has been shown that the parasites need lympoid cells near the point of penetration to initiate their development. In the case of *Trypanosoma cruzi*, the pruritus that frequently accompanies the immediate-type hypersensitivity may be of critical importance for the autoinoculation of the flagellate by scratching. In all these instances, and possibly in many others, the cellular composition around the inoculation site, or local physiological reactions such as pruritus, with their corresponding influences on the fate of the infection, are dependent on the immune response to the arthropod bite and on the stage of its evolution.

Protective Reactions   On several occasions, it has been observed that the repeated attack of an individual by an ectoparasite often results in a reduction of the susceptibility of the subject to future attacks by the same species of arthropod. Studies with mice have shown that, after 2 or more weeks of intense and permanent infestation with lice (*Polyplax serrata*), the insects tend to leave the host as if it were no longer an adequate source of food. The resistance in this case appears to be specific, because the same mice are attacked and sustain the life of ticks normally. Similar but more limited observations have been gathered with respect to the cattle louse, *Haematopinus eurysternus*. In the case of the louse-fly of sheep (*Melophagus ovinus*), the density of the ectoparasite population begins to decrease by the third month of infestation. In contrast to the lice, the reduction in the number of *Melophagus* is mainly due to increased mortality rather than to migration away from the host.

In both cases, lice and louse-fly, the presence of specific antibodies has been demonstrated in resistant animals, and the resistance has been abolished by administration of immunosuppressive agents. There thus seem to be abundant arguments to regard these phenomena as immunological in nature. Also, in both cases the reduced susceptibility to the insects has been related to an arteriolar vasoconstriction and a cellular infiltration induced by the bites, which would prevent the attainment of the blood meal by the arthropod. The skin infiltration of mice resistant to *P. serrata* is characterized by the presence of neutrophils, eosinophils, and lymphocytes between the second and the fifth week of infestation. Between the sixth and the 13th week, the infiltrate contains lymphocytes, mast cells, and fibroblasts. Reduction of the blood flow to the skin reaches a peak in the fifth week and persists reduced until the 11th week. Certain parallelism between the cellular and vascular phenomena in the site of bites and the time at which the resistance appears has encouraged the conjecture that this latter is due to the reduction of the local blood irrigation, which impedes the feeding of the insects. Some authors have speculated that the ischemia is biphasic—caused initially by a reaction of immediate hypersensitivity to the bites, and later by a nonspecific reaction of the tissues to the chronic irritation from the bites.

Resistance to *M. ovinus* is similar superficially, but the mechanisms entailed appear to be different. In this case the histological picture is characterized by the abundance of neutrophils, particularly in the periphery of blood vessels, eosinophils, and mast cells, and by the presence of edema and fibrinoid degeneration of the arterioles. These morphological findings and the fact that the insects die when they feed on resistant hosts have suggested the existence of an Arthus phenomenon. It is believed that antigen-antibody complexes formed at the site of the bite would fix complement and produce the lesions described. When the complexes are ingested by the insects, they would be toxic for them.

Acquired resistance to ticks has received preferential attention from the investigators. Numerous observations have demonstrated that the prolonged or repeated infestation of cattle or of laboratory animals results in a degree of resistance that is expressed as a longer period of feeding by the arthropods and by a reduction in the amount of blood drawn and of eggs produced. On occasion, a delay in the completion of the life cycle and an increased mortality of the larval stages have been reported.

Early observations attributed this resistance to the formation of a mass of inflammatory cells, believed to be a manifestation of cell-mediated immunity, which occurred around the site of feeding and impeded the attainment of the meal by the arthropod. Subsequent examination of the infiltrate revealed the presence of abundant basophils and eosinophils. This histological picture is characteristic of a new type of hypersensitivity recently identified (*cutaneous basophil hypersensitivity*), which appears to be a novel variety of delayed-type allergy. Some authors (Bossard, 1976) have been able to correlate the presence of resistance to ticks with the existence of circulating antibodies.

Experiments of transfer of resistance from immune animals to susceptible subjects have not contributed much to the disclosure of the immunological mechanisms entailed in resistance. Almost 40 years ago,

Trager reported transfer of partial protection to *Dermacentor variabilis* by inoculation of serum. Later studies verified his results with other species of ticks. This procedure constitutes a satisfactory proof that circulating antibodies play a role in resistance.

In other studies, however, resistance to *Haemaphysalis* showed no relationship with the presence of skin-sensitizing antibodies, and resistance to *Ixodes* was transferred with cells but not with serum. The former results may be explained by prior findings that skin-sensitizing antibodies protected against a toxin present in the eggs of ticks, whereas only precipitating antibodies defended against the hematophagous arthropod itself.

Recent communications have claimed that resistence to *Dermacentor* can be transmitted by inoculation of lymph node cells of immune animals. This method does not totally exclude the possibility that the inoculum contained antibody-forming cells, but constitutes an additional argument of the participation of cell-mediated immunity in the protection against the attack by ticks.

Taking into consideration clinical, histological, and experimental evidence, it seems that immune protection against ticks is a complex phenomenon that may include manifestations of immediate and delayed hypersensitivity and other events mediated by circulating antibodies, such as the Arthus reaction. It is even possible that various modalities of the immune response may be protective at different times, depending on the species or the biology of the particular tick, or the peculiar conditions of reactivity of the host. At any rate, resistance to ticks is a subject that still requires further research.

Artificial Modification of Immunity

In addition to its use for experimental purposes, artificial modification of immunity to hematophagous arthropods is usually meant to inhibit allergic reactions that affect the health of the host, or to stimulate the protective reactions to the attack by the arthropod. Ideally, the attainment of the desired purposes in such situations entails a precise knowledge of the immunological phenomena responsible for the allergic or protective reactions. Unfortunately, the information currently available is too limited yet to apply rational principles at large, so a great proportion of the investigation in this area has been mostly empirical.

The participation of cell-mediated immunity in the allergic responses is already fairly well defined, but its influence on the protective reactions is still far from being completely understood. Knowledge with respect to humoral immunity is even vaguer: not even the classes of antibodies produced in diverse situations are known. The possibility of transfering immediate hypersensitivity to the bites of mosquitoes, fleas, bed-bugs, and ticks by inoculation of serum from sensitized individuals was taken at the beginning as a demonstration of the presence of IgE antibodies. The subsequent finding of a subclass of IgG that is able to sensitize the skin of homologous species has obliged researchers to review this conclusion.

At least in the case of immediate hypersensitivity to bites of *Pediculus* and *Culicoides*, the respective antibodies are heat sensitive, which argues in favor of their IgE nature. Immediate hypersensitivity to bites of mosquitoes and fleas has been transferred by inoculation of serum between different vertebrate species, which suggests the participation of heterocytotropic IgG antibodies. Laboratory animals attacked repeatedly by mosquitoes (*Aedes*), kissing-bugs (*Rhodnius*), or ticks (*Hyalomma* or *Rhipicephalus*) produce specific antibodies that form precipitates with the proper antigens in semisolid media: antibodies with this characteristic commonly belong to the IgG class. Similar animals infested with lice (*Pediculus humanus* or *Bovicola bovis*), however, did not produce precipitating antibodies.

Desensitization   Prolonged inhibition of allergic reactivity by serial inoculation of small quantities of specific antigens (called *desensitization, hyposensitization, immunotherapy*, or *immunodeviation* by various authors) has been attempted with variable results in a number of allergies to arthropod bites. In human hypersensitivity to mosquito bites, this treatment has produced temporary cure of all symptoms in some cases, but has relieved only the manifestations of the delayed hypersensitivity without affecting the immediate-type allergy in others. At least in one case, it was communicated that desensitization to the bites of *Aedes* mosquitoes also reduced the reaction to the bites of *Culex* mosquitoes. Mosquitoes share common antigens extensively, so it is possible that the allergic response is elicited by materials at least partially identical in both genera.

Spontaneous desensitization has been observed in patients previously allergic to the bites of body lice (*P. humanus corporis*), sandflies (*Phlebotomus*), fleas *(Pulex)*, bed-bugs (*Cimex*), or mosquitoes. These reports suggest that therapeutic desensitization against these insects in clinical practice must be feasible.

Hyposensitization of domestic carnivores to flea bites, by three serial inoculations of a commercial antigenic preparation, has been reported to produce excellent results in 75% of the cases and to fail only with 5% of the patients. However, the complete treatment for flea dermatitis includes elimination of the fleas, so that part of the success in reducing the allergic condition may be due to the removal of the insulting antigen rather than to an effective desensitization. The frequent observation that flea dermatitis rarely affects stray dogs with massive infestations but usually affects dogs with rather few fleas suggests, however, that desensitization occurs consistently in nature. Experiments with laboratory animals have demonstrated that desensitization to flea bites is rather specific, since it did not change a concurrent hypersensitivity to mosquito bites.

Possibly a reason that conspires against the consistent success of the desensitizing procedures to arthropod bites is the quality of the antigenic preparations in current use. These are complex extracts of the arthropod that, besides the specific allergen(s), contain a number of other antigens that may be allergenic by themselves or may compete with the desensitizing activity of the relevant allergen. Evidently, desensitization techniques require a sound knowledge of the antigens that participate in the allergic reaction, which is far from available at present in the case of the arthropods.

The most widely accepted mechanism to explain the occurrence of desensitization is that periodic inoculations of antigens induce the formation of circulating IgG antibodies. These antibodies remain in the circulation and combine with new doses of antigens as soon as these are introduced, so that the allergen does not have a chance to stimulate the primed T lymphocyte effectors of cell-mediated immunity or the mast cells sensitized with specific IgE antibodies. In addition, it is now known that conventional IgG antibodies may compete with IgE antibodies for the reaginic receptors on mast cells, but they do not induce degranulation of the cells on binding of the corresponding antigen as IgE antibodies do. Some investigators believe that serial inoculation of the relevant antigen produce specific immunological tolerance to it. Recent findings on immune tolerance do not support the production of this phenomenon in adult subjects injected with small but not insignificant doses of antigens.

Protection   The meager protection produced by the natural infestations by hematophagous arthropods and the failure of some early attempts to produce resistance by inoculation of extracts of the arthropods have diminished the interest in the possibilities of vaccination against these organisms and have discouraged most researchers. Besides, subjectively it would appear that only those ectoparasites that remain in the host for periods long enough to allow the action of the host immune response were appropriate subjects for immunological control. This assumption is erroneous, as is discussed below.

The literature registers several attempts to produce resistance against ticks artificially. Although in most cases the results have been disheartening, Bossard (1976) was able to induce considerable resistance by inoculating 2-day-old calves with salivary glands of *Boophilus microplus*, followed by artificial infestations with a small number of tick larvae.

Schlein and Lewis (1976) recently reported an experiment that may have great repercussions: *Stomoxys calcitrans* flies that fed daily on rabbits immunized with tissue of the same species of fly experienced a mortality rate twice as high in 15 days as that of flies feeding on nonimmunized rabbits. Tsetse flies (*Glossina morsitans*) that fed on the same animals underwent a similar fate, which indicates that the phenomenon is not strictly specific. In both cases the insects showed general symptoms, such as motor paralysis and alterations in the development of the tegument, that were attributed to the activity of the antifly antibodies ingested. Although the immunization of the host does not prevent the bites, this method may constitute an additional technique for the biological control of populations of hematophagous arthropods, with the added benefit that the bites on immunized bait animals probably would maintain the immunity at an effective level.

Another recent report of great interest indicated that the sera of rabbits immunized with extracts of *Aedes* mosquitoes are able to neutralize in vitro the replication of viruses transmitted by the same species of mosquito. The interpretation given to this phenomenon was that antimosquito antibodies reacted with mosquito material present on the viral particles, killing the microorganisms indirectly. This finding may be of extraordinary importance in devising new methods for the control of arthropod-borne infections, especially because it has been recently demonstrated that *Trypanosoma vivax, Trypanosoma lewisi*, and possibly *T. cruzi* adsorb host proteins on their own external membranes.

Immunodiagnosis

In the case of hematophagous arthropods, immunological tests are commonly used to diagnose allergies to a particular species of arthropod in patients, to identify the origin of the meal or the predators of arthropods captured in the field, or to evaluate the possible relationships among these organisms by comparing their antigenic composition.

The most common techniques employed in the diagnosis of allergies to arthropods are the skin tests. With variable frequency, it has been found that individuals repeatedly attacked by hematophagous arthropods may also exhibit positive results with the tests of precipitation, hemagglutination, complement fixation, fluorescent antibodies, and others. However, these tests are not positive consistently and their correlation with the immunoallergic state of the host is much less clear than in the case of the skin tests.

The relative crudeness and inspecificity of the antigenic preparations currently available and the fact that the allergic patient often relates his affliction to a particular species of arthropod have considerably reduced the usefulness of these diagnostic methods in medical practice. These methods are discussed again in regard to bee stings.

Epidemiologists are often interested in knowing the common hosts of a vector arthropod in order to evaluate its potential as transmitter of the infection to humans or domestic animals. Determination of host preferences of nocturnal or flying arthropods by mere observation of their behavior may be extremely difficult and inaccurate. Serological methods, however, because of their specificity and sensitivity, provide a much more exact and reliable method. In such studies, the intestinal content of the arthropod under study (containing fluids of the unknown host) is utilized as antigen and reacted with laboratory-raised antisera to blood proteins of all probable hosts. Strong positive reactions will occur only when the host blood in the arthropod intestinal content meets the serum against the same host species. By using these techniques in Yugoslavia, it was found that *Phlebotomus papatasi* feeds preferentially on humans and horses, rarely on rats and cattle, and not on birds.

Although precipitation in gel is the routine method utilized, other more sensitive techniques are equally adequate. Evidently, the blood ingested by the arthropod undergoes processes of digestion that make it inappropriate as an antigen after some time. Studies with intestinal content of mosquitoes have demonstrated that the specificity of the ingested blood is preserved for 24 hr; it then begins to yield nonspecific reactions, and after 96 hr it loses its antigenic properties.

A modification of this procedure has been used to identify the natural predators of arthropods of medical importance for the purpose of designing campaigns of biological control. In this case, the intestinal content of potential predators is used as antigen and reacted against laboratory-produced antiserum to extracts of the arthropod under study. A study with this technique showed that about 43% of the examined predaceous diving beetles (*Dytiscidae*) fed on larvae or pupae of *Aedes cantans*, about 24% of predaceous diptera consumed adult mosquitoes of the same species at the moment they left the puparium, and about 22% of spiders ingested adult specimens of *A. cantans*.

The exquisite specificity of the serological reactions has permitted the establishment of very precise comparisons among the antigens of diverse species of arthropods, which presumably would indicate the degree of evolutionary relationship among them. Comparison of the results of reactions of mosquito extracts with homologous and heterologous antisera have permitted the differentiation of species of *Anopheles*, *Aedes*, and *Culex* and enabled workers to distinguish *Culex pipiens pipiens* from *C. pipiens fatigans* and *C. pipiens molestus*. Even the hybrids of *C. p. fatigans* + *C. p. pipiens* can be separated from both progenitors.

Antigens

The oral secretions of numerous species of hematophagous arthropods have been studied on several occasions, but the minute attainable volume and their complex composition have been a deterrent to the exact definition of all their components in most cases. Despite this, the zeal of the investigators has identified the presence of vasoactive amines (histamine and analogs), anesthesics, anticoagulants, hemolysins, hemagglutinins, irritants, toxins, enzymes, necrotizing compounds, and a number of low molecular weight materials.

Among the known enzymes, there are decarboxylases, which may act on host peptides to generate vasoactive amines; hyaluronidase, which operates as a diffusing agent by degradating the hyaluronic acid of the intercellular matrix; and various proteolytic and cytolytic enzymes.

Several of these compounds are antigenic, so they may induce the corresponding immune reactions when they are introduced into the host. In the case of fleas, it has been demonstrated that the allergen that produces immediate hypersensitivity is a hapten of about 600 daltons that requires conjugation with the collagen of the host skin to produce an immune response. An allergen of 36,000 daltons, structurally related to collagen, has been identified in *Aedes aegypti*. The immediate hypersensitivity to the bite of the tick *B. microplus* has been associated with a protein of about 60,000 daltons that appears to exhibit esterase activity.

Investigations on the presence of precipitating antibodies in animals repeatedly attacked by hematophagous arthropods have demonstrated three precipitation bands in the oral secretions of *Aedes* mosquitoes, *Rhodnius* kissing-bugs, and *Hyalomma* ticks, and two in *Boophilus* ticks. The potential number of antigens might be higher, however, because the inoculation of extracts of salivary glands of *Hyalomma* or of *Boophilus* produced antisera that formed nine and seven bands of precipitation in gel, respectively.

The first experiments directed to identify the existence of cross-reactivity among antigens of various arthropod species reported the presence of a considerable degree of specificity. Modern studies, however, have found that cross-reactions are rather frequent.

Some authors have demonstrated that patients sensitized to the bite of the common bed-bug (*Cimex lectularius*) also react to the bites of other species of the same genus and even of other genera of the same family. In the case of fleas, subjects sensitive to the bite of *Cten-*

ocephalides felis also respond to bites by *Pulex* and *Xenopsilla* fleas but not to bites of mosquitoes or bed-bugs. Among mosquitoes, abundant cross-reactivity has been reported among species of the same subfamily and even of the same family. A study of the antigenic relationships among extracts of the entire body of mosquitoes (*Culex pipiens pallens, C. p. molestus,* and *C. p. fatigans*) revealed up to 26 antigens in each extract; only two of them were exclusive to *C. p. molestus* and only one was specific for each one of the other species. It must be mentioned that many of these antigens must be somatic material and, therefore, not relevant to cross-reactivity to the bites.

Experiments with ticks could not demonstrate cross-reactions between precipitating antibodies to extracts of salivary glands of *Hyalomma* and *Rhipicephalus*, but other reports found cross-reactivity between *Dermacentor* and *Haemaphysalis*. It must be kept in mind, however, that precipitating antibodies usually belong to the IgG class, so these results do not exclude cross-reactions of immediate hypersensitivity, normally associated to IgE antibodies, or of delayed hypersensitivity, which is a manifestation of cell-mediated immunity.

It is likely that the occurrence of cross-reactions depends to a certain extent on the length or frequency of the bites. Short or infrequent bites must inject quantities of oral secretion that are too limited, so that only the most potent antigens in them will elicit immune responses. A large number of secondary antigens would not have a chance to act as such in these cases. On theoretical grounds, one can speculate that the strongest antigens tend to make the host-parasite association rather uncertain and should be selected against during evolution (see "Imitation of Host Antigens," Chapter 7). The acquisition of these antigens must be rather recent in the evolutionary sense, then, and may have occurred when the speciation of the various groups of arthropods was already well advanced. Therefore, stronger antigens would have a lower statistical possibility of having persisted from the common ancestral stock when compared to the weaker antigens. These latter might then have a particularly important participation in the reactions of cross-reactivity, since they must have originated in more ancient forms.

THE INVASIVE ARTHROPODS

Under this heading I have grouped the larvae of myiasis-producing flies, mange-producing mites, and some mites that affect the human lungs. The arthropods included in this section also constitute a rather heterogenous group whose categorization does not have well-defined boundaries (Figure 17).

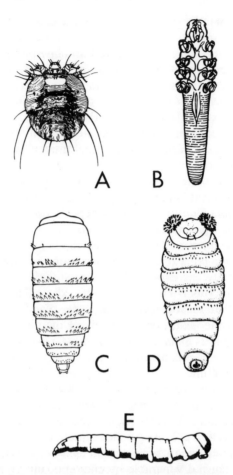

Figure 17.   Some invasive arthropods. **A,** *Sarcoptes scabiei,* the itch mite of humans and other mammals; **B,** *Demodex* species, the hair follicle mite of humans and lower mammals; **C,** *Gastrophilus intestinalis,* the common horse bot; **D,** *Dermatobia hominis,* the skin bot of "man" that affects numerous homoiotherms; **E,** muscoid larva (maggot) typical of a large number of myiasis-producing flies. Reprinted by permission from *Arthropods, Reptiles, Birds, and Mammals of Public Health Significance.* Publication No. 1955. Public Health Service, Washington, D.C. (1969).

Among the myiasis-producing larvae there are organisms that remain stationary in the host skin for all the extent of their parasitic lives (*Cochliomyia, Dermatobia*), others that perform limited migrations with only brief or transient contact with the internal tissues of the host (*Oestrus, Gastrophilus, Cuterebra*), and others that undergo extensive wanderings in the host tissues (*Hypoderma*). Some persist in the host for periods of 8 to 10 months (*Hypoderma, Oestrus, Gastrophilus*), others for 1 or 2 months (*Dermatobia, Cuterebra*), and others for only a week (*Cochliomyia*).

Their host specificity is also variable: *Gastrophilus* and *Oestrus* are exclusive parasites of horses and sheep, respectively, affecting humans only on rare occasions; *Hypoderma* is specific for cattle, although is found in horses and humans infrequently; *Cuterebra* is a parasite of rodents and lagomorphs that infests domestic carnivores and humans sporadically; *Cochliomyia hominivorax (Callitroga americana)* and *Dermatobia* affect humans and a large number of species of mammals and birds.

Among the mange-producing mites, there are organisms that live on the surface of the skin perforating or grinding the cornified cells to reach deeper layers from where they obtain lymph and, occasionally, blood (*Psoroptes, Chorioptes, Otodectes*). The separation of these organisms from the hematophagous mites (e.g., *Dermanyssus, Liponyssus*) is mostly conventional and based mainly on the length of the stay on the host and the clinical picture produced. Other mites, however, bore into the superficial layers of the skin and live in the tunnels so formed (*Sarcoptes, Notoedres, Knemidocoptes*). This latter group constitutes an intermediate zone between endo- and ectoparasites. The members of the genus *Demodex* live in the dermis of their hosts and, in the case of *Demodex canis* at least, they are occasionally found in lymph nodes or spleen.

Immunological studies require the precise identification of the insulting parasitic species in order to determine reactive or protective specificity. Unfortunately, some mange-producing mites currently appear to be in an active process of speciation, and the populations that affect diverse hosts seem to be subspecies or physiological strains rather than true species separable by conventional criteria. *Sarcoptes scabies*, for instance, may be a single species with varieties *suis, equi, bovis, canis, hominis*, and so on that affect the corresponding hosts. Varieties *equi* and *canis* are able to produce transient disease in humans. *Psoroptes comunis (P. equi)* also has a number of varieties with specificity for different species of hosts, but that are indistinguishable morphologically.

Because the exact identification of these mites often is based on the host species from which they are recovered, it is frequently impossible to determine cases of cross-infestation or of cross-resistance, unless the infestation is provoked artificially with arthropods of previously known origin.

Finally, there are a considerable number of free-living mites that eventually bite or are inhaled by humans and other mammals. Lately, a number of studies have been conducted with *Dermatophagoides* species mites, which are one of the components of the "house dust," and appear to have a role in the production of respiratory allergies.

Natural Resistance

Evidently, the susceptibility to myiasis depends, in part, on the factors that attract the adult female fly (or the vector of the eggs, in the case of *Dermatobia*) toward a particular host to deposit its eggs (see "The Hematophagous Arthropods," above). In the particular case of *Cochliomyia*, the condition of parasite appears to be due to the habits of the adult fly rather than to some restriction in the physiology of the larvae, because these latter develop normally in dead meat if it is kept at 37°C. The adult fly is attracted by exudates of wounds and physiological secretions of the host and requires a preexisting solution of continuity of the skin for the larvae to initiate invasion. This latter requisite is easily fulfilled, however, since wounds as small as those produced by mosquito bites suffice. The fact that *Cochliomyia* exhibits little host specificity but affects farm animals much more frequently than humans suggests that the mechanical reactions to the sight or contact of the fly (presumably more effective in humans) play a preponderant role in preventing the initiation of parasitism.

Gastrophilus and *Oestrus* have similar host restrictions: in both cases the larvae normally affect only horses and sheep, respectively, and on the rare occasions in which they parasitize humans, the larvae usually wander in the thickness of the skin for days (*Oestrus*) or months (*Gastrophilus*) without developing beyond the first stage, and finally die. This behavior suggests that humans do not provide the larvae with the necessary biochemical stimuli to moult and continue development, although they seem to be an adequate source of food. Larvae of *Hypoderma*, on the contrary, although not common in humans or horses, can undergo their usual migrations and attain the third larval stage in these hosts, as well as in cattle. This phenomenon indicates that the infrequent hosts must contain the necessary elements for the nutrition and development of the larvae, so their relative refractoriness must be related to mechanisms that precede the penetration of the larvae in the host. What these mechanisms are is unknown at present.

Several reports have indicated that *Hypoderma*, *Dermatobia*, and *Sarcoptes* affect children more often than adults. It is not known yet whether this characteristic is related to behavior, to skin structure, or to the relative efficiency of some physiological mechanism in both age groups. Several years ago, it was speculated that scratching might help reduce the severity of scabies in humans. Recent studies by Carslaw in Glasgow have lent support to this notion, since it was observed that the mites proliferated unrestrictly in paraplegic patients. The important participation of the mechanical activities of the host in limiting the multiplication of ectoparasites has also been verified with lice in mice whose forelimbs had been amputated, or with the mite *Ornithonyssus sylviarum* in chickens whose beaks had been removed.

In the case of mange, it seems clear that some not yet identified factors of the host are essential to maintain the infestation: some mites are able to proliferate on inappropriate hosts (*Sarcoptes* of horses and dogs in humans, *Knemidocoptes* of birds in young rabbits and guinea pigs), but they usually disappear spontaneously in the course of a few weeks.

The preferential distribution of mites on diverse regions of the host body suggests that microecological conditions of the skin also exert some influence on the susceptibility to parasitism.

The general conditions of the physiology of the host also appear to be of great importance in the initiation of ectoparasitism; numerous attempts to infest healthy and well cared for dogs with their specific *Sarcoptes* mites have failed, whereas the infestation is achieved when the animals are subjected to stress situations. However, the infestation disappears spontaneously when the animals are returned to satisfactory environmental conditions. A large number of clinical and epidemiological reports have emphasized that mange affects neglected animals preferentially, especially during the period of restricted alimentation in winter. Similarly, human scabies predominantes during times of war or public disasters; a recent study in India has shown a strong correlation between malnutrition and the presence of scabies in human populations.

Recent findings indicate that not only the presence of the parasite, but also its capacity to cause clinical disease are greatly influenced by the previous stage of health of the host; various studies have revealed that a large proportion of clinically healthy sheep, pigs, and dogs are carriers of *Chorioptes, Sarcoptes,* and *Demodex* mites, respectively. It has also been known for some time that human *Demodex* commonly behave as commensals of human skin. At present, however, there is no solid evidence to determine whether the asymptomatic presence of these ectoparasites in their hosts corresponds to a phenomenon of natural resistance or is an expression of acquired resistance.

Acquired Resistance

Acquired resistance against invasive arthropods has received almost as little attention from investigators as natural resistance. Much of the existing information is fragmentary and based on field observations, which rarely are susceptible to proper controls. As in the case of the hematophagous arthropods, immunity in this case is clinically expressed either by allergic reactions that are deleterious for the host, or by protective reactions that are deleterious for the parasite. Kim (1977) has recently reviewed the immunity to pulmonary *Pneumonyssus* and *Rhinophaga* mites in Old World monkeys.

Allergic Reactions   Although there do not seem to be systematic studies on the allergies produced by myiasis, it is common knowledge among clinicians that trituration of larvae of *Hypoderma* and of *Dermatobia* (and probably of other species as well) in the skin of the host, or their spontaneous death in situ, may produce manifestations of anaphylaxis (Euzeby, 1976). It has also been possible to produce local or systemic anaphylaxis in horses infested with *Gastrophilus* by depositing extracts of the larvae in the eye or by injecting the extracts parenterally.

Clinical and experimental evidence of allergy is more conclusive in the case of human scabies. A primary infestation with the mite usually remains asymptomatic during the first month, after which the characteristic pruritus begins. On occasion, a generalized exanthema develops toward the sixth week of the initial contact with the parasite. In reinfestations, pruritus appears already during the first or second day and it does not correlate with the number of mites present. These characteristics are consistent with the need for a lag period for antibody formation in the primary infestation, and with a rapid reaction of the parasite antigens with preformed antibodies in the reinfestations.

Intraepidermic injection of extracts of *Sarcoptes* in patients produces reactions typical of immediate-type hypersensitivity, which appear for the first time about 6 months after the primary infestation and persist for a few months after the definitive cure (possibly sustained by parasitic material retained in the thickness of the skin). The histological aspect of scabies lesions, either in humans or in swine, is also very suggestive of immediate hypersensitivity. Similar reactions probably take place in manges of domestic animals.

Numerous investigations directed to assess the participation of the various mites of "house dust," especially *Dermatophagoides* species, in the genesis of respiratory allergies of humans have been reported lately (Wharton, 1976). In one study, it was found that 45% of 122 subjects showed positive cutaneous tests to extracts of *Acarus siro* and 37% of 106 were positive to *Dermatophagoides*; of these latter 106, 25% had cross-reactions to both mites. Nevertheless, only 6 individuals presented respiratory symptoms. On the other hand, an experiment in France showed that 78% of 176 asthmatic patients were sensitive to house dust as well as to extracts of *Dermatophagoides*, and a study in Papua, New Guinea, demonstrated that 90% of the asthmatic patients had IgE antibodies against this mite, whereas only 5% of the asymptomatic population revealed the same antibodies.

In an attempt to resolve these inconsistencies, a group of 32 asthmatic patients who had shown sensitivity to *Dermatophagoides* were maintained for 6 weeks in an environment free of the mite. Despite this, none of the patients showed remission of the symptoms. This experiment is not conclusive, however, since the mites or their products might persist in the pulmonary tissue for more than that span of time. Studies of hyposensitization by serial inoculation of extracts of *Dermatophagoides* have produced inconsistent results, but this may be partly attributable to the crudeness of the antigenic preparations (see "The Hematophagous Arthropods").

At present, there does not appear to be solid evidence to affirm or deny the participation of mites in respiratory allergies. It is even possible that the chitin (a probable stimulant of the production of IgE) acts as a precipitating immunological factor only in those individuals with a prior predisposition to respiratory allergies.

Protective Reactions   Field observations have demonstrated that cattle become less susceptible to *Hypoderma* with age, until they are about 3 years old. They become more susceptible, however, after campaigns of control that reduce greatly the population of adult flies. These facts were taken as indications that cattle develop some degree of protective immunity to the repeated attack by the parasite, protection that wanes rapidly in the absence of reinfections.

Later studies showed that the experimental inoculation of collagenase of *Hypoderma* larvae produced neutralizing antibodies and considerably reduced the numbers of larvae that completed their development in immunized hosts. Based on the symptoms induced by the death of the larvae in the host, on the presence of circulating antibodies, and on the histopathology of the lesions, some authors have speculated that *Hypoderma* infection causes immediate hypersensitivity and an Arthus phenomenon (Euzeby, 1976).

Experimental infections of *Peromyscus* mice with *Cuterebra* larvae have revealed that 80% of the larvae develop after a primary infection, but only 50% of them grow when the infection is repeated 5 weeks later. In a study in which rabbits were immunized with extracts of *Dermatobia* in such a manner as to stimulate a response predominantly humoral in some and preponderantly cellular in others, only the latter were able to restrict the number and vitality of the larvae that developed and the intensity of the symptoms of the infection. *Cochliomyia* infections with more than 3 larvae per 100 g of body weight are usually lethal for guinea pigs, apparently because of toxins produced by the parasite. Some authors have reported that the infection elicits effective immunity against the toxin but not against the larvae, whereas others have claimed that immunized guinea pigs allow the growth of only half as many larvae as nonimmunized controls. From the existing evidence, it appears that myiases produce a degree of protective immunity in the host, but its efficacy and mechanisms are still a matter for research.

Several investigators have observed that the prevalence of human scabies experiences periodic increases every 15 to 20 years, although the circumstances surrounding many outbreaks did not seem to favor transmission more than during the intervening periods. These cycles have been interpreted as evidence that human populations exhibit a degree of acquired resistance to the infestation; this resistance would wane with time, creating the proper set-up for an outbreak, which, in turn, would stimulate "herd" immunity in the population for another 15–20 year period.

There is a good amount of indirect evidence in support of the existence of acquired resistance in human scabies. Observation of numerous patients has shown that the average patient carries 25 mites 50 days after a primary infection and 500 by the 100th day of infection, and that the population of arthropods diminishes markedly thereafter, with only a few mites present in chronic cases or in reinfections. Considering that a female *Sarcoptes* lays about 30 eggs per generation and that the intergeneration period is only about 2 weeks, around 24 million mites would be expected by day 105 of infections if the parasites multiplied without restrictions. This unchecked proliferation is seen in an infrequent form of the infection called "Norwegian scabies." The absence of pruritus in this form had already stimulated speculations that the production of this variety might be related to an immunological defect inherent in the host. This hypothesis has received considerable support recently from reports of induction of Norwegian scabies in patients receiving massive steroidal treatments (known to be potent immunosuppressors) or immunosuppressor treatments as a consequence of renal transplants.

Recent reports have consistently correlated the presence of generalized demodectic mange in dogs with a hyporeactivity of the mechanisms of cell-mediated immunity in the same patients. It has not been possible to decide yet whether the immunological defect is preexisting or induced by the parasites, but serum from these patients reduces the cellular reactivity when injected into normal dogs, which indicates the existence of an immunodepressor factor in the circulation of the diseased animals. In goats, demodectic mange is particularly prevalent in animals during gestation, which is suggestive since it has been satisfactorily demonstrated that gestation interferes with the expression of cell-mediated immunity. The limited studies available appear to indicate production of immunological protection against mange, probably mediated by reactions of cellular immunity.

Antigens and Humoral Responses

There is no doubt that the compositions of the oral secretions of invasive arthropods must be as complex as that of hematophagous arthropods, but their study is often impeded by the minute size of many of the former organisms and by the ever-present possibility of contamination with native or altered host materials. Investigations with larvae of *Hypoderma* have revealed the presence of collagenases and proteolases (especially abundant in the first larval stage), substances that must also exist in other arthropods that tunnel through the host tissues.

In the case of the invasive arthropods, the secretions and detritus are particularly important as antigens, since they are often deposited directly in the intimacy of the host tissues. Studies of the composition of the body of these organisms has shown the presence of histamine and other vasoactive substances, of enzymes such as tyrosinase and phosphatase, of proteins typical of arthropods (arthropodin and sclerotin), of waxes (which may act as immunological adjuvants), and of polysaccharides. Most of these materials are effective antigens, and there is some evidence that the arthropod cuticle may be particularly important for the production of immediate-type hypersensitivity. In the case of *Dermatophagoides*, a protein of 25,000 to 125,000 daltons has been identified as an important allergen.

Infection of cattle and rabbits with *Hypoderma* has demonstrated the production of hemagglutinating antibodies that increase rapidly for the first 2 months of infection, remain stable for a few months, begin to decline by the seventh month of infection, and disappear in the ninth month. Reactions of precipitation in gel with larval extracts and serum of infected rabbits have revealed one precipitin band with the collagenase of larvae between days 30 and 60 of infection; two bands against secretions and excretions between days 60 and 90; and one band against somatic antigens between days 90 and 100 of infection. The antibodies disappeared after that date. This sequence probably represents the most prevalent antigens at each point in time in the infection: collagenase must be produced in quantities during the initial migration of the larvae to their resting location; excretions and secretions must be particularly abundant during the larval growth and development; and some mortality (with production of somatic antigens) must occur when the larvae reinitiate their wandering in the tissues of a host that has had 3 months to develop some immune resistance.

The proteolytic secretions of *Hypoderma* have been employed in the diagnosis of the human infection by immunoelectrophoresis: the serum of 10 of 13 patients yielded positive reactions, whereas the serum of 131 healthy controls did not react. Positive reactions disappeared 9 days after parasitological cure.

Inoculation of extracts of cuticle or of internal tissues of *Dermatobia* larvae in rabbits produced four precipitin bands with each antigenic preparation, but only one of them was shared by the cuticle and the internal tissues.

Horses infected with *Gastrophilus intestinales* and *Gastrophilus nasalis* and studied by hemagglutination with homologous extracts demonstrated the presence of antibodies against the former species from the third week of infection, and against the latter from the seventh week: in both cases, the antibodies reached a peak during the eighth week and descended posteriorly. Complement-fixing antibodies were found occasionally between the seventh and the 11th weeks of infection, but precipitating antibodies were never detected. The class to which these antibodies belong has not been determined yet, but it is suggestive that reactions of hemagglutination and complement fixation (in which IgM is particularly reactive) were positive, whereas the precipitin reaction (usually produced by IgG antibodies) was consistently negative.

The cutaneous reactions of humans and swine to *Sarcoptes* infection are very suggestive of immediate-type hypersensitivity. In humans, scabies was diagnosed in 22 of 23 patients by skin tests with extracts of *Notoedres alepis*, the mange mite of rabbits and rats, in order to avoid nonspecific reactions against contaminating host components.

Pigs infected with *Sarcoptes* showed an increase of the total serum proteins and of the β- and γ-globulin fractions, but hemagglutinating antibodies could not be demonstrated. On the contrary, rabbits infected with *Psoroptes* had precipitating antibodies directed against at least two antigens in the extracts of the mites. This is rather surprising because *Sarcoptes* has a more intimate association with the host in the galleries of the epidermis than *Psoroptes* on the surface of the skin. Other factors besides the strict location of the mites may have important roles in the stimulation of immunity by the various species. This fact advises against premature generalizations regarding immunology of different manges.

Studies of individuals with pulmonary infection by *A. siro* or by *Dermatophagoides* species have demonstrated the presence of skin-sensitizing and precipitating antibodies that have been identified as IgE and IgG, respectively. Monkeys with pulmonary acariasis have been shown to have IgM and IgE antibodies and cell-mediated immunity to the mite extracts. By reactions of hemagglutination, it has been found that the serum of these monkeys reacts strongly with extracts of free-living mites, especially of *Dermanyssus farina,* and weakly with extracts of ticks and mosquitoes. Cross-reactivity appears to be extensive among the crude extracts of mites assayed so far. This may be of benefit for the diagnosis of invasive acariasis, since culture of cross-reactive free-living mites will provide large amounts of antigens uncontaminated with host material much more readily than collection of the corresponding parasitic organisms.

THE POISONOUS ARTHROPODS

In a review done by Parrish (1963) on 460 fatalities due to poisonous animals that occurred in the U.S. in the decade 1950–1959, 50% were attributed to hymenoptera stings, 30% to snake bites, 14% to spider bites, and about 2% to scorpion stings. These proportions do not necessarily hold true for other countries; in Mexico for example, scorpions are responsible for more than 80% of the deaths due to venomous animals. It therefore seems important to know the poisonous fauna of one's own area in order to be able to predict the most common occurrences.

For the purposes of immunology, it is convenient to separate the insects from the arachnids. The venom of the former harms the victim mainly through the production of allergic conditions, whereas the poison of the latter causes disease mostly by direct damage to the structures of the victim. The immunological procedures in current use are exclusively preventive in the first case and essentially curative in the second.

Among the most important poisonous insects (bees, wasps, ants, buffalo-gnats) (Figure 18), the study of the toxic action of the oral secretions of the buffalo-gnats (*Simuliidae*) has been comparatively neglected, although they may have serious consequences for the health of domestic animals and humans. The scarce work done indicates that they contain a hemolytic toxin that induces the formation of protective and, occasionally, hypersensitizing antibodies. The following discussion refers to the Hymenoptera, which is the group by far the most researched. Among the most important poisonous arachnids (Figure 18), I comment on the spiders and scorpions and mention briefly the little known phenomenon of tick paralysis or tick toxicosis.

Poisonous Hymenoptera

Current estimations state that between 0.4% and 0.8% of the U.S. population (this is roughly 1 to 2 million people) has had systemic manifestations of hymenoptera allergy; possibly 8 million have any degree of sensitivity to the stings, and 50 persons are known to die of them every year. The number of cases of individuals with severe local reactions to the sting who may develop systemic sensitization later on is unknown, but undoubtedly numbers in the millions.

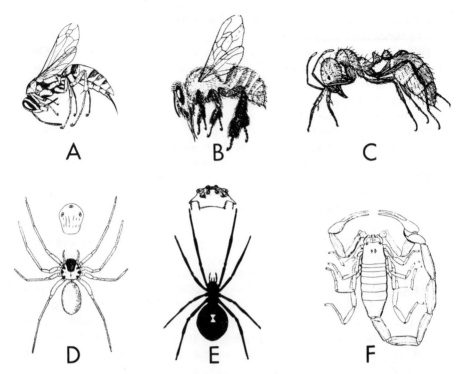

A   B   C

D   E   F

Figure 18. Some poisonous arthropods. **A,** *Polistes* species, a social wasp; **B,** *Apis mellifera,* the honey bee; **C,** *Solenopsis* species, a fire-ant; **D,** *Loxosceles reclusa,* the brown recluse spider (the distribution of the eyes in three pairs, shown above the spider, is an important characteristic in the identification of the genus); **E,** *Latrodectus mactans,* the black widow spider (the distribution of the eyes is depicted above the spider); **F,** *Centruroides sculpturatus,* the deadly sculptured scorpion of the U.S. Reprinted by permission from *Arthropods, Reptiles, Birds, and Mammals of Public Health Significance.* Publication No. 1955, Public Health Service, Washington, D.C. (1969).

The concern about insect sting allergy at present is evidenced by the large number of papers written on the subject lately (many of them published in the *Journal of Allergy and Clinical Immunology*) and by the conference on Emergency Treatment of Insect Sting Allergy convened by the U.S. National Institute of Allergy and Infectious Diseases at Bethesda, Maryland, in September, 1978.

Antigens   Most of the studies of hymenoptera allergy have been done with honeybees (genus *Apis*), paper wasps (genus *Polistes*), yellow-jackets, and hornets (genus *Vespula*). Less work has been published about ants (especially fire-ants—*Solenopsis* and *Pogonomyrmex*). Examination of the venom of honeybees has shown the presence of biogenic amines (histamine, dopamine, and noradrenaline), toxic polypeptides (melittin, apamin, mast cell–degranulating peptide, and minimine), and enzymes (hyaluronidase and phospholipases A and B). The venoms of wasps, hornets, and ants have a similar but not identical composition (Habermann, 1972; Cavagnol, 1977). Although the pharmacological, physicochemical, or enzymatic activities of these compounds have well-defined deleterious effects on the victim, most of the medical problems related to the sting of hymenoptera are due to the production of Type I allergy, since a large amount of venom has to be injected to produce evidences of toxicity.

The main allergens in the venom are phospholipase A and hyaluronidase, with a smaller proportion of the patients showing sensitivity to melittin. Two new allergens (B and C), which react with the serum of 98% of the individuals allergic to bee stings examined, have been identified recently (Hoffman et al., 1977).

Clinical Data   In a study of 100 lethal cases of hymenoptera stings (Barnard, 1973) 69% of the deaths were attributed to respiratory pathology, 12% to vascular disturbances, 12% to anaphylactic shock, and 7% to neurological alterations. Only 59 of 400 subjects were known to have been sensitive to hymenoptera prior to the fatal accident and 26 had been stung before without exhibiting reactions. In 87% of the respiratory cases and in 59% of all cases, death occurred within 1 hr of the sting. This illustrates how quickly the clinician must act and how cautiously he must evaluate remote information when facing a case of hymenoptera sting. Emergency treatment with epinephrine seemed to be more effective than treatment with isoproterenol in the same study.

Immunodiagnosis   The diagnosis of hypersensitivity to hymenoptera stings is often made by questioning the patient, but immunological procedures are frequently necessary to identify the species of insect involved and are always necessary to evaluate the degree of reactivity of the subject before initiating immunotherapy. Three tests are currently used to assess the allergic state of the patient: the immediate-type skin test, the assay of histamine release from peripheral leukocytes, and the radioallergosorbent test (RAST) (see Chapter 2). The skin test is the simplest and most accurate assay, since it measures variables that may be absent in an in vitro system, such as the particular reactivity of the individual to vasoactive amines. Unfortunately, it may also trigger systemic reactions in strongly sensitized subjects. In a recent investigation, the results of the skin tests showed correlation with the results of histamine release and RAST in cases of hypersensitivity to the stings of honeybees, yellow-jackets, and white-faced hornets, but not in cases of hypersensitivity to paper wasp or yellow-hornet stings.

Until very recently the immunodiagnosis and the immunotherapy of these conditions were performed with whole body extracts of the insects; it was reported in 1965 that 95% of patients so treated had less

intense reactions when restung. Later work has demonstrated that inoculation of whole body extracts for diagnostic purposes gave origin
to numerous false-positive reactions, and its injection for therapeutic
purposes failed to produce effective desensitization in many cases.

There is a strong trend currently to employ the venom rather than
the whole body extract: intradermal injections of a preparation containing 1 μg of venom per ml gave positive reactions in all 30 patients
and in only 1 of 150 controls; preparations with 100 μg per ml (and on
occasions with 10 μg per ml) produced a wheal and flare in the controls
that were indistinguishable from those in the patients. Concentrations
of 0.1 μg per ml yielded positive reactions in only 75% of the patients
(Hunt et al., 1976). Earlier experiments had failed to show cross-reactivity among the venoms of different hymenoptera by in vitro tests,
although cross-reactions were observed in vivo with the same venoms.
Recently, cross-reactivity among all the common members of the genus
Vespula and with *Polistes* has also been demonstrated in vitro. It is
possible that these cross-reactions are also valid in regard to
desensitization.

Immunotherapy   The immunotherapy (*desensitization*) for hymenoptera hypersensitivity follows a number of schemes: one of them,
which has been successful, consists of injecting subcutaneously three
doses of 0.1 μg, 1.0 μg, and 10.0 μg of venom in half-hour intervals
during the first day. These are followed by weekly injections doubling
the dose each time until a single dose of 100 μg (equivalent to two
stings) is achieved. In a trial with this procedure with 28 patients treated
with a total of 455 μg of venom on average per patient, for 6 to 10
weeks, only one reacted to a subsequent sting (Hunt et al., 1978). Skin
testing to diagnose the allergy and to adjust the dose of the first injection
is required prior to the treatment. Intense local, or any degree of systemic, reaction to the inoculation of venom during therapy calls for
conservation or even reduction of the dose for the next week and may
necessitate treatment. Maintenance of the hyposensitization requires
spaced reinoculations, about a month apart.

Effective desensitization coincides with a decrease of the IgE antibodies to the venom or to the main allergen (if identified for that
particular patient) with an increase of the IgG blocking antibodies. The
use of these criteria permits investigators to follow the course of the
therapy and to predict with certain approximation the reactivity of the
patient at a given time. The same relationship of low IgE antibody
content with high IgG antibody titers has been found in beekeepers
who do not respond to the stings anymore. At least in the case of
hymenoptera hypersensitivity, there is abundant evidence of the development of blocking IgG antibodies.

An accelerated schedule of desensitization with increasing doses of venom every hour to complete the treatment in 2 or 3 days has also been assayed, but the occurrence of too many strong local or systemic anaphylactic reactions during the treatment makes it inadvisable.

Although there is consensus that treatment with the venom of the particular insulting species is more adequate, successful desensitization can be achieved with mixtures of venoms of several hymenoptera species. The skin cross-reactivity to the inoculation of venoms of different hymenoptera suggests that treatment with venom of one species will reduce the hypersensitivity to the stings of related species. This area requires further evaluation, however.

Poisonous Arachnida

The most important poisonous arachnida in the Americas are spiders of the species *Loxosceles reclusa*, (brown recluse), *Loxosceles laeta* ("araña de los rincones"), and *Latrodectus mactans* and related species (black widow), some scorpions of the genus *Centruroides* in parts of the U.S., in Mexico, and in Central America, and of the genus *Tityus* in Brazil and other parts of South America, and various ticks (particularly *Dermacentor* in the U.S.) that produce paralysis.

Spider bites or scorpion stings normally produce clinical symptoms too acute for the immune system of the victim to play any effective role in the course of the current disease, and their occurrence is too infrequent for the immunity built against one episode to be protective against the next one. The use of immunology in these cases is restricted to the employment of antivenoms for therapeutic purposes. On occasion, however, the venom may act as an allergen rather than as a toxin: 2 of 8 deaths by scorpion stings in the U.S. were attributed to anaphylactic shock rather than to envenomization in one study. The same general criteria discussed in relationship to allergy to hymenoptera sting must be applicable to these cases.

Loxoceles **Spider**   *Loxosceles* venom has shown to be highly active in mice, guinea pigs, rabbits, and dogs, moderately effective in hamsters, birds, and toads, and of little or no effect in rats, frogs, and fishes. For obvious reasons, human sensitivity has not been tested extensively, but, of 133 cases diagnosed, 86% were only cutaneous, 14% were cutaneovisceral, and 4% were lethal. Recent studies on this venom have been done by Futrell and Morgan (1977).

Electrophoretic separations of the venom have shown seven or eight major and three or four minor protein components, although up to 22 components have been demonstrated by gel electrophocusing. The skin-necrotizing and hemolytic activities are apparently associated with only one of them. At least six or seven of these materials, including the toxic factor, are antigenic when inoculated in rabbits. Other workers have reported the presence of two toxins: T_1, which is lethal to mice and rabbits and neutralized by antiserum; and T_2, which is lethal only to rabbits and not neutralized by antiserum.

Early research had suggested that the toxic action of *Loxosceles* venom was mediated by some host factor, since it did not cause damage when injected in dead guinea pigs but produced extensive lesions in live animals. Later it was demonstrated that the toxin attaches to red blood cells, activates complement, and produces hemolysis and skin necrosis by a complement-mediated mechanism; the latter phenomenon is rather similar to the Arthus reaction. Based on the histological picture of the lesion, some authors have speculated that there may be participation of delayed hypersensitivity manifestations.

Mixture of the venom with serum of rabbits extensively immunized against it abolishes all these activities, as well as the action of the spreading agent, hyaluronidase, present in the venom. Inoculation of the antiserum in the skin of normal rabbits up to 18 hr prior to injection of the venom inhibits the toxic action, whereas inoculation immediately after administration of the venom only reduces it. It remains to be determined whether the binding of the toxin to the cells is irreversible, or it requires a large concentration of antibody to be broken. Alternately, it may be that the complement-activating action is independent of the cell-binding property and the lesions in this latter case were produced by the irreversible activation of complement factors.

Although the toxin of *Loxosceles* appears to be a protein of about 34,000 daltons, its immunogenicity is rather disappointing. Rabbits may require six to 14 bites before developing a protective immunity; inoculation of 10 μg of venom-protein every 3 weeks produced hemagglutinating antibodies in some rabbits after the first injection and increasing numbers of precipitating antibodies after 2 weeks from the second injection. Severe local lesions were observed with the first three injections of venom, however, which suggest that the antibodies present at that time were insufficient in quantity or quality to neutralize all the toxic activity.

Study of the antigenic patterns of the venom of three species of *Loxosceles* revealed differences that warn against assuming cross-protection within the genus too readily.

Since the bite of *Loxosceles* is often not identified as such by the victim and the symptoms are uncharacteristic for the first few hours or days, an immunodiagnostic test has been recently developed. It consists of arranging an indirect hemagglutination test with venom-sensitized erythrocytes and antivenom serum, and trying to inhibit the reaction with the venom contained in the fluid taken from the suspected lesion. It was demonstrated experimentally that 5 of 6 guinea pigs bitten by the spider had enough venom in their lesions after 24 hr to cause the inhibition of the indirect hemagglutinating reaction.

The Instituto Butantan in Sao Paulo, Brazil, has recently prepared a therapeutic anti-*Loxosceles* serum, but its clinical evaluation against the conventional treatment has not been completed yet.

Latrodectus **Spider**   *Latrodectus* have been known to be poisonous to humans for considerably longer than *Loxosceles*, and most people think of them when talking about arachnidism. From my own experience and because *Latrodectus* are outdoor spiders as compared to the more domestic *Loxosceles*, I suspect that humans are bitten by *Latrodectus* considerably less frequently than by *Loxosceles*. *Latrodectus* venom causes sytemic neurologic symptoms in all mammals: horses are highly sensitive to its activity and dogs are much less so; cold-blooded vertebrates are quite resistant.

The venom of *Latrodectus* contains at least 11 different proteins and maybe as many as 17. One of them, with a molecular weight of about 130,000, is toxic for the neuromuscular junctions of mammals and is different from one or more other substances that are neurotoxic for arthropods (Frontali et al., 1976). Earlier reports had attributed mammal toxicity to a protein fraction of 5000 daltons. It is possible that more than one toxin exists or that the heavier species corresponds to a polymer.

The mammal toxin is highly antigenic and immunization with it produces high-titered antisera in short periods. Current evidence has shown that anti-*Latrodectus* serum produced with venom of one species will protect against the bites of other species of the same genus.

Merck, Sharp and Dohme Laboratories manufacture an antiserum ("Antivenim-*Latrodectus mactans*") that is recommended for victims younger than 16 or older than 60, or suffering from hypertension. Usually an intramuscular or intravenous dose relieves the symptoms in 1 to 3 hr. Since the antiserum is raised in horses, investigation of the patient for hypersensitivity to horse protein, by skin test or by depositing a drop of the suspension in the eye, is essential before administering the medication.

Scorpions   Scorpion stings appear to be a rather frequent accident in Mexico; current estimations hold that upward of 2000 cases occur every year, with a mortality of about 5%. According to the doses of antiserum used, there seem to be 40–60 cases annually in the U.S.

The studies on scorpion venom have increased greatly lately: work with the secretions of *Tityus serrulatus* from Brazil has recently demonstrated up to 16 protein components, five of which exhibited neurotoxic activity. These toxins are generally of rather low molecular weight (between 7000 and 9000), which may contribute to their reduced immunogenicity. Work with the American species *Centruroides sculpturatus* has given comparable results.

Antisera are prepared in Mexico and in Brazil, in the Americas. The antiserum against Mexican species of *Centruroides* is effective against the stings by American species of the same genus. Although protection across genera is frequent among scorpions, this cannot be taken as a general criterion, since several exceptions are known.

The experience of the Institute Pasteur of Algier indicates that production of antiserum against scorpion venom is a lengthy process. It may take the periodic inoculation of the venom of 400 to 500 scorpions over a period of at least 8 months before a satisfactory antiserum is obtained in horses. Even then, the protection is never complete, since the animals always show evidence of pain at the site of the injection and of colics following the inoculation. Even the best serum rarely is able to neutralize more than 0.4 to 1.0 mg of venom per ml (according to the species of scorpion) and its relative inefficiency is illustrated by the fact that 20 to 30 ml of antiserum are often necessary for a patient. Theoretically, this dose should neutralize 8 to 30 mg of venom, which is at least 2 to 8 times the amount of venom obtainable from the best-producing scorpion. Although it can be argued that irreversible damage of the neuronal membranes may have already occurred in the patient receiving the serum, the same argument is not valid for horses, which have been immunized for long periods and still become symptomatic on injection of venom. In North Africa, the mortality of serum-treated patients is still 3%, but untreated stung children may exhibit 50% mortality.

At any rate, the dynamics of the neutralization of scorpion toxins by the corresponding antivenom is an area that deserves study.

Tick Paralysis   The engorgement of some ticks occasionally produces an ascending flaccid paralysis in the host that can progress to death if the arthropod is not removed in time. This condition affects humans, a number of domestic and wild mammals, and birds. Humans, dogs, and sheep appear to be particularly susceptible, because one single arthropod may cause a fatal accident in them. Tick paralysis in the U.S. accounts for serious livestock losses in some areas and for about 150 reported human cases, these latter with 10–12% mortality. In humans, most cases affect children below 7 years of age, possibly because of their higher exposure to tick infestation and their lower corporeal mass.

The problem in North America is generally connected with parasitism by species of *Dermacentor*, but *Ixodes, Rhipicephalus, Hyalomma, Rhipicentor, Haemaphysalis, Amblyomma* and other tick species have been held responsible elsewhere and, occasionally, in the U.S. In most instances the disease is caused by engorging females, but the larval stages of *Argas persicus* and of *Ixodes holocyclus* can also produce it.

The agent of tick paralysis is believed to be a toxin present in the oral secretions of only some members of a species, during particular periods of their life cycle. The toxin has not been isolated yet, nor have the factors that induce its production been identified.

The symptoms of the condition begin 4 to 7 days after attachment of the tick; they then develop rapidly, and death may ensue in 48 hr or less in highly susceptible species (9 days in cattle). If the insulting arthropod is removed before the bulbar structures become compromised, recovery is equally rapid: it usually begins in a few hours and it is completed after 48 hr in dogs and sheep, after about 10 days in cattle, and often after a longer period in humans.

It has been recorded that the toxicity of the ticks is inversely related to the age and weight of the host and directly related to the number of ticks attacking it: it diminishes with storage of the arthropods and it is more severe with ticks that engorge slowly. The lag period before appearance of symptoms makes one wonder whether the oral secretions have a direct effect on the neuromuscular junctions of the host or they act through the production of mediators derived from the host. It has been impossible so far to transfer the paralysis from sick animals to healthy ones by inoculation of serum: this suggests that either the toxin must concentrate very slowly in the neuromuscular junctions until reaching a pathogenic threshold (which would be reversible and surprisingly discriminative, taking into account the prolonged lag period as compared to the brief course and recovery) or it exhibits an exquisite specificity that reminds one of the immune system.

There have been claims that tick paralysis induces immunity, and a recent work in Australia reported that 12 affected calves left untreated died, whereas only 3 of 7 affected calves receiving 10–30 ml of canine antitick serum met the same fate. This latter result, however, may have been due to an effect of the immune serum on the engorging ticks that impaired their feeding, as well as an effect on the toxin itself. Along the same lines, Cebu cattle are more resistant to tick paralysis in nature than the conventional breeds, but, under controlled conditions in which the number of engorging ticks is related to the mass of the host, they have not shown any particular resistance over the other breeds. It is believed that the resistance observed in nature is connected to the fact that Cebu cattle are less susceptible to tick attack than the European breeds.

Other arguments somewhat negate the development of protective immunity to tick paralysis: the course of the condition in calves infested with *I. holocyclus* (7 days to beginning of symptoms, and 9 days to death or 10 days to complete recovery) certainly does not indicate any evidence of acquired resistance being produced, and the fact that sheep may succumb to the disease within 21 days after recovery from a previous episode suggests lack of acquired protection.

Besides the classic tick paralysis, there are at least four other less frequent neurological syndromes attributed to the bites of *Ixodidae* ticks. A paralytic syndrome caused occasionally by the bites of *Argasidae* ticks (*Argas, Ornithodorus*, and *Otobius*) is less severe and believed to be different from the *Ixodidae* tick paralysis. All these conditions are thought to be due to diverse toxins about which virtually nothing is known.

SOURCES OF INFORMATION

Barnard, J. H. 1973. Studies of 400 hymenoptera sting deaths in the United States. J. Allergy Clin. Immunol. 52:259–264.

Benjamini, E., and Feingold, B. F. 1970. Immunity to arthropods. In: G. J. Jackson, R. Herman, and I. Singer (eds.), Immunity to Parasitic Animals. Vol. II, pp. 1061–1134. Appleton-Century-Crofts, New York.

Bossard, M. 1976. Relations immunologiques entre bovins et tiques, plus particulierement entre bovins et *Boophilus microplus*. Acta Trop. 33:15–36.

Cavagnol, R. M. 1977. The pharmacological effects of hymenoptera venoms. Annu. Rev. Pharm. Toxicol. 17:479–498.

Euzeby, J. 1976. Traitement et prophylaxie de l'hypodermose des bovins: Donnes actualles. Rev. Med. Vet. 127:187–235.

Feingold, B. F., Benjamini, E., and Michaeli, D. 1968. The allergic responses to insect bites. Annu. Rev. Entomol. 13:137–158.

Friend, W. G., and Smith, J. J. 1977. Factors affecting feeding by blood sucking insects. Annu. Rev. Entomol. 22:309–331.

Frontali, N., et al. 1976. Purification from Black Widow spider venom of a protein factor causing the depletion of synaptic vesicles at neuromuscular junctions. J. Cell Biol. 68:462–479.

Futrell, J. M., and Morgan, P. N. 1977. Identification and neutralization of biological activities associated with venom from the Brown Recluse spider, *Loxosceles reclusa*. Am. J. Trop. Med. Hyg. 26:1206–1211.

Gaafar, S. M. 1972. Immune response to arthropods. In: E. J. L. Soulsby (ed.), Immunity to Animal Parasties, pp. 273–285. Academic Press, Inc., New York.

Habermann, E. 1972. Bee and wasp venoms. Science 177:314–322.

Hoffman, D. R., Shipman, W. H., and Babin, D. 1977. Allergens in bee venom. J. Allergy Clin. Immunol. 59:147–153.

Hunt, K. J., Valentine, M. D., Sobotka, A. K., and Lichtenstein, L. M. 1976. Diagnosis of allergy to stinging insects by skin testing with hymenoptera venoms. Ann. Int. Med. 85:56–59.

Hunt, K. J., et al. 1978. A controlled trial of immunotherapy in insect hypersensitivity. N. Engl. J. Med. 299:157–162.

Kim, J. 1977. Pulmonary acariasis in Old World monkeys. Vet. Bull. 47:249–255.

Nelson, W. A., Bell, J. F., Clifford, C. M., and Keirans, J. E. 1977. Interaction of ectoparasites and their hosts. J. Med. Entomol. 13:389–428.

Parrish, H. M. 1963. Analysis of 460 fatalities from venomous animals in the United States. Am. J. Med. Sci. 245:129–141.

Russell, F. E. (ed.). 1977. Animal Toxins. Pergamon Press, Oxford.

Schlein, Y., and Lewis, C. T. 1976 Lesions in haematophagous flies after feeding on rabbits immunized with fly tissues. Physiol. Ent. 1:55–59.

Wharton, G. W. 1976. House dust mites. J. Med. Entomol. 12:577–621.

Chapter 7

Evasion of
the Immune Response
by Parasites

In the preceding pages a number of manifestations of the host's immunity that, factually or potentially, can damage the parasites in it have been reviewed. The indisputable truth is that, despite all of these, most parasites find a means of subsisting and multiplying in hosts that appear to be fully immunocompetent. *Eppur si muove*!

The host-parasite relationship must be almost as old as life itself (the organisms that could not synthesize their own nutrients probably invaded those that could), and the parasites that we know today are the victors in a long evolutionary process that only permitted the survival of those associations whose members learned to tolerate each other; at least until the parasite reached the phase of reproduction and dissemination. An extreme virulence on the part of the parasite would have eliminated the host species, with the consequent extinction of the parasitic species; on the other hand, a totally effective defense reaction on the part of the host would have exterminated the parasitic species. In this sense, the parasites we know today are a highly selected population, and it should not be a surprise to find that they have acquired peculiar mechanisms along their evolution that are present only rarely in nonparasitic organisms.

The evolution of the host-parasite relationship is a seductive field for speculation, and the ability of the parasite to persist in its host is particularly fascinating since it represents the defeat (if we yield to a pinch of teleology) of a highly organized physiological system whose specific function is precisely to eliminate parasites in their broadest sense.

It is quite possible that the currently known mechanisms of evasion of the host's immune response by the parasite are only a part of the potential of these organisms for survival in the host. On the other hand, our contemporary picture is abundantly sprinkled with speculations. Therefore, the major principles of parasite subsistence in the immunocompetent host are discussed here as they are now understood.

Recent reviews by Cohen (1976) and Ogilvie and Wilson (1976) and a full symposium by the Ciba Foundation (1974) have been devoted to this subject.

LOCALIZATION IN IMMUNOLOGICALLY PRIVILEGED SITES

Anatomical structures such as the fetus, the inner eye and the tissues of the cerebrum, the testicles, and the thymus are relatively isolated from the immune system because of peculiar morphophysiological barriers. While the integrity of these barriers is maintained, the structures located beyond their boundaries are mostly out of the reach of the lymphoid apparatus. They have been called, perhaps whimsically, *immunologically privileged sites*. It seems logical to expect that the parasites lodged in these sites will also partake of the protection against immunity: in support of this assumption, larvae of *Toxocara canis* in mouse brain, larvae of *Taenia solium* in human brain, and a larva of a nematode followed for 10 years in a human eye have been reported not to produce the tissue reactions that the same infections would trigger in other anatomical locations. The often peracute course of Chagas' disease and of toxoplasmosis in congenital infections may also be expressions of parasitic proliferation out of the control of the immune system.

The intracellular parasites (such as the reproductive forms of *Toxoplasma gondii*, *Trypanosoma cruzi*, *Leishmania*, *Babesia*, and *Plasmodium* in the vertebrates) have constituted their own immunologically privileged sites inside the cells of the host, since all the current evidence indicates that the host's immunity is unable to reach the interior of its own cells. In the case of *Babesia* and *Plasmodium*, the intracellular environment is not particularly effective as a mechanism to circumvent the attack by the host because their exclusive or predominant existence in dissociated cells permits the destruction of the host cell along with the parasite. In this respect, it might not be a coincidence that these species also exhibit another effective deterrent of the host immunity, such as antigenic variation (see below). It is surprising, on the other hand, that *T. gondii*, *T. cruzi*, and *Leishmania* can invade and multiply in macrophages, which are on occasion obtained from hosts that have demonstrated resistance to the acute homologous infection. It has been verified for *T. gondii*, and there are some preliminary indications for *T. cruzi*, that the live parasite in the macrophage is able to inhibit the fusion of the phagocytic vacuole with the lysosomes in order to prevent its own digestion.

In the case of parasites surrounded by cystic membranes of the host, such as *Trichinella spiralis* and hydatid cysts, it has been assumed that these membranes are effective barriers against immune reactions. The actual persistence of the parasites despite abundant evidence of specific immunity tends to support this notion. Also, in trichinellosis antigens obtained from muscle parasites elicit protection, and the freed parasites themselves have been reported to be affected by the immunity in some experiments. The question remains, however, of whether the cystic membrane is really protective (or necessary) or the insusceptibility to the immune defenses is simply a result of the intracellular environment of the parasite. In the case of hydatid cysts (and probably of all larval cestodes), the host-derived cyst membrane could only be effective in stopping the effector cells of cell-mediated immunity (not demonstrated yet to be protective) because antibodies and complement are found inside the parasite. At present, it is not known why the antibody-complement combination, which is lethal for the cestode in vitro, does not operate in vivo. It has been found, however, that substances produced by the parasites activate the complement nonspecifically, and it is believed that this event may elude the action of complement on the parasitic structures by exhausting it in the immediate vicinity of the parasite. Disguise with a coat of host protein may be another protective mechanism for larval cestodes (see below).

A great deal of indirect evidence and some speculations based on well-documented phenomena suggest that the physical seclusion of the parasites from the orbit of action of the immune system is indeed an efficient method of avoiding the responses that are adverse to the invader; in all fairness, however, there is little direct proof in support of this presumption.

IMITATION OF HOST ANTIGENS

In Chapter 2 it was mentioned that sharing of antigens between the parasite and its host, or among diverse species of parasites, was a rather common event. This phenomenon has been the subject of a number of speculations. Sprent (1963) hypothesized that an "adaptation tolerance" or adaptive host's immunological unresponsiveness should be a condition of successful parasitism: this condition could have arisen from a selective convergent evolution of antigen structure between host and parasite, or from mechanisms by which the parasite could disguise or minimize its functional antigens. Damian (1964) referred to the phenomenon of parasitic antigenic determinants evolving to become identical to those in the host as "molecular mimicry," and called the antigens involved "eclipsed antigens." In his conception, the final goal is the attainment of parasitic substances that will not elicit immunity on the part of the host. Dineen (1963) took the view that the host's immunity was a part of the total environment of the parasite and that the parasite may therefore have taken advantage of it for its own necessities of self-regulation. He proposed that the disparity between the functional antigens of the parasites and the host's antigens must have diminished during evolution only to such a degree that the parasitic antigens would trigger the corresponding protective reactions solely when they passed a threshold concentration ("fitness antigens"). In this manner, the parasite itself controls the size of the parasitic population in the host. This ingenious theory, which has had some experimental confirmation, explains the chronicity of many parasitic diseases, the persistence of small populations of parasites in immune hosts, the classic phenomenon of self-cure, and the generally low pathogenicity of most parasitoses in the undisturbed nature.

For Damian, the existence of identical antigenic determinants in different parasitic species constitutes an evidence of convergent evolution, in which various parasites are evolving to imitate the molecular pattern of the host. Schad (1966) extended Dineen's concepts by hypothesizing that a given parasite may utilize the host's immunity to limit the competition by another parasite with which it shares antigens: in this case, the first parasite that arrived in a particular host would elicit protective responses that would manifest themselves against subsequent competing species, but not against the original organism. This notion might be particularly applicable to helminths that possess common antigenic substances in their preadult stages; this could explain why the postulated immunity is active against invading parasites but not against established specimens.

The best known and more studied instance of imitation of host antigens is the case of Schistosoma. The first clue to this phenomenon was the observation that mouse Schistosoma mansoni transferred to normal monkeys survived, but died if the recipients were immunized previously with mouse red blood cells. Subsequent studies have shown that schistosomes adsorb red cell antigens and, to a lesser extent, serum proteins from the host, at least in cases of mouse, monkey, and human parasitism. The presence of these host materials on the parasite surface impedes the combination of its superficial antigens with the complement-fixing and opsonizing antischistosoma antibodies, but does not prevent the priming of helper T cells. It thus seems that the host antigens protect the parasite against the efferent branch of immunity but do not stop the afferent limb. Since the new schitosomula acquire host materials only 4–5 days after the infection, they are affected by the immunity elicited by a previous infection during the first days of their life in the mammalian host. By comparison with a similar phenomenon in tumor immunology, the immune protection against reinfections, in the absence of an effect on the already established parasites, has been called "concomitant immunity." Most studies until now have demonstrated that schistosomes adsorb the "disguising" substances from the host, but some evidence has been produced that the parasite is also able to synthesize similar materials.

Host substances have also been investigated on adult *Fasciola hepatica*, but to no avail so far. It has been found that *Taenia taeniformis* cysts survive when transferred to immune hosts, but they die if treated with trypsin before implantation. This report suggests that a proteinic coating, presumably of host origin, exerts a protective action against the effect of the host's immune response. The presence of host materials has also been demonstrated on other helminths and on some African trypanosomes (*T. vivax* and *T. gambiense*), and it is suspected in *T. cruzi*.

It would be wrong, however, to assume automatically that host substances on a parasite would always be a deterrent for protective immunity: in a study with *Ascaris suum* in pigs it was found that the parasitism was as prevalent in animals with blood group A as in those with blood group O, although the third-stage larvae of *A. suum* react with anti-A and with anti-B, but not with anti-O, antibody. In this case, sharing of antigens between host and parasite does not seem to have a detectable effect on the infection.

PRODUCTION OF SOLUBLE ANTIGENS

Studies in the last few years have found parasitic antigens in the circulation or in nonparasitized tissues of subjects infected with an increasing number of species of parasites. Thus far, this phenomenon has been reported in visceral leishmaniasis, toxoplasmosis, babesiosis, malaria, African trypanosomiasis, haemonchosis, trichinellosis, dirofilariasis, and schistosomiasis. The production of complement-activating substances by cestodes (see above) may fit here also. New reports are likely to add other parasitoses as the investigations continue.

Work in nonparasitic systems has shown that the simultaneous injection of soluble antigens by the intraperitoneal or intravenous route depresses the cell-mediated response to the same antigens given in particulated form or administered in the tissues. Several investigations have also found immunity-blocking factors in the serum of cancer patients; these factors appear to be tumor antigens or antigen-antibody complexes. It is perfectly conceivable, therefore, that the soluble antigens of the parasite can also interfere with the recognition of other more relevant antigens, compete with them, or neutralize the effector branch of the immunity before it reaches the parasite. In any case, they would provide an effective system of eluding the protective responses of the host. Systematic work in this area is urgently needed. Wilson (1978) has commented on this subject recently.

VARIATION OF ANTIGENS IN THE COURSE OF AN INFECTION

One of the most striking discoveries in parasitology in recent years has been the finding that the functional antigens of some parasites may change in the course of a single infection: in this way, the new antigenic variants are unaffected by the immunity to previous variants and, by the time that the response to the new antigen reaches effective levels, a still newer variant is being produced. This mechanism keeps the parasite a step in front of the host's protective response and allows its survival regardless of the effectiveness of the immunity.

The best-known example of antigenic variation is the African trypanosomes. The fluctuation of the parasitemia of patients of sleeping sickness, at approximately weekly intervals, was already known and suspected to be related to the host's immunity as long as 60 years ago. Numerous modern works (see Doyle, 1977) have revealed the gross details, but much research is still to be done. The genome of the salivarian trypanosomes codes for a glycoproteinic coat on the cellular membrane of these protozoa. At present up to 22 different glycoproteins or antigenic variants have been detected in the same strain (although undoubtedly more exist), but only one of them is expressed at a given time. Each variant differs from the others in amino acid composition and in antigenicity. On infection, the parasites of the first wave of parasitemia exhibit a predominant variant, to which agglutinating and lytic IgM antibodies that abate the parasitemia are soon produced. A new wave of parasitemia with parasites that possess a different glycoprotein begins then, and again the corresponding antibodies are formed. For all that is now known, these cycles of parasitemia followed by parasiticidal antibodies may continue indefinitely, but observations in the field indicate that infected cattle can recover and remain resistant to reinfections. Since functional antigens common to all variants have not been found, but a few cases of repetition of the same variant are known, it is possible that only a limited number of variants actually exist and that, after several relapses of parasitemia, the affected animal has developed protective antibodies against most of the variants of the particular strain. This circumstance will be detected most often as resistance to all the variants of a strain, within a particular parasite species.

Ingestion and development of the trypanosomes in the vector fly makes them revert to a "basic" antigen: unfortunately, this is different for each strain within a trypanosome species, and the numbers of strains are so many that vaccination at present does not seem practical. Antigen variation in trypanosomes is undoubtedly related to the host's immunity, because it does not occur in immunosuppressed hosts, but it is not clear whether the immune response induces the change or only selects preexisting populations of parasites. In the stercorarian trypanosomes (e.g., *T. lewisi, T. cruzi*), strains that differ antigenically have been described, but there is no indication that a strain is able to change its antigens in the course of an infection.

The first and most extensive experiments of antigenic variation in malaria have been done with *Plasmodium knowlesi* (Brown, 1977). Recent studies indicate that it also occurs in *Plasmodium berghei, Plasmodium cynomolgi,* and *Plasmodium falciparum.* Many authors feel that antigenic variation is an inherent property of the genus *Plasmodium.* The existence of immune responses to antigens exclusive of a given relapse has been demonstrated by schizont-infected cell agglutination tests (SICAT), by opsonization tests (OT), and by merozoite inhibition tests; since the latter two tests indicate protection, there is no doubt that the variable antigens are functional. Brown has found that the SICAT for a given variant (which does not indicate protection) becomes positive considerably sooner than the OT at the beginning of an infection, but they appear at the same time, or their order is reversed, when the infection has already stabilized at a low level (thus revealing higher efficacy of the protective mechanism). On this basis, it has been proposed that the immune response to the blood forms of malaria occurs in two phases: a first wave of variation-inducing antibodies, and a later wave of parasiticidal (opsonizing and merozoite-inhibiting) antibodies. As the infection progresses, the difference in time between these phases would diminish until disappearing, so that most parasites would be destroyed before having an opportunity to change antigens. In contrast to the African trypanosomes, the plasmodia will undergo antigenic variation even in the absence of parasiticidal immunity, which suggests that the mechanism of variation is not the selection of preexisting populations of parasites.

Despite the existence of variable antigens in malaria, it appears that all variants share some common antigenic determinant, because the occurrence of repeated relapses extends the specificity of the immune protection until it becomes effective against all the variants of that particular strain. It is even possible that different strains of the same species have some weak functional antigens in common, because vaccination of owl monkeys with *P. knowlesi* or *P. falciparum* merozoites in Freund's complete adjuvant produced resistance to the infection with heterologous strains of the corresponding species.

Antigenic variants have been detected by agglutination of parasitized erythrocytes in relapses of *Babesia argentina* in cattle, and by tests of protection in relapses of *Babesia rhodaini* in rodents. As with the African trypanosomes, babesias revert to a basic antigen that is characteristic for each parasite when developed in the tick. No antigenic variation was detected when *Babesia* was transferred rapidly among splenectomized calves, which suggest that a host's immune response is needed for the protozoan to express its ability to change antigens.

Some authors suspect that antigenic variation also occurs in *Entamoeba* and *Toxoplasma*, but appropriate confirmation is still necessary. Recently some evidence has been reported that antigenic variation may occur in the nematode *Nippostrongylus brasiliensis* and perhaps also in *Onchocerca volvulus*. If verified, this phenomenon would be known to exist in viruses, spirochetas, protozoa, and nematodes, which suggests that it is a biological mechanism more general than has been thought until now.

DEPRESSION OF THE HOST'S IMMUNE RESPONSIVENESS

Epidemiological observations in zones of malarial hyperendemia in Africa had associated the infection with a reduced efficacy of the vaccines of childhood, with the infrequent presence of autoimmune diseases, and with an uncommonly high prevalence of Burkitt's lymphoma. Since these findings suggested that malaria was related to a decreased reactivity of the immune system, the problem was soon taken to the laboratory for appropriate investigation. Later studies showed that mice infected with *P. berghei* or children infected with *P. falciparum* produced less antibody to unrelated antigens when these were injected during the acute infection; the cell-mediated immunity was not modified in these circumstances, but it became affected in mice as the infection turned chronic. In those cases in which it was investigated, the immunodepression persisted despite successful treatment of the malaric condition.

Studies on the cause of immune depression in malaria have revealed a reduction of the weight of the thymus and of the number of mature T and B cells in the lymph nodes of infected animals. Other authors have reported alterations in the function of the macrophages in infected people and mice. These findings may indicate that *Plasmodium* interferes with the preliminary processing of the antigen by the macrophages and with the subsequent necessary proliferation of the lymphoid cells. How the parasite does this remains to be revealed.

Work with trypanosomes of rodents has shown that infections with these protozoa also inhibit the response to unrelated antigens, but, unlike the case of malaria, the immune depression disappears with the successful treatment of the condition. Murray et al. (1974) found that trypanosome infections exert a nonspecific stimulation ("mitogen effect") on B lymphocytes that might pre-empty or smother the populations of cells required to respond to subsequent antigens. This experimental finding agrees well with the nonspecific hypergammaglobulinemia commonly observed in African trypanosomiasis of man.

Depression of the host's immune responsiveness has also been documented in infections by *Toxoplasma* and by *Leishmania*, in terminal cases of *Entamoeba histolytica* in humans and in babesiosis, American trypanosomiasis, and *T. vaginalis* inoculations in mice.

In recent years, immune depression has been also reported in connection with a number of experimental helminth infections (trichinellosis, ascariasis, nematospiroidosis or heligmosomoidiasis, *Brugia pahangi* filariasis, canine ancylostomiasis, possibly haemonchosis, schistosomiasis, *Taenia crassiceps* cysticercosis) and, possibly, in canine demodectic mange. The infection most studied from this standpoint at the moment is trichinellosis. During the second to the eighth week of a *T. spiralis* infection, laboratory mice show a markedly inhibited ability to produce antibodies (Barriga, 1978a) or manifestations of cell-mediated immunity (Barriga, 1978b) to nonrelated antigens. This has been demonstrated to be mediated by suppressor T cells and transferable with splenocytes.

In the case of *Demodex canis* infection of dogs, it has been found that the peripheral lymphocytes of a high proportion of animals with the generalized form of the disease do not respond well to T cell mitogens, and that this phenomenon is induced by incubation of normal T cells with serum of diseased dogs. It is not known yet whether the suppressive substance in the serum is derived from the parasite or is the result of some host reaction.

The possibility that large quantities of parasitic antigens cause specific immunotolerance of the immature lymphoid system of very young hosts has been often raised in the literature, but no systematic investigations have yet been attempted.

Current information seems to indicate that depression of the host's immune response is a common event in a number of parasitic infections, but the mechanism utilized to achieve the final result appears to vary from one parasite to the next. This should not be surprising, however, since the consequences of possessing this ability are so crucial for the survival of the parasitic species that it would be expected that parasites had used any means at their disposal to attain this advantage. Besides, the immune system has built-in regulatory mechanisms to set a limit to the immunological responses (otherwise, once triggered, they would continue forever); for organisms that preceded immunity in time, it might not have been too difficult to incorporate these mechanisms into their own natural histories by a process of selective evolution.

A disturbing fact is that parasites elicit an array of immune responses to their own antigens despite their ability to produce immune depression to unrelated antigens; in some cases (such as toxoplasmosis and trichinellosis) they even stimulate nonspecific responses that are usually protective (such as phagocytosis). It is possible that parasites also selectively suppress the response to some of their own functional antigens, maybe those most relevant to the protective responses. If this is the case, the demonstration of this phenomenon will be technically difficult, since the verification of antigenic activity (protective or not) is based on the ability of the host to react to the corresponding substance.

Besides the critical importance of immune depression for the life of the parasitic species, it probably has an important impact on the health of the parasitized population. Inhibition of the responses to functional antigens of the insulting parasite itself will certainly contribute to the establishment of peracute or chronic infections, according to the degree of pathogenicity that the particular organism is able to exhibit. Depression of the response to heterologous antigens may stimulate the expression of concurrent subclinical infections, facilitate the development of intercurrent diseases, and interfere with immunization procedures. It is possible that this phenomenon is partially responsible for the chronic "poor health" often seen in parasitized infants, whose lymphoid systems might be particularly affected by the parasite-induced depression because of their relative immaturity.

METABOLIC INHIBITION of parasite

In the last few years it has been repeatedly demonstrated that a number of parasites are able to remain dormant in the host for significant periods of their parasitic life. The intracellular resting stages of *Toxoplasma* and of *Trypanosoma cruzi*, the hepatic forms of the recurrent human malarias, the arrested larvae of many nematodes, and possibly the third-stage larvae of long-lasting myiasis may be appropriate examples.

The quiet existence of hypobiotic parasites implies a reduced metabolism that, in turn, must result in the diminished production of functional antigens to stimulate the host's immune system. This idea is supported, among other things, by evidence that the immune response produced by the acute infection wanes in animals carrying arrested larvae of *Haemonchus* or *Ostertagia* and by the marked drop of antibody levels in patients with chronic toxoplasmosis or Chagas' disease or in the late phases of *Gastrophilus* or *Hypoderma* infections.

Even if the host's immunity participates in the production of hypobiosis, the metabolic response of the parasites to it tends to create opportunistic chronic infections that favor their persistence in the stable host environment until the conditions for successful transmission return. This appears to be an effective mechanism of parasite survival that has been little studied so far.

SOURCES OF INFORMATION

Barriga, O. O. 1978a. Modification of immune competence by parasitic infections. I. Responses to mitogens and antigens in mice treated with *Trichinella spiralis* extract. J. Parasitol. 64:638–644.

Barriga, O. O. 1978b. Depression of cell-mediated immunity following inoculation of *Trichinella spiralis* extract in the mouse. Immunology 34:167–173.

Brown, K. N. 1977. Antigenic variation in malaria. Adv. Exp. Med. Biol. 93:5–25.

Ciba Foundation. 1974. Parasites in the Immunized Host: Mechanisms of Survival. Symposium 25 (new series). Elsevier, Amsterdam.

Cohen, S. 1976. Survival of parasites in the immunized host. In: S. Cohen and E. H. Sadun (eds.), Immunology of Parasitic Infections, pp. 35–46. Blackwell Scientific Publications, Oxford.

Damian, R. T. 1964. Molecular mimicry: Antigen sharing by parasite and host and its consequences. Am. Natural. 48:129–149.

Dineen, J. K. 1963. Immunological aspects of parasitism. Nature 197:268–269.

Doyle, J. J. 1977. Antigenic variation in the salivarian trypanosomes. Adv. Exp. Med. Biol. 93:31–63.

Murray, P. K., Jennings, F. W., Murray, M., and Urquhart, G. M. 1974. The nature of immunosuppression in *Trypanosoma brucei*. Immunology 27:815–840.

Ogilvie, B. M., and Wilson, R. J. M. 1976. Evasion of the immune response by parasites. Br. Med. Bull. 32:177–181.

Schad, G. A. 1966. Immunity, competition, and natural regulation of helminth populations. Am. Natural. 100:359–364.

Sprent, J. F. A. 1963. Parasitism. Williams & Wilkins Company, Baltimore.

Wilson, R. J. M. 1978. Circulating antigens of parasites. In: Immunity to Parasitic Diseases. Colloque INSERM-INRA, pp. 87–101. Editions INSERM, Paris.

Chapter 8

Principles of
Immunoprophylaxis in Parasitic Infections

On a world basis, parasitic infections exert a tremendous toll on human health and food supplies. This is particularly severe in tropical regions that, because of their present stage of economic and social development, are the least prepared to prevent and compensate for their effects. The modification of the natural environment by humans, in order to produce the large conglomerates of human beings and domestic animals characteristic of our civilization, altered the existing balance between host and parasite and favored the unrestricted proliferation of the parasitic species. Whereas the infective stage of a paleolithic parasite must have waited uncertainly for the eventual return of its nomadic host, its modern counterpart now can count on the abundant presence of adequate hosts in its immediate environment. The evidence shows that our feeble attempts at epidemiological control and our chemotherapeutic paraphernalia have been unable to regenerate a peaceful coexistence between parasite and host within the new order.

Since the characteristics of our present life style preclude the return to a low population density (which will largely prevent the transmission of the parasites) and antiparasitic drugs are effective only temporarily (when used prophylactically) and do not correct the damage already produced (when used therapeutically), many efforts have been devoted to artificially increasing the resistance of the host to the infection.

Vaccination against animal parasites has not been nearly as successful as vaccination against bacteria and viruses. Several factors contribute to this difference: on one hand, the widely held belief, until relatively recent years, that immunity to parasites was inherently different from immunity to other exogenous invaders discouraged intense research in this field; on the other hand, the involved life cycle of many parasites demands the individual study and correlation of the immunological phenomena that take place in various biological phases of the parasitic infection, which may sometimes be technically difficult to achieve and always obscures the total picture of the protective immunity to the infection. Also, parasites are more complex organisms than bacteria or viruses and the identification of functional antigens is correspondingly more difficult. Finally, the numerous possibilities to which parasites may resort in order to elude or depress the host's immunity (and that have become known only recently; see Chapter 7) are an essential issue in antiparasite vaccination, and a difficult one to circumvent. In addition, immune competence against animal parasites may appear rather late in ontogeny: lambs do not respond to infection by *Haemonchus contortus* or *Trichostrongylus* until they are 4–6 months old or to vaccination with *Dictyocaulus filaria* until after 8–16 weeks of age; the more severe course of many parasitoses in young individuals than in adult subjects suggests that this might be a rather general phenomenon.

References to attempts at vaccination against specific parasitoses have been given in the corresponding sections. Recent general reviews have been made by Cohen (1975), Soulsby (1975), Rowe (1978), and Cox (1978). Although there is still a long way to go before vaccination becomes a major tool in the fight against parasitic infections, a few important advances have been already achieved, and many results encourage further research. The major approaches to immunoprophylaxis of parasitism that have been utilized so far are discussed below.

CONTROLLED INFECTIONS

The observation that many natural parasitic infections result in the production of effective resistance to later challenges has stimulated attempts to produce the infection artificially in order to elicit immunity to natural occurrences. This method is being used successfully for human cutaneous leishmaniasis, in which the primary purpose is to avoid scars in exposed areas of the body. In most other cases, however, the infection is too dangerous to permit its normal course and it is preferable to treat it as soon as enough time to develop a protective response has elapsed; this technique is currently used in bovine babesiosis. A modification of this approach consists of administering subtherapeutic doses of antiparasitic drugs when the danger of infection is imminent; the infective parasites that survive on introduction in the host are not sufficient to cause disease, but do elicit some degree of resistance that builds up with the reinfections. This procedure is currently in use for avian and rabbit coccidiosis. A similar approach is used as chemoprophylaxis in canine heartworm disease, but this parasite does not appear to elicit effective immunity in these conditions. Sterilizing treatment of infections of *Ascaris suum* in rats and of *Echin-*

ococcus granulosus in dogs, after the infection has had the opportunity to develop for a few days, has also resulted in moderate protection. Similar results have been obtained with the subcutaneous injection of infective eggs of *A. suum* in guinea pigs or the intravenous or intramuscular inoculation of activated oncospheres of *E. granulosus* in dogs. These stages develop only for a limited time in these abnormal habitats, but produce functional antigens, and the corresponding resistance, in the interim.

Theoretically, the same procedure is applicable to human parasitoses, but the difficulties of predicting the course of the infection in each particular case and the existence of remote sequelae (sometimes associated with lack of an effective treatment) in many of the infections in which the use of this method would be desirable (e.g., Chagas' disease, toxoplasmosis) advise against its utilization. Similarly, its use in veterinary medicine is severely limited by the cost of giving individual attention to each member of the herd. A major inconvenience of this procedure is that often it does little to prevent the contamination of the environment, and therefore it does not help to eradicate the infection. Use of subliminal and prolonged chemotherapy, on the other hand, facilitates the selection of drug-resistant parasites.

KILLED PARASITES AND CRUDE PARASITIC EXTRACTS

Possibly in imitation of many successful bacterial vaccines, the inoculation of whole, dead, or ground parasites was one of the first methods used in the attempts to produce acquired resistance. A detailed review of these attempts has been written by Clegg and Smith (1978). Most studies of vaccination with dead organisms or with their crude extracts have produced disappointing results: the protection produced has often been nil or negligible, and waned rapidly. Nowadays, it is known that only the antigens that elicit responses that interfere with vital functions of the parasite, that alter its structures, or that drastically modify its environment will be able to produce reactions of resistance. Experience indicates that most antigens that behave in this way are metabolic products of the parasite; the whole organism or its extracts, on the contrary, contain a tremendous predominance of somatic parasitic material that, in addition to being inert for the purposes of protection, triggers the normal mechanisms of control of the immune response before the less potent antigens have a reasonable chance to act. Most likely, the reduced protection elicited by the crude extracts is attributable to small quantities of metabolic products present in them; the general need for strong adjuvants in these cases also suggests that the functional antigens are in subliminal quantities.

In general, vaccination with crude extracts of larval cestodes or of protozoa has been somewhat more successful than the immunization with extracts of other parasites. At least in the case of the protozoa, this may be attributed to the smaller proportion of "inert" structures, as compared to actively metabolizing organelles, that they possess.

PURIFIED ANTIGENS

The obvious next step after the use of crude homogenates of parasites was the selection of parasitic substances that showed particular ability to produce protective reactions in the host. It was soon discovered that many of these substances were secreted or excreted by the parasite in culture, which made it easier to collect them from the culture media instead of attempting their separation in minute quantities from complex crude extracts. Soluble antigens in the plasma of animals infected with *Babesia* or of mice infected with *Plasmodium berghei* (presumably metabolites of the parasites) have also proved to produce acquired resistance in the laboratory. The granules of the stichosome of *Trichinella spiralis* and *Trichuris muris* have also produced protective immunity in laboratory studies. Perhaps the most exciting recent development in vaccination with purified antigens has been the claims of successful immunization of cattle against *Taenia saginata* and of sheep against *Taenia ovis* by inoculation of in vitro secretions of the corresponding parasites.

The use of purified parasitic substances for immunization should greatly facilitate the standardization, control, transportation, storage, and administration of the vaccine. The widespread use of bacterial and viral vaccines frequently makes us forget how formidable these problems are in the case of live vaccines, the only ones commercially available today in parasitology. Besides, this technique would permit the selection of the proper core antigen to eliminate those parasites capable of antigenic variation before they put their potential to play. On the negative side, this method requires the adequate identification of the appropriate antigens, and their production in satisfactory quantities, neither of which techniques is well developed today. In addition, there is evidence in several cases that protective immunity may depend on the activity of a specific branch of the lymphoid system (humoral or cellular), or on the action of a particular class of antibodies, or on the operation of the immune elements at a given location; complete knowledge of these peculiarities and of the procedures to replicate them is necessary if success is to be expected. Finally, immunity against soluble antigens is usually transient, which may force frequent revaccinations; this is obviously an inconvenience when dealing with large human populations and economically unfeasible with food animals;

methods to prolong the effect of the immunization (or to secure natural infection during the protected period) must be found.

In summary, immunization with selected antigens seems to be an ideal procedure that appears to be within the reach of our present technology, but a number of questions must be answered in the laboratory before their rational application becomes a reality.

ATTENUATED PARASITES

Because the identification and purification of functional antigens presented some technical difficulties, a logical substitute approach was to let the parasite itself produce the relevant antigens in the body of the host. In order to avoid the pathology of the natural infection without the trouble or uncertainty of individual treatment, it was necessary to select apathogenic strains or to attenuate virulent lines, making sure that the protective antigenicity remained. Natural strains of reduced pathogenicity are known for *Toxoplasma gondii* and for *Trypanosoma cruzi* and are under active investigation for *Leishmania*; successful vaccination of laboratory animals has been already achieved with the first two species. Presumably, the spontaneous production of strains with variable pathogenicity is more likely to occur in protozoa than in helminths, since the faster generational turnover of the former affords them more chances for selection and mutation in a given period of time. At any rate, this procedure is risky to use in humans because no animal model can accurately predict the behavior of a particular parasitic strain in the human being. On the other hand, there are no warranties that a strain that was originally apathogenic will not revert with time to an enhanced virulence. These considerations greatly limit the use of mild strains in humans, or in animals that can act as reservoirs of the infection.

Several methods of artificial attenuation of the pathogenicity of parasites have been utilized to obtain a better control of their behavior in the host and to escape from the dependency of fortuituous findings of avirulent strains. In vitro culture, pretreatment with drugs, passage in abnormal hosts, and so on have been used at various opportunities. The most favored system, nevertheless, is the irradiation of the infective forms of the parasite. The International Atomic Energy Agency (1968) has published a report that reviews numerous experimental assays with this method. Adequate doses of irradiation prevent the reproduction of the parasite but apparently do not affect its metabolism in any other important way: parasites so treated are incapable of multiplying and producing important colonization of the host's tissues on infection, and eventually will die, but in the interim they secrete antigens that frequently are functional. In the case of the protozoa, irradiated vaccines of *T. gondii, T. cruzi, Plasmodium,* and *Babesia* have given preliminary results that encourage further research. Similar methods have been considerably less effective with trematodes and have not been investigated extensively with cestodes. Among the nematodes, irradiated vaccines have often been successful in experimental trials and have led to the commercial production of vaccines against *Dictyocaulus viviparus* and *Ancylostoma caninum.*

A major drawback of the irradiated vaccines is their brief shelf life, which poses important problems of distribution: irradiated *D. viviparus* larvae survive only 2–4 weeks in artifical media and irradiated cercariae of *Schistosoma* lose their activity after 24 hours.

HETEROLOGOUS PARASITES

Despite the fact that the functional antigens usually exhibit an exquisite specificity, cases are known in which a given species can protect against the infection by a different parasite. On occasion, the species that elicits immunity is less pathogenic than the parasite affected by it, in which case the former organism can be advantageously used as a vaccine against the latter one.

Several cases of cross-protection between different species have been verified in the last years; some of them were mentioned in the preceding chapters. A few have generated particular interest for their possibilities of practical applications: immunization of sheep with nematodes of insects or of the soil has produced strong but transient protection against *Dictyocaulus filaria*; vaccination of monkeys with the zoophilic strain of *Schistosoma japonicum* or with schistosomes of ruminants protected against subsequent infections with the anthropophilic strain of *S. japonicum* or with *Schistosoma mansoni*; infection of cattle or sheep with *Taenia hydatigena* produced resistance to further infections with *Taenia saginata* or with *Fasciola hepatica.*

Heterologous immunization appears to have only a limited application, however, since it depends on the fortuituous finding of slightly pathogenic or nonpathogenic organisms that can elicit effective immunity against agents of disease. A number of attempts have failed to confer resistance (e.g., vaccination with *Toxocara canis, Haemonchus contortus,* or *Caenorhabditis briggsae* against *Ascaris suum* in guinea pigs), and others have intensified the disease (immunization of sheep with *Nippostrongylus brasiliensis* or *Ascaris suum* followed by infection with *Dictyocaulus filaria*). The finding of the appropriate immunizing agent requires that the actual experiment of immunization and infection be performed, since serological cross-reactivity has not been related to cross-protection in the cases in which both have been studied.

NONSPECIFIC IMMUNIZATION

There are reports that the previous inoculation of BCG in rodents protects them, to variable degrees, against subsequent infections with *Babesia, Plasmodium, Leishmania, Echinococcus,* and *Schistosoma. Corynebacterium parvum (Propionibacterium acnes)* produces some resistance against *Babesia, Plasmodium, Trypanosoma cruzi,* and avirulent *Toxoplasma gondii.* Several similar substances have been used with partial success in experimental malaria.

Most of these materials are immunological adjuvants when injected with antigenic preparations, and nonspecific stimulants of cell-mediated immunity and phagocytosis when administered alone. It is not clear which of these activities is particularly relevant to this case: since these materials remain in the body for some time, they can potentiate the host's response to the antigens released by a parasite even if this is administered some days later; on the other hand, the parasites may be affected by the hypermacrophagia elicited nonspecifically by the BCG or a similar inoculum. In this latter regard, it might not be a coincidence that nonspecific immunization has proved to be particularly effective against parasites normally eliminated by phagocytosis. At any rate, this is a subject that deserves further research.

CONCLUSIONS

The achievements that have thus far been obtained in vaccination against parasites have not been especially outstanding when compared to the accomplishments of other branches of immunology. For different reasons, parasitologists have lagged behind the pack in this particular area. The gathering of basic information and the structuring of the ideas in the field have been extremely fruitful in recent years, however, and the possibility of developing effective immunoprophylactic schemes against parasitic diseases is perceived as a real eventuality in a not-too-distant future.

The final success of vaccination in parasitology will depend on the ability of the specialists to take full advantage of the latest concepts and techniques and on their ingenuity in producing immunizing preparations that are safe, effective, inexpensive, and easy to administer. Conditions for which there is no adequate therapy available, such as American trypanosomiasis, toxoplasmosis, and cysticercosis, or in which the parasite successfully circumvects the resistance naturally acquired, such as malaria, babesiosis, and African trypanosomiasis, appear to be natural candidates for the most urgent research efforts.

The greatest challenge comes from those infections that are controllable already by relatively cheap and effective chemical treatments (e.g., avian coccidiosis, some ectoparasites). Immunoprophylaxis, in association with other ecological procedures, is directed to regaining the biological balance that nature originally provided. Since conditions of life are different now than in primeval times, allowances must be made and improvements on the original design must be furnished. Restoration of the ecological host-parasite equilibrium without altering drastically our present way of living is a scientific feat and, most likely, a long and laborious process. Thus, it is tempting to close our eyes to the mounting chemical pollution of our planet, trust that our capability to synthesize new active compounds is truly inexhaustible, and hope that we will beat the parasites at the game of becoming drug resistant. The long-term prognosis is open to anybody, but the prospects that the ingenuity of man will defeat a billion years of genetic information without nature's help appear to be despairingly flimsy.

SOURCE OF INFORMATION

Clegg, J. A., and Smith, M. A. 1978. Prospects for the development of dead vaccines against helminths. Adv. Parasitol. 16:165–218.

Cohen, S. 1975. Immunoprophylaxis of protozoal diseases. In: P. G. H. Gell, R. R. A. Coombs, and P. J. Lachmann (eds.), Clinical Aspects of Immunology. 3rd Ed., pp. 1649–1680. Blackwell Scientific Publications, Oxford.

Cox, F. E. G. 1978. Specific and nonspecific immunization against parasitic infections. Nature 273:623–626.

International Atomic Energy Agency. 1968. Isotopes and Radiation in Parasitology. I.A.E.A., Vienna.

Rowe, D. S. 1978. Vaccines and other immunological approaches to the control of parasitic diseases: Needs and prospects. In: Immunity to Parasitic Diseases. Colloque INSERM-INRA, pp. 307–310. Editions INSERM, Paris.

Silverman, P. H. 1970. Vaccination: Progress and problems. In: G. J. Jackson, R. Herman, and I. Singer (eds.), Immunity to Parasitic Animals. Vol. II, pp. 1165–1185. Appleton-Century-Crofts, New York.

Soulsby, E. J. L. 1975. Immunoprophylaxis of helminth infections. In: R. G. H. Gells, R. R. A. Coombs, and P. J. Lackmann (eds.), Clinical Aspects of Immunology. 3rd Ed., pp. 1681–1689. Blackwell Scientific Publications, Oxford.

Index

DATE DUE

MAY 11 1996

DEMCO, INC. 38-3012